Nanostructured Materials for Sustainable Energy and Environmental Remediation

Online at: https://doi.org/10.1088/978-0-7503-5143-0

Nanostructured Materials for Sustainable Energy and Environmental Remediation

Edited by
R V Mangalaraja
Faculty of Engineering and Sciences, Universidad Adolfo Ibáñez, Santiago, Chile

R Udayabhaskar
Institute of Scientific and Technological Research, Universidad de Atacama, Copiapó, Chile

P Sathishkumar
Department of Chemistry, Vellore Institute of Technology (VIT), Vellore, India

Joydeep Dutta
Applied Physics Department, Functional Materials Group, KTH Royal Institute of Technology, Stockholm, Sweden

IOP Publishing, Bristol, UK

ISBN 978-0-7503-5143-0 (ebook)
ISBN 978-0-7503-5140-9 (print)
ISBN 978-0-7503-5141-6 (myPrint)
ISBN 978-0-7503-5144-7 (mobi)

DOI 10.1088/978-0-7503-5143-0

Version: 20221201

IOP ebooks

British Library Cataloguing-in-Publication Data: A catalogue record for this book is available from the British Library.

Published by IOP Publishing, wholly owned by The Institute of Physics, London

IOP Publishing, No.2 The Distillery, Glassfields, Avon Street, Bristol, BS2 0GR, UK

US Office: IOP Publishing, Inc., 190 North Independence Mall West, Suite 601, Philadelphia, PA 19106, USA

Contents

Editor biographies

Mangalaraja Ramalinga Viswanathan

Professor Mangalaraja Ramalinga Viswanathan is a Full Professor in the Faculty of Engineering and Sciences, Universidad Adolfo Ibáñez, Peñalolén, Santiago, Chile. He worked as the Director of International Relations at the University of Concepción for the years 2018 and 2019 and as the Director for the Department of Materials Engineering at University of Concepcion for six years from 2010 to 2016. Further, he also served as the President of the Chilean Metallurgy and Materials Society (SOCHIMM) for the period from 2015 to 2017. Professor Mangal is well recognized in the scientific community and he also serves in many professional bodies like Life-Member of Indian Ceramic Society (ICS), Fellow of Indian Institute of Ceramics (IIC), Member of The Mineral, Metals and Materials Society (TMS), Singapore Materials Research Society (MRS-S) during 2017-2018; member of American (ACerS) and PAN-American Ceramic Societies.

Professor Mangalaraja earned his BSc degree in Physics from Madurai Kamaraj University and MSc degree in Materials Science, MTech in Ceramic Technology. Later, he continued his PhD in Materials Technology (Advanced Ceramics) from Anna University-Chennai, India. Prior to joining Universidad Adolfo Ibáñez, Peñalolén, he served as Professor for 17 years at the University of Concepcion, he was a Postdoctoral researcher at AIST-Nagoya, Japan, Lulea University of Technology-Lulea, Sweden and Anna University-Chennai, India. His primary research focuses on the areas of nanotechnology-driven advanced ceramics (functional ceramics for energy and environmental applications) and micro-nano-structure-property relationship of functional and structural ceramics and metallic composite materials.

Professor Mangal has published over 260 refereed international journal publications, 27 book chapters, three patents, edited two books, and more than 185 research presentations/invited talks in several international conferences. Further, he has participated as Responsible and Co-Investigator of 40 research projects funded by ANID-National Agency for Research Development (formerly CONICYT—National Commission of Scientific and Technological Research), Government of Chile and two projects funded by Universidad Adolfo Ibáñez, Peñalolén, Santiago, Chile. For his credits, he has guided 30 Postdoctoral researchers and 13 doctoral candidates in the area of materials science and technology. Also, he acted as a member in the expert committee of FONDECYT-ANID twice (four years) in the area of materials science and engineering, and now extending his expertise as a reviewer of various projects in the area of materials science and engineering.

Dr Udayabhaskar Rednam

Dr R Udayabhakar is an Assistant Professor at Institute of Scientific and Technological Research (IDICTEC), University of Atacama, Copiapó, Chile. He worked as the Assistant Professor in Department of Physics, Vignan's University, Vadlamudi, Guntur, Andhra Pradesh, India. He earned his BSc and MSc degrees from Andhra Uinversity, Vishakapatnam, Andhra Pradesh, India. He received his PhD (Physics) from National Institute of Technology, Tiruchirapalli (NIT Trichy), Tamilnadu, India. He has postdoctoral experience from University of Concepcion, Chile. He was a member of The Mineral, Metals and Materials Society (TMS) during 2016–17, Society of Photo-Optical Instrumentation Engineers (SPIE) and the Optical Society (OSA) during 2010–2014. His primary research interests are focused on the development of functional nanomaterials for energy harvesting and environmental remediation. He is the author or co-author of 38 scientific publications in reputed international journals, and one book chapter and has presented his research outcomes at various international conferences. He has participated as a principal responsible investigator and co-investigator for three research projects that are funded by ANID-National Agency for Research Development (formerly CONICYT—National Commission of Scientific and Technological Research), Government of Chile. Further, he is extending his expertise as a reviewer for the FONDECYT-ANID in the materials science and engineering discipline.

Dr Sathishkumar Panneerselvam

Dr P Sathishkumar presently serves as Assistant professor in the Department of Chemistry, Vellore Institute of Technology (VIT), Vellore, India. Previously he served as academician at the University of Concepcion, Chile, Periyar Maniammai University, India and Aksum University, Ethiopia. He received his BSc and MSc degrees from Bharathidasan University, India and earned his PhD from National Institute of Technology (Chemistry), Tiruchirappalli, India. He is a life member of Indian Society for Technical Education (ISTE). He has postdoctoral experience from University of Concepcion, Chile and he visited the University of Melbourne as visiting researcher. His research interest focused on nanomaterials synthesis and its application in environmental remediation technologies. He is the author or co-author of over 40 refereed international journal publications and four book chapters and more than 10 research presentations at several international conferences. Further, he has participated as Responsible Investigator of three research projects funded by ANID-National Agency for Research Development (formerly CONICYT—National Commission of Scientific and Technological Research), Government of Chile, University of Concepcion, Chile and Aksum University, Ethiopia and now extending his expertise as a reviewer of various projects in the area of materials science and engineering.

Professor Joydeep Dutta

Professor Joydeep Dutta is the Chair of Functional Materials at the KTH Royal Institute of Technology, Stockholm, Sweden. He was the Chair Professor in Nanotechnology for Water Desalination and other applications in Sultan Qaboos University from October 2011–September 2015. Earlier he served as the Vice President (Academic Affairs), Director of the Center of Excellence in Nanotechnology and Professor in Nanotechnology at the Asian Institute of Technology (AIT), Bangkok, Thailand, whose faculty he joined in April 2003. He completed his PhD in 1990 from the Indian Association for the Cultivation of Science, India (Calcutta University). In 1991 and 1992 he did Post Doctoral work at the Electrotechnical Laboratory (ETL, Japan) and at Ecole Polytechnique (France) before moving to Switzerland in 1993 where he was associated with the Swiss Federal Institute of Technology (EPFL), Lausanne, Switzerland until 2003. His broad research interests encompass development of nanomaterials for perovskite and quantum dot solar cells, hydrogen energy, enhanced treatment of impaired water, desalination, catalysis, photocatalysis, amongst others. From 1997–2001 he worked in technical and managerial qualities in high technology industries in Switzerland before returning back to academia in 2002. He is chairman of the board of Stockholm Water Technology AB and has been the member of the board of two other companies working in high technology electronics and environmental consulting, respectively.

Professor Dutta is an award winning author (Choice award for Outstanding Academic title of 2010 from American Library Association) of the book *Fundamentals of Nanotechnology*. He has also written two other text books entitled *Introduction to Nanoscience* and *Introduction to Nanoscience and Nanotechnology* (CRC Press of Taylor and Francis Group LLC). Professor Dutta has over 200 research publications, with more than 10,000+ citations (h-index 50), 11 chapters in science and technology reference books, five patents (11 ongoing applications) and has delivered over 100 invited and keynote lectures. He is on the editorial board of a few journals and regularly referees articles in international journals.

List of contributors

M Abinaya
Alagappa University, Karaikudi, India

Radhamanohar Aepuru
Universidad Tecnológica Metropolitana, Santiago, Chile

A Arulraj
Universidad Adolfo Ibáñez, Peñalolén, Chile
Universidad de Atacama, Copiapo, Chile

Mohammad Ashfaq
Advanced Ceramics and Nanotechnology Laboratory, Department of Materials Engineering, Faculty of Engineering, University of Concepción, Concepción, Chile
and
University Centre for Research & Development (UCRD), Chandigarh University, Gharaun, Mohali, Punjab, India
and
Department of Biotechnology, Chandigarh University, Gharaun, Mohali, Punjab, India

Gnana Sundara Raj Balasubramaniam
University of Concepción, Chile

F Béron
State University of Campinas (UNICAMP), Brazil

Cristian H Campos
Universidad de Concepción, Chile

A M G Carvalho
State University of Maringá, Brazil

Divya Chauhan
University of South Florida, United States

J C Denardin
University of Santiago, Chile

Saaed Farhang
Universidad Adolfo Ibáñez, Peñalolén, Santiago

M A Gracia-Pinilla
Universidad Autónoma de Nuevo León, Mexico

Bhalchandra Kakade
SRM Institute of Science and Technology, India

B Karthikeyan
National Institute of Technology, India

Moorthi Lokanathan
Universidad Adolfo Ibáñez, Peñalolén, Chile

R V Mangalaraja
Universidad Adolfo Ibáñez, Peñalolén, Chile

Héctor Valdés Morales
Universidad Católica de la Santísima Concepción, Chile

Gnanaprakasam Periyasamy
Saveetha Engineering College, Chennai, India

S Rajalakshmi
Anna University, MIT Campus, India

C Rajkumar
Centre for High Energy Systems and Sciences, DRDO, Hyderabad

Suresh Renganthan
SRM Valliammai Engineering College (Autonomous), India

K Shanmugaraj
Universidad de Concepción, Chile

R Shin Juang
Chang Gung University, Taiwan

P Sundaresan
Chang Gung University, Taiwan

E Sundaravadivel
SRM Institute of Science and Technology, India

R Suresh
Easwari Engineering College (Autonomous), Chennai, India

Neetu Talreja
Advanced Ceramics and Nanotechnology Laboratory, Department of Materials Engineering, Faculty of Engineering, University of Concepción, Concepción, Chile
and
Department of Science, Faculty of Science and Technology, Alliance University, Anekal, Bengaluru, Karnataka, India

N Thachnatharen
National Defence University of Malaysia, Malaysia

Prabhakaran Thandapani
VIT-AP University, Amaravati, Andhra Pradesh, India

Pandiyarajan Thangaraj
University of Concepcion, Chile

R Udaybhaskar
Universidad de Atacama, Chile

B Vaseeharan
Alagappa University, Karaikudi, India

IOP Publishing

Chapter 1

Introduction to engineering nanostructured materials for energy and environmental applications

R V Mangalaraja and R Udayabhaskar

Miniaturization and compatibility are two significant trends that are becoming more pervasive throughout the development of nanotechnology using nanostructured materials. At an annual meeting of the American Physical Society at Caltech in 1959, Richard Feynman gave a talk called 'There's plenty of room at the bottom' and laid the conceptual foundation for the field of nanotechnology. Over the past few decades, the global research community has been actively realizing intense research activity to conceptualize the benefits of the nanostructured materials obtained by various scientific methodologies to succeed in benefiting the energy and environmental sectors for the betterment of human living conditions. This book caters to the needs of enthusiastic forthcoming researchers to overcome the obstacles to their efficient research output. The utilization of multifunctional nanostructure materials for energy harvesting, storage, and conversion includes devices such as solar cells, batteries, supercapacitors etc. The scope of applications of nanostructured materials for environmental applications comprises green fuel harvesting, pollutant removal, refrigerants, and sensors for the detection of toxic and biomolecular components.

To meet global energy demands and to overcome the issues related to fossil fuels, researchers across the globe have taken advantage of harvesting the abundant 'free' solar radiation. Photovoltaic solar cells provide the most common alternative energy at different scales. Currently, silicon (Si) based photovoltaic technology is dominant in solar energy harvesting. Solar cell based energy harvesting using alternative materials is a promising approach to overcome the energy supply–demand crisis. Among the various materials, chalcogenide and graphene based materials exhibit

doi:10.1088/978-0-7503-5143-0ch1

the most promising results. Chalcogenides are known to be analogous to graphene owing to their layered structure. The chalcogenides with their unique physicochemical properties offer a huge volume of reactive/active catalytic sites and facilitate the ease of movement through pores and channels. Graphene, a two-dimensional (2D) carbon-based material with remarkable mechanical, thermal, electrical, and optical properties, has been used in various technological applications. In many instances only a few milliwatts of power are needed to power, for example, a wireless sensor node, therefore smaller scale solutions are needed. Thus low-temperature fuel cells, specifically proton exchange membrane fuel cells (PEMFCs), are considered a viable solution in small-scale energy harvesting due to their high efficiency, facile operation, and rapid start-up/shut-down cycle under ambient conditions.

For the optimal utilization of the green and sustainable energy harvested from nanostructures, the energy harvester module should be combined with a storage device, such as a battery or supercapacitor, to deliver continual power while in non-operating conditions. The flexible energy storage devices have unique characteristics, such as high flexibility, a higher charging–discharging rate, and economic viability, that make them ideal candidates for energy storage devices, in particular for wearable devices. Currently, numerous 2D nanolayered materials (porous carbon-based, graphene, hybrid transition metal oxide, and chalcogenide nanostructures) are under consideration as nanostructured materials for energy storage (battery or supercapacitor) applications. These materials prevail over the aggregate structured materials which can restrict direct access to the optimal charge storage surface, leading to ionic resistance within the electrodes. Also, employing metal chalcogenides as an additive lubricant to suppress the energy loss by heat dissipation in various micro-scale mechanical devices significantly contributes to the reduction of energy demand. In addition, the utilization of graphene nanostructures as an alternative for photothermal reagents highlights the suitability of nanostructured materials as an *energy conversion* medium for the advancement of biomedical applications.

Over the decades, pollution has occurred in air, land, and water ecosystems, and has became a serious problem at the global level. The tremendous usage and depletion of fossil fuels and the negative impact caused by toxic releases in the environment have propelled the global research community to innovate in the field of alternative green energy sources and remedies for restoring and maintaining healthy ecosystems. Even though wind and solar power are the main focus in the fight against climate change, hydrogen is gaining momentum as a promising green fuel. Hydrogen gas production using electricity from renewable sources has emerged as a sustainable solution for the green energy–green fuel cycle to reduce environmental contamination. This clean fuel is experiencing great demand in the industrial, transport, power, and energy sectors. In its nascent stage, the most established method of hydrogen production is to split water into hydrogen and oxygen through electrolysis. Hetero-atom doped graphene nanostructures are emerging as an excellent alternative for expensive Pt, IrO_2, and noble metals through their synergic catalytic activity. Several advancements have been reported for capturing carbon effectively, and for the reduction of emissive carbon dioxide (CO_2) using value-added chemicals and fuels. It should be noted that heterostructured chalcogenide based materials display unusual activity towards the reduction and effective conversion of CO_2 due to their vast surface areas and large electrocatalytic active sites.

In addition to the existing water pollution removal techniques (ion-exchange, membrane processes, activated sludge processes, reverse osmosis, precipitation methods), typical nanostructured material based advanced adsorption, catalysis, photocatalysis, and electro-catalysis processes are also being developed for wastewater treatment. The incorporation of solar energy to degrade/remove pollutants by photocatalysis is an economic and sustainable alternative. In particular for treating organic pollutants, a few metal oxides (ZnO, TiO_2, etc) and hybrid carbon–metal oxide composites are known to exhibit excellent photocatalytic activity for antibiotic wastewater treatment. The detection of toxic chemicals and molecules plays a vital role in primary assessment and in the selection of suitable treatment techniques for wastewater. The emerging fluorophores, classified as fluorescent metal nanoclusters (NCs), have attracted the research community for sensor applications to support environmental remedial activities. A variant of fluorescent materials, known as functionalized noble metal NCs, are effectively ratified for the detection/sensing of heavy metal ions, pesticide residues, and biomolecules. Also, for the effective removal of colorants and heavy metal ions, Fe_2O_3 based nanostructures exhibit versatility through different approaches (adsorption, catalysis, and disinfection techniques). At the global level, among the various home appliances, cooling devices (refrigerators, air-conditioners) consume a great share of energy and also contribute to global warming. Thus, as a replacement for existing conventional toxic refrigerants, spinel ferrites as solid-state refrigerants have attracted researchers due to their magnetism based heating and cooling behavior for use in green magnetic refrigeration.

As a concluding remark, the current research trends are toward either improving the efficiency of existing conventional devices or miniaturizing devices to minimize the energy consumption and environmental contamination.

Chapter 2

Engineering 2D chalcogenides for energy and environmental remediation

A Arulraj, N Thachnatharen, C Rajkumar and R V Mangalaraja

Energy and environmental remediation involve a broad range of sectors, including solar energy conversion (solar cells), energy storage (batteries), energy transmission, etc, for the former, and addressing the greenhouse effect, CO_2 capture, catalysis (dye degradation, water-splitting), etc, for the latter. In addition to their broad range, a long-standing aim for these applications is the synthesis/fabrication of suitable materials with high efficiency and long-term stability at low cost. The realization of such a dream depends on the properties of the materials, experiment design, prototype development, architecture, processing, etc. In this regard, two-dimensional (2D) chalcogenide materials have received great interest in recent decades for a diverse range of applications. This chapter will examine the effect of 2D chalcogenide materials in energy and environmental remedial applications. Details on how materials are classified as chalcogenides and the best approaches for fabricating such materials are provided. The chapter focuses on the impact of chalcogenides on emerging solar cells (thin films), supercapacitors, lubricants, carbon dioxide (CO_2) capture, and water-splitting. Finally, a summary of the features of this emerging material along with its future potential and challenges are provided.

2.1 Introduction

Chalcogenides are chemical compounds consisting of at least one chalcogen (anion) that corresponds to the VI A group of the periodic table (S, Se, and Te). They are also referred to as analogs of graphene due to their layered structure. In general, the chalcogenides form a stable layered crystalline structure with the transition metals of the IV–VII B group (except for manganese) [1]. Originating the existence levels between the anionic sp and cationic d the chalcogen atoms (X) will

doi:10.1088/978-0-7503-5143-0ch2

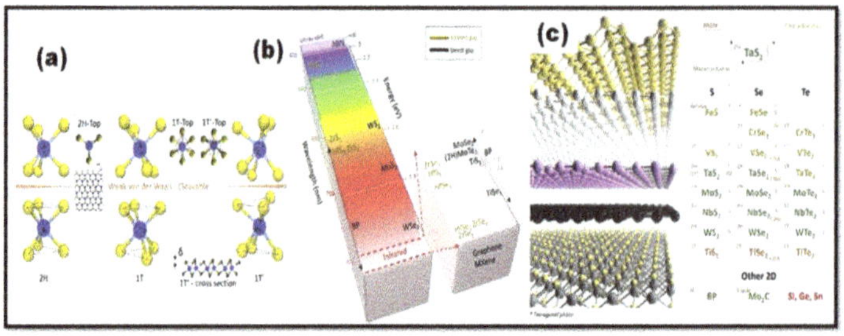

Figure 2.1. (a) Cleavable 2H, 1T, and 1T′ structures of layered TMDs. (b) The bandgap of layered materials starting from graphene to hBN (zero to wide). (c) Various physical properties and crystal structures of 2D TMDs and other 2D materials. (Reproduced with permission from [3]. Copyright 2017 Elsevier.)

react with transition metal atoms (M) to form an X–M–X configuration with a stoichiometry of MX_2 [2]. In the two-dimensional (2D) configuration system of transition metal chalcogenides (TMCs) or transition metal dichalcogenides (TMDs), a strong covalent bond forms between the hexagonal M (transition metal) layer and two X (chalcogen) layers and the adjacent sheets stack via weak van der Waals interaction forces. Depending on the atomic arrangement, chalcogenides structures can be characterized as trigonal prismatic (hexagonal, H), octahedral (tetragonal, T), and distorted phase (T′), giving rise to the dramatic anisotropy and different material properties (figure 2.1(a)) [3].

In accordance with the oxidation state of the metal atoms and coordination chemistry, TMDs can cover wide spectrum of properties ranging from insulators such as HfS_2, through semiconductors such as MoS_2, WS_2, $MoSe_2$, and WSe_2, to metals such as NbS_2 and VSe_2 [2, 4]. Among these the semiconducting chalcogenides are crucial in the systematic choice of materials for diverse applications owing to their intriguing chemical, electrical, optical, thermal, and mechanical properties. In addition, the bandgap engineering of TMDs plays a vital role in energy and environmental applications and has been investigated in depth. The bandgap of the TMDs ranges from 0 eV (semimetal) to 2 eV (semiconductor) depending on the number of layers, defects, etc. Figure 2.1(b) shows a wide range of bandgaps in chalcogenide materials, covering the entire regime of visible and infrared electromagnetic radiation. As aforementioned, the stacking of layers (the number of layers) in chalcogenide materials plays a role in determining the nature of the bandgap. Single-layered TMDs mostly exhibit direct bandgap energy, but as the stacking of layers increases this transforms into indirect bandgap energy. The structure and properties, such as charge density wave, superconductivity, and magnetism, are given in figure 2.1(c) [3, 5, 6].

This chapter encapsulates the state-of-the-art of nanostructured TMDs and their impact on energy and environmental remedial applications. It is well known that nanomaterials differ in their properties compared to bulk materials; the same applies to chalcogenide materials (due to their layered structure). Hence, the different synthesis approaches for TMDs, their fundamental properties, and their impact on their corresponding applications are discussed first.

2.2 Synthesis approaches for chalcogenides

Nanostructured TMDs demonstrate fascinating properties specific to their dimensions and chemical compositions, thereby offering solutions to diverse research issues, including fundamental and practical applications. The properties of TMDs are strongly dependent on their synthesis method, shape, size, the nature of surface, crystallinity, defects, etc. For example, a reduction in the size of the materials may lead to quantum confinement effects (for quantum dots), resulting in enhanced optical and optoelectronic properties [7]. A variety of synthesis approaches have been reported for the fabrication of nanostructured TMDs, including hydrothermal, solvothermal, hot-injection, template-assisted, liquid exfoliation, thermal evaporation sputtering, mixed solvent, cationic exchange, electro-spinning, Kirkendall-effect-induced, etc [8]. As considerable review and book chapters exist on the synthesis of TMDs, only a few crucial TMD nanostructures engineered via different fascinating routes are discussed in this section.

The hydrothermal technique is one of the more common routes of chemical synthesis for fabricating different nanostructured materials. In this regard, the controlled synthesis of TMDs and special intrinsic properties (ferromagnetism) was developed using a modified hydrothermal route [9]. In this report, microrod structured TMDs (marcasite (FeS_2) and ferrosulfide mineral greigite (Fe_3S_4)) were synthesized using an *in situ* magnetic field-assisted hydrothermal route. The autoclave is the main device used for hydrothermal synthesis, and in this modified approach two magnetic rings (circular disks) were placed over both sides (top and bottom) of a stainless steel autoclave. During the synthesis, a magnet placed over the autoclave provided a magnetic intensity of 450 G (1 G = 104 Wb m^2), which assisted in the growth and direction of the TMDs, resulting in the formation of pearl-necklace-like microrod structure. The building blocks of the microrod structure were dependent on the concentration of the sulfur source used in the synthesis. Irregular microsphere structures were obtained for the same material in the absence of a magnetic field, demonstrating that the magnetic field is one of the vital tools in the synthesis parameters for developing a controlled architecture [9].

Controlled shape, size, and phases of materials can be achieved using the solvothermal synthesis route. This synthesis route is quite similar to the hydrothermal route with the difference lying in the source of the medium; in the solvothermal route, organic solvents are often used. Both these techniques can also be considered low-temperature synthesis routes. During synthesis the reaction temperature will usually lie between 100 °C and 250 °C. For the synthesis of TMDs, sulfur, selenides, tellurides, thiourea, etc, are used as sources of chalcogen. The nanostructure of the materials also depends on the precursor source. Yao *et al* demonstrated the solvothermal synthesis route for one-dimensional nanowire ZnS/amine composites using organic amines (ethylamine, *n*-butylamine, and tetraethylenepentamine) as solvents. The presence of amines (the solvent) contributes to the uni-directional growth of zinc atoms along the *c*-axis (0 0 2) faces, as shown in figure 2.2(e) [10]. Similarly, nanobelt structured β-In_2S_3, graphene-like MoS_2

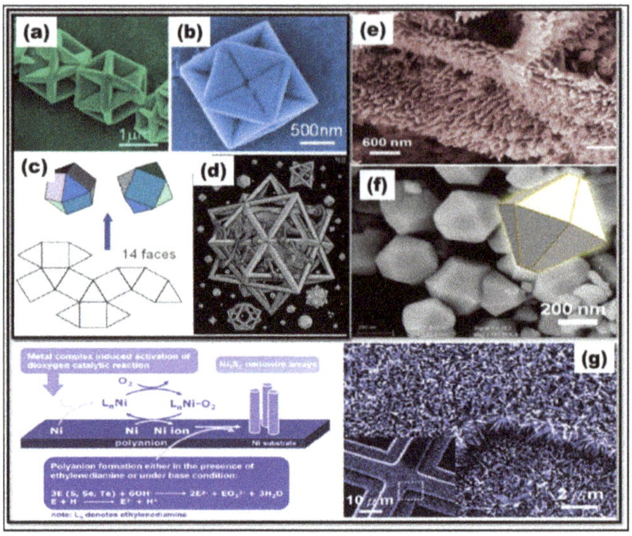

Figure 2.2. (a)–(d) Caved cuboctahedral CuS crystals, reproduced with permission from [13], copyright (2006) American Chemical Society; SEM migrographs of (e) and (f) ZnS nanowire arrays and Fe_7Se_8 polyhedrons [10, 15]; and (g) schematic growth mechanism of Ni_3S_2 nanowire arrays with their morphology, reproduced with permission from [17]. Copyright 2009 The Royal Society of Chemistry.)

nanoplates, and concave cuboctahedral CuS TMDs have been synthesized using different organic solvents such as pyridine, xylene, and ethylene glycol (figures 2.2(a)–(d)) [11–13]. In the case of the astonishing CuS cuboctahedral structured nanocrystals, each face has four identical hexagonal flakes creating 14 concave cavities in total. The formation of these unique architectural nanostructured CuS TMDs is due to the use of EG as a solvent [13].

Combining two or more solvents for synthesizing TMDs also results in a yield of nanostructured materials. For example, ethylenediamine and glycol in a volumetric ratio of 3:1 with $NiCl_2{\cdot}H_2O$ results in the formation of monodispersity dodecahedron structured NiS_2 with a diameter of 6 μm [14]. In addition to combining two or more different solvents, the correct ratio of organic solvents with water may also result in oriented nanostructured materials. For example, polyhedral Fe_7Se_8 with twelve {0 1 2} facets and two {0 0 1} facets has been synthesized with the aid of diethyl triamine (DETA) and water using $FeSO_4{\cdot}H_2O$ (figure 2.2(f)). The proper volume of mixed solvent (2:1) DETA:water resulting in the formation of high index facets, suggests it has some 'magic power' in the formation of polyhedral structured TMDs [15]. In most cases, organic solvents of small organic molecules cannot assist in the direction orientated growth of nanomaterials (TMDs), however, they can assist in the formation of hybrid and/or nanocomposites [16].

The oriented growth of nanostructures plays a vital role in determining the performance of their corresponding applications. The growth of materials such as TMDs over desired substrates is very attractive for energy conversion and storage devices, such as solar cells, batteries, sensors, supercapacitors, etc. In this regard, the small organic molecules such as amines, ligands, etc, contribute to the surface interaction between the substrates and the deposited materials (TMDs). For example, the growth of one-dimensional nanowire Ni_3S_2 over a nickel (Ni) grid at a large scale was achieved using ethylenediamine and hydrazine as the solvent along with the source materials under ambient conditions for 20 h (figure 2.2(g)) [17]. Likewise, several similar approaches have been followed to synthesize nanostructured TMDs, nanocomposites of TMDs, and hybrid and/or doped TMDs with metals, metal oxides, etc.

2.3 Applications of chalcogenides

Unlike oxides, chalcogenides form a complex network structure through the building blocks of chalcogen–chalcogen bonds and new structures through phase homologies. Chalcogenides with a high degree of covalence and a tunable bandgap in turn have a propensity to form semiconducting nanostructures and facilitate ease of movement through pores and channels in energy and environmental remedial applications. The hierarchical structures of chalcogenides with their unique physicochemical properties offer a huge volume of reactive/active catalytic sites in energy sector applications. The tuneable stoichiometric and chemical compositions of the chalcogenides enable their inherent properties to be adjusted for applications. The following section will describe the different energy (solar cells, supercapacitors, lubricants) and environmental remedial (water-splitting, CO_2 capture) applications of chalcogenides.

2.3.1 Energy

The pioneering work on chalcogenides in energy applications was carried out starting in the 1970s. The promising properties of chalcogenides have led to the further exploration of the potential of these materials in energy conversion and storage systems. An interesting feature of chalcogenides that distinguishes them from metal oxides is their higher melting point, making them a suitable candidate for the energy sector. In this section, the significant applications of chalcogenides in solar cells (specifically thin-film solar cells), supercapacitors, and lubricants are covered in detail.

2.3.1.1 Conversion (solar cells)

Energy supply is a hot topic and there is an emerging energy crisis due to the demand for energy, which needs to be resolved sustainably. In this regard, renewable energy is the key tool for tackling the challenging energy crisis sustainably. Among the different renewable energy sources, solar energy, the wide range of electromagnetic radiation emitted by the Sun, will dramatically help supply energy demands. The generation of power (electricity) from solar energy can be achieved using photovoltaic (PV) technology. To date, silicon (Si) has dominated the energy market owing to its higher performance, but the cost and high demand for Si have motivated the need to develop alternative types of solar cells to produce energy.

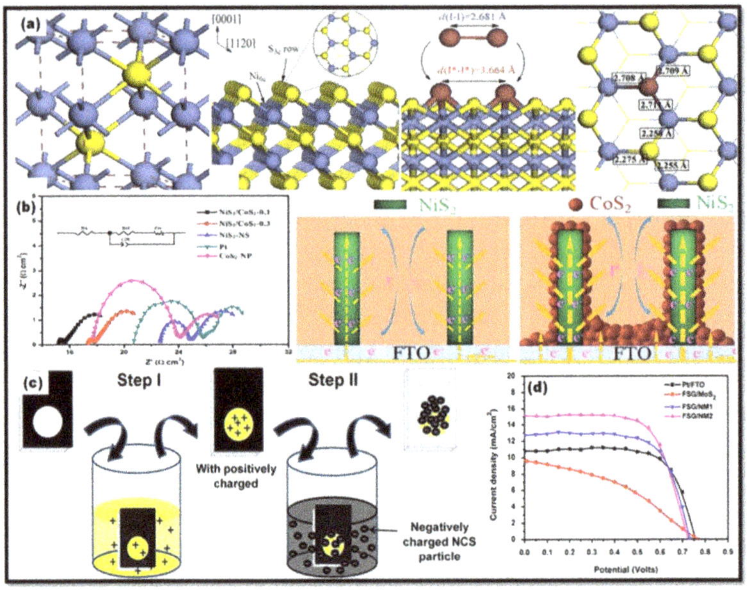

Figure 2.4. (a) Optimized bulk structure of NiS with rows of three coordinated S ions exposed and direct dissociation of the I_2 molecule upon adsorption at the S-vacancy dimer with adsorption configuration [31]. (b) Nyquist plot with the structure diagram of NiS_2 and CoS_2–NiS_2 CEs [32]. (c) Illustration of the dip-coating process of NCS CEs [33]. (d) JV characteristics curve of NiMoS–FSG [34]. ((a) Reproduced with permission from [31]. Copyright 2014 Royal Society of Chemistry. (b) Reproduced with from permission [32]. Copyright 2018 Elsevier. (c) Reproduced with permission from [33]. Copyright 2016 Elsevier. (d) Reproduced with permission from [34]. Copyright 2018 Elsevier.)

coordinated Ni^{2+} and S^{2-} create a stable configuration in the oriented facets which terminate with three coordinated sulfur ions (S_3C) along with its oriented growth of direction (figure 2.4(a)). The iodine (I_2) molecule needs to be activated to adsorb molecules during the electrochemical reactions to enrich the electrocatalytic activity. The iodine molecules/atoms preferably fit into the sulfur (S) vacancy sites, thereby forming Ni–I bonds at a distance of 2.709 Å [31].

Li *et al* reported NiS_2–CoS_2 composites synthesized via an *in situ* chemical bath deposition and sulfurization process and used as a counter electrode (CEs) in DSSCs. In this composite, CoS_2 nanoparticles wrap over the NiS_2, thereby reducing the sheet resistance and diffusion coefficient of the CE materials. Moreover, the formed NiS_2–CoS_2 interface modulates the composite structure's electronic structure, thereby contributing to tuning the performance of the device (8.22%), which is 24.36% higher than that of conventional devices. Since the NiS_2 was grown vertically on the fluorine doped tin oxide (FTO) interface, the contact between them is very small, as shown in figure 2.4(b), leading to an increase in sheet resistance (R_s), and when the CoS_2 was loaded onto the NiS_2 the R_s value was reduced through filling up the empty areas on the FTO (figure 2.4(b)) [32]. As research has advanced, ternary/quaternary chalcogenides have been developed extensively for the fabrication of solar cells due to their outstanding electrical conductivity and electrocatalytic activity [35]. For example, a scalable $NiCo_2S_4$ solution was developed via the solvothermal method. Subsequently, it was immobilized over a conducting transparent substrate (FTO) and flexible polyethylene naphthalate (ITO/PEN) substrates by a series of dip-coating processes, as shown in figure 2.4(c). The flexible $NiCo_2S_4$ based DSSCs had a PCE performance of 8.62%, which is equivalent to that of conventional Pt CEs (8.51%) [33]. Similarly, NiMoS was electrochemically deposited over flexible free-standing graphene paper as an alternative for Pt and expensive FTO substrates. The bimetallic (NiMoS–FSG) chalcogenides performed effective electrocatalytic activity with a PCE of ~7% (figure 2.4(d)) [34].

2.3.1.2 Storage (supercapacitors)

Energy storage devices, particularly wearable devices (flexible supercapacitors), which can cater to our daily energy needs are developing rapidly. To produce wearable supercapacitors the scientific community has pursued the development of nanostructured materials with long-term stability and higher-order cyclic performance. In this regard, a large number of chalcogenides has been employed as active electrode materials for electrochemical supercapacitor devices. These chalcogenides have a layered structure identical to that of graphene, which has good ionic conductivity with a large surface area. The large surface area of chalcogenides supports accommodating a large volume of ions or atoms between the layers, enabling them to store energy either by pseudocapacitive or electric double-layer mechanisms [36, 37].

The hierarchical architecture of nanostructures enabled the rapid development of highly stable flexible supercapacitor devices. For example, copper–cobalt selenides ($Cu_xCo_{1-x}Se_2$) were synthesized via the simplest hydrothermal method followed by a selenization process. The fabricated chalcogenides with an optimized percentage of chemical composition ($Cu_{0.5}Co_{0.5}Se_2$) boost the electrochemical performance of the devices. A flexible chalcogenide supercapacitor provides an ultrahigh specific capacitance of 1695 F g^{-1} at a current density of 1 A g^{-1} with a retention rate of ~95%. The enhanced performance of the devices is mainly attributed to their hierarchical architecture and porous network with outstanding intrinsic conductivity along with the synergistic effects of their composition (figure 2.5(a)). Asymmetric supercapacitors (ASCs) employing nanosheets such as structured $Cu_{0.5}Co_{0.5}Se_2$ as the positive electrode deliver high energy and a power density of ~85 Wh kg^{-1} and ~15 kW kg^{-1} at an operational voltage of 1.6 V, respectively. Moreover, the fabricated ASC device reveals an exceptional cyclic stability of 91% over 10 000 charge–discharge life cycles [38–40].

Flexible hybrid ASCs were developed using multivalent metal selenides (Zn–Co–Se and Zn–Mn–Se) over nickel foam via an electrochemical deposition approach. During the synthesis process, the Zn–Co–Se chalcogenides develop good contact with the nickel foam, start growing in the form of a bunch of nanoparticles, and end up as nanotube-like structures compared to Zn–Mn–Se. The Zn–Co–Se architecture manifests a strong coupling effect which accelerates the movement of electron transportation with the proliferation of electrolyte ions, thereby enriching the

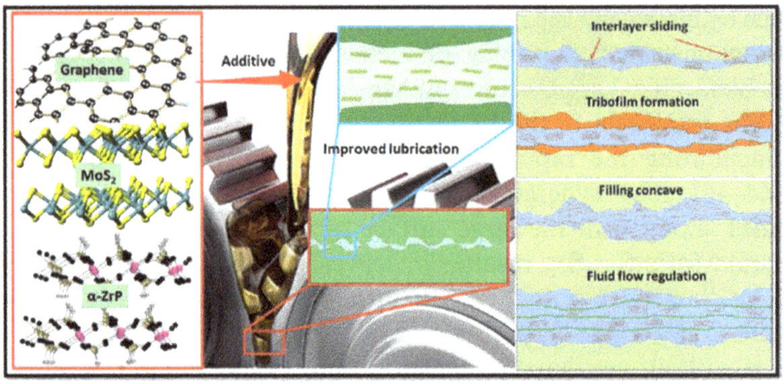

Figure 2.6. Graphical representation of chalcogenides serving as lubricant additives. (Reproduced with permission from [56]. Copyright 2017 Elsevier.)

The advantage of employing metal chalcogenides (MoS_2, WS_2, etc) as additive lubricants is their anti-impact and high thermal stability properties, which allows them to function in extreme conditions such as high load, contact pressure, and speed. MoS_2 and WS_2 employed as additives form a chemical tribofilm during the rubbing of materials against a metal surface [57]. In addition to their thermal stability, most research reports suggested that the chalcogenides' outstanding performance is mainly due to their morphology and size. However, some contradictory reports also exist. For example, Dassenoy *et al* reported the effects of the morphology and size of MoS_2 as a lubricant additive and studied its frictional behavior using a polyalphaolefin base oil. The results suggest that the size of the chalcogenides (MoS_2) has little impact on friction activity, and morphology has a negligible influence. A similar study was performed and reported by Kogovsek *et al*, with contradictory results on the effects of size and morphology of chalcogenides in their tribological performance [58–60]. Compared to morphology and size, structural defects induced during the exfoliation or synthesis of chalcogenides have synergetic effects in reducing the frictional rate [58, 60].

Thus the frictional coefficient (COF) can be decreased by varying the composition modulation, structure modification, post-treatment, and adaptive tribological coatings. However, there are some limitations to chalcogenide based additives that prevent further practical approaches. For example, WS_2 is limited by its high-density causing aggregation in the base oil. In some cases, a greater amount of chalcogenides need to be added in the base oil; increasing the concentration of the additives results in the softening of materials and impairs the mechanical properties of the lubricants [61].

2.3.2 Environmental remedial applications

The need for environmental protection has grown in recent decades due to the rapid development of industrialization and urbanization. The growth of industries has a huge impact, causing environmental issues such as water contamination, the release of hazardous gases, soil contamination, and so on. Researchers have developed solutions to overcome these detrimental environmental impacts under the term 'environmental remediation'. Environmental remedial solutions make use of semiconductor chalcogenides, particularly in the order of nanometer size, to break down toxic molecules into useful by-products. For example, the breakdown of toxic dye molecules into carbon dioxide and water molecules. TMDs play a major role in environmental remediation due to their higher catalytic activity and thermal stability, allowing them to be suitable candidates for diverse applications such as CO_2 capture, photocatalysis, water-splitting, etc [62–65].

2.3.2.1 CO_2 conversion/reduction

The reduction of emissive carbon dioxide (CO_2) into value-added chemical feedstock is emerging in recent decades as an approach to resolve environmental problems. The production of fuels from the conversion of CO_2 has gained much interest in the field of research and in industrial sectors. The technology and/or materials used for CO_2 conversion should be cost-effective and highly efficient. To date, vigorous research has been carried out to develop novel materials (catalysts) to increase their proficiency [66]. However, the existing electrocatalysts are too efficient in practice due to their weak bonding with reaction intermediates, leading to overpotential or a poor kinetic electron transfer rate leading to low current density. These metrics mainly depend on the intrinsic electronic properties of the catalytic materials and the effect of the solvent used during the synthesis [67–70]. In this regard, TMDs have displayed unusual activity in the reduction of CO_2 owing to their large surface area and highly electrocatalytic active sites. Several TMDs and their composites have been developed as catalyst materials for the effective conversion of CO_2. In this section, heterostructured chalcogenide materials, chalcogenides in ionic liquids, and chalcogenides with an organic complex as electrocatalysts for environmental remediation are discussed briefly.

As mentioned, heterostructured bimetallic co-doping with chalcogenides has been developed using a microwave-assisted solvothermal approach. It is a two-step process, first ruthenium (Ru) and Cu are co-doped with zinc sulfide nanopowders and then anchored with hybrid chalcogenide (MoS_2)–graphene. The developed novel hybrid chalcogenide nanomaterials enhance the photocatalytic activity of CO_2 reduction by hindering the recombination rate (electron–hole pairs) (figure 2.7(a)). The reduction in the recombination rate of the materials can be achieved by optimizing the hybrid nanomaterials' concentration, which results in tuning its bandgap. The optimized heterostructured hybrid TMDs performed CO_2 reduction activity of 58.6 $\mu mol\ h^{-1}\ g^{-1}$, even in a concentrated NaOH solution. Ru along with the TMDs support better photocatalytic activity in remedial applications [71].

The S and Se bonding with the inorganic/organic complexes are known before in the CO_2 reduction, but hitherto the mechanisms involved in the complex are not clear. In this regard, the synthesis of an isocyclam nickel (Ni) complex with chalcogenides containing macrocycles Ni–LS and Ni–LSe demonstrated by Gerschel and co-workers provides insight into the mechanism of the CO_2 reduction activity (figure 2.7(b)). The electrochemical results of the isocyclam complex

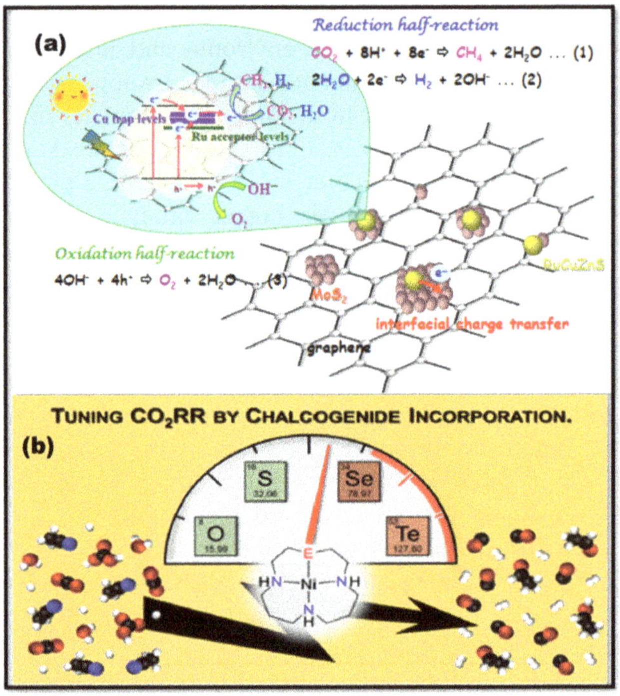

Figure 2.7. (a) CO_2 reduction mechanism of heterostructured hybrid chalcogenide electrocatalysts [71]; and (b) graphical illustration of CO_2 reduction by chalcogenide incorporation [72].

demonstrate that the incorporation of different chalcogenide (S and Se) atoms has major effects in tuning their properties. Depending on their stoichiometric and specific precursors, the active chalcogen species (S) becomes more feasible along with its increase in electron density facilitating the CO_2 reduction at a mild potential (0.3 V). Notably, the Se based isocyclam Ni complex (Ni–LSe) exhibits poor CO_2 reduction but shows promising activity in the evolution of hydrogen. Moreover, the synthesis reaction route of the isocyclam Ni chalcogenide complex also provides new insights into generating acetaldehyde from acetonitrile [72–74].

The most common concern in CO_2 reduction during photoelectrochemical conversion is its inertness. To address this issue, nanoarchitecture TMDs have been dispersed in an ionic liquid for effective conversion of CO_2 into carbon monoxide (CO). A suitable solvent with high electrocatalytic active sites results in a low overpotential reduction rate of CO_2. Asadi and co-workers initially developed three-dimensional (3D) structure bulk MoS_2 as a catalyst for the reduction of CO_2 at a low overpotential of 54 mV in an ionic liquid [75]. Later, the same research group developed nanoflake structured tungsten diselenides (WSe_2) for CO_2 reduction, which exhibited 24% faradaic efficiency (CO) with 18.95 mA cm^{-2} current density (figure 2.7(c)). The turnover frequency (TOF) measured for the WSe_2 electrocatalysts per-site was found to be 0.28 s^{-1}

(producing CO). The measured total organic carbon (TOC) was comparatively better than the measured bulk MoS_2 ($0.016\ s^{-1}$) and metallic nanoparticle Ag (zero). Moreover, the developed electrocatalysts (WSe_2) recorded a current density of 330 mA cm^{-2} at a potential of -0.76 V and the same concurrently oxidized water molecules in the absence of an external voltage. A relative current density versus potential graph for different electrocatalytic (chalcogenides) materials carried out under various potentials is provided in figure 2.7(d) [76].

Yang *et al* demonstrated the synthesis of copper selenide ($Cu_{2-x}Se_y$) nanocrystals via a solvent coordination molecular template method for the reduction of CO_2. The researchers investigated the effect of the morphology and size of the $Cu_{2-x}Se_y$ nanocrystals and reported the reaction pathway for CO_2 reduction in both the theoretical (DFT) and experimental aspects. The possible reaction pathway includes enhancement of CO_2 concentration in the ionic liquid (electrolyte), and binding of CO_2 radicals on the surface of the $Cu_{2-x}Se_y$ nanocrystal catalysts, accelerating the formation of CO species. Further, the active sites of Cu in the nanocrystals enhance the facilitation of adsorbed –CO species by rapid movement of protons and electrons to form –CHO. Based on the possible reactive intermediates for facilitating the production of methanol, a schematic illustration is provided in figure 2.8(a) [77–79]. To support the reaction pathway of the $Cu_{2-x}Se_y$ nanocrystals, DFT calculations were conducted on multiple reaction steps, as shown in figure 2.8(a). The optimized $Cu_{1.6}Se_{0.33}$ has a more stable configuration in the activity of CO_2 reduction by means of intermediate formation with *COOH. The Cu forms bonds with C and O (Cu–C and Cu–O) with lower free energy and accelerates *CO species adsorption. Further, the *CO species facilitates the moderate binding energy, benefitting the transformation of CO_2 [78, 80]. Based on the Brønsted–Evans–Polanyi (BEP) relation, the reduction of *CHO from *CO species is found to be an endothermic reaction owing to its relative energy potential (0.56 eV) [81, 82]. Hence the origin of the strong binding energy with the species of $Cu_{1.63}Se_{0.33}$ nanocatalysts yields an ~42 mA cm^{-2} current density with a Faradaic efficiency of ~78% at -2.1 V [83].

Similarly, the effects of the active site edges of the chalcogenides (MoS_2, $MoSe_2$, and Ni-doped MoS_2) in the reduction activity of CO_2 was investigated theoretically by Chen *et al.*

The results, using DFT, indicate that a relationship scale exists between the binding energies of the species (COOH, COH, CO) with various edges of chalcogenides. Mostly the CHO and COOH species form stable configurations in the orientation of (111) and (211), facilitating the adsorption of COOH and CHO to S or Se atoms and CO to the metal atom (Mo, W, etc). COOH binding at the S edge of MoS_2 with $\theta_H = 0.5$ exposes weak bonding, resulting in a poor reduction rate (figure 2.8(b)). *OCHO was considered in CO_2 reduction into formic acid but it was weaker than COOH* on all the edges and are negligible. The lowest free-energy paths concerning CH_4 reaction intermediates are via CH_2O, *CH_2OH, CH_2, and CH_3. The binding of *OCHO at the edges can effectively reduce the CO_2, as illustrated in figure 2.8(b). C–C coupling of CHO* can reduce further to produce higher-order hydrocarbons [84–86]. Thus the theoretical DFT calculation supports

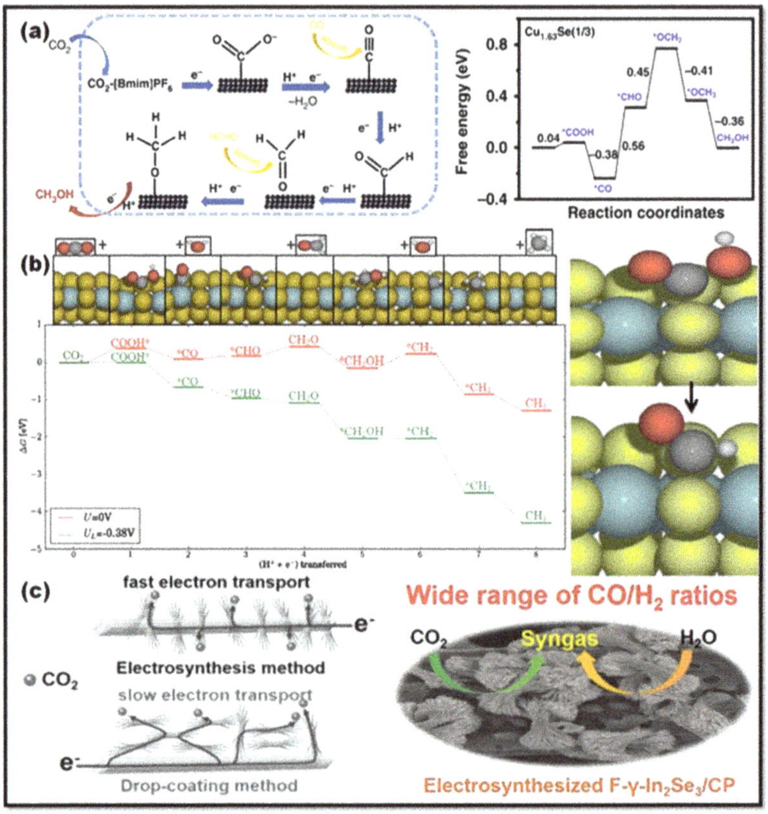

Figure 2.8. (a) Proposed CO_2 reduction mechanism of a $Cu_{2-x}Se_y$ electrode with its free-energy diagram [83]. (b) Free-energy diagram for the reduction of CO_2 to CH_4 on the Mo edge of $MoSe_2$ with its binding configurations of COOH, CHO to the Mo edge of MoS_2 through S species [87]. (c) Electron transfer schematics of F-γ-In_2Se_3 electrodes prepared via electrosynthesis and a drop-coating method with its morphological structures [88]. ((a) Reproduced with permission from [83]. Copyright 2019 Nature. (b) Reproduced with permission from [87]. Copyright 2014 Wiley. (c) Reproduced with permission from [88]. Copyright 2020 Wiley.)

the research to understand the reaction mechanism by bridging the adsorption species with edges of chalcogenides [87].

Among the various value-added products of CO_2, the production of syngas (a mixture of CO and H_2) is the most promising. However, the production of highly efficient syngas is more difficult and various nanostructured TMDs have been developed to this end. In this regard, nanoflower shaped indium selenides were coated over carbon paper (γ-In_2Se_3–CP) by an electrosynthesis and dip-coating approach, as shown in figure 2.8(c). The developed defect rich electrode (γ-In_2Se_3–CP) yields CO and H_2 by tuning its potential range in the electrochemical tests. By varying its potential, the γ-In_2Se_3–CP delivers a performance of ~90 mA cm^{-2} current density at a 1:1 ratio of CO: H_2. Moreover, the faradaic efficiency of CO reaches 96.5% at a lower overpotential of 220 mV [88, 89]. Although high faradaic efficacy, high current density, and low overpotential are achieved for novel TMDs and their composites, the practical applications of TMDs are still limited due to their poor stability, selectivity, and activation barrier, and induced defects, etc. Thus developing novel architecture TMDs addressing the issues mentioned are key aims for budding researchers interested in the field of CO_2 reduction.

2.3.2.2 Water-splitting

Water-splitting is a outstanding tool that is regarded as an 'artificial synthesis' method for harvesting renewable energy and reducing the greenhouse effect. In water-splitting by electrolysis, the water molecules break into hydrogen (H_2) and oxygen (O_2), and the production of H_2 from water is receiving great interest. H_2 fuel can be generated upon the redox reaction with the electrons/holes of water molecules during electrolytic reactions [90, 91]. In this regard, effective catalytic materials have been developed for water-splitting applications. TMDs, owing to their atomic-scale large surface catalytic active sites, were found to be excellent candidates for water-splitting applications. Good reviews exist on the commonly used TMDs, such as MoS_2 and WS_2 for water-splitting. In this section a few instances of ternary and monolayered Janus TMDs and their effects in electrolytic water-splitting are discussed.

The performance and stability of the TMDs can be enhanced by selective doping of metals, rare earths, etc. For instance, cobalt (Co) was synthesized with different doping ratios in Ni_xSe ($Co_xNi_{0.85-x}Se$) and their hydrogen evolution reaction (HER) properties were examined. Depending on the doping percentage of Co, the catalysts' chemical composition is altered, leading to tunable electrochemical properties. The synergetic effects existing between the metals Co and Ni enhance the electrocatalytic performance. Among the different doping ratios, the $Co_{0.1}Ni_{0.75}Se$ electrocatalysts exhibit better HER activity at a -10 mA cm^{-2} current density. Further, synthesized electrocatalysts with an optimized composition ($Co_{0.1}Ni_{0.75}Se$) were introduced to reduced graphene oxide (rGO) forming the hybrid architecture of $Co_{0.1}Ni_{0.75}Se$–rGO, as shown in figure 2.9(a). rGO assists in preventing the aggregation of $Co_{0.1}Ni_{0.75}Se$ during the electrolytic reaction, leading to more active sites and exposure to deliver higher HER performances at a low overpotential of 103 mV (figure 2.9(a)). The overpotential recorded for the hybrid electrocatalysts ($Co_{0.1}Ni_{0.75}Se$–rGO) was found to equivalent to the conventional Pt–C electrode [92–95].

The electronic band structure of the catalysts plays a crucial role in water-splitting applications. A detailed investigation of the electronic transition of TMDs (Ba_2ZnSe_3) was carried out with the aid of DFT. The highly symmetric Ba_2ZnSe_3 explores the band dispersion with a considerable anisotropic optical response, supporting the occurrence of adsorption edges in the visible region. The optical bandgap of the TMDs was estimated to be ~2.70 eV and the corresponding EPc and EPv electrode potentials were -0.145 V and $+2.605$ (versus normal hydrogen electrode), respectively. The conduction (CB) and valence (VB) band positions of the Ba_2ZnSe_3 facilitates in accommodating a redox capacity with high reduction potential at the CB position. The effective mass difference between holes and

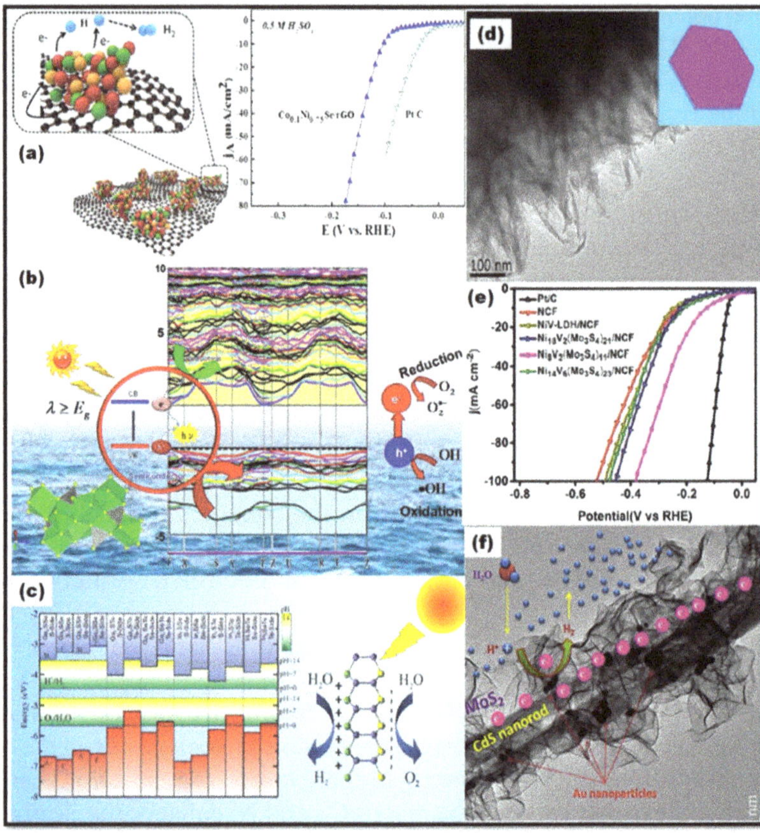

Figure 2.9. (a) Schematic illustration of $Co_xNi_{0.85-x}Se$ electrocatalysts with their corresponding LSV curve for water-splitting [92]. (b) Schematics with the calculated partial density of states for Ba_2ZnSe_3 chalcogenides [98]. (c) Illustration of the band edges for all Janus materials with a redox potential of water with respect to the pH value [99]. (d) and (e) morphology and LSV curve of $Ni_8V_2(Mo_3S_4)_{11}$ electrocatalysts at 1.0 M KOH [100]. (f) High-resolution TEM micrograph of CdS–Au/MoS_2 electrocatalysts [101]. ((a) Reproduced with permission from [92]. Copyright 2018 American Chemical Society. (b) Reproduced with permission from [98]. Copyright 2018 Elsevier. (c) Reproduced with permission from [99]. Copyright 2019 American Chemical Society. (d) and (e) Reproduced with permission from [100]. Copyright 2019 American Chemical Society. (f) Reproduced with permission from [101]. Copyright 2018 American Chemical Society.)

electrons boosts the carrier mobility, easing the migration of electrons/holes leading to enhanced photoexcitation performance. At a zero set condition, the maximum VB and minimum CB positions result in an optical bandgap of 2.7 eV, causing an increase in participation in the photocatalysis reaction [96, 97]. Moreover, the presence of $BaSe_7$ polyhedra and ZnO_4 tetrahedra groups gives the TMDs a large surface to volume ratio that promotes high photon absorption. In the water-splitting scheme, the Ba_2ZnSe_3 oxygen is absorbed on the surface of the catalytic species forming a superoxide (due to the electron) and meanwhile holes support generating hydroxyl. A schematic illustration of the photocatalytic and charge transfer mechanism of Ba_2ZnSe_3 is shown in figure 2.9(b) [98].

Janus monolayers identical to pristine structures composed of TMDs, such as Ga_2XY and In_2XY (X and Y are S, Se, or Te), exhibit stable mechanical properties. The electronic, structural, and catalytic properties are investigated using DFT. A theoretical study reveals the band alignment and electronic structure of TMDs in photocatalytic water-splitting activity. In this report, the Janus monolayer is characterized by an electric dipole moment by locating different chalcogen atoms on adjacent sides. Depending on the size of the chalcogen atoms, the electron band alignment can be tuned, and most of the Janus crystals exhibit direct bandgap energy (except for Ga_2SSe). Figure 2.9(c) shows the band edge alignments of different TMDs over the pH range of 0 to 14 at the water's redox potentials. It is worth noting that most of the different compositions of TMDs deliver good photocatalytic activity in reducing water molecules. The pH variation carried out in the study suggests the green–yellow (figure 2.9(c)) regime provides better redox potential. Finally, the internal intrinsic electric field can be tuned based on the size resulting in the improvement of exciton dissociation promoting electron–hole separation, ensuring enhanced activity in water-splitting reactions [99].

Recently, metal-free electrocatalysts have been developed for next-generation practical water electrolysis applications. The developed materials should have robust preparation and abundance, and be environmentally friendly and cost-effective. In this regard, 3D nanostructured nickel–vanadium (Ni–V) double hydroxide nanosheets were synthesized and exchanged with $(MoS_4)^{2-}$ ions to form Ni–VMoS. The nanosheet-like structures (figure 2.9(d)) of the $Ni_8V_2(Mo_3S_4)_{11}$ with low adsorption free-energy benefit rapid electronic transport movement reducing the water molecules at an overpotential of 129 mV (HER) and 330 mV (oxygen evolution reaction) at a current density of 10 mA cm^{-2} (figure 2.9(e)). The amount of O_2 and H_2 generated in an alkaline medium are about 342 μmol h^{-1} and 517 μmol h^{-1}, which are comparatively better than non-noble electrodes. Moreover, the developed electrode performs prolonged life cycling stability in both the OER and HER conditions [100]. Likewise, several novel catalyst materials such as CdS–Au–MoS_2 composites (figure 2.9(f)), MnS–In_2S_3, CdS–CdTe heterodimer, $Co_3FeS_{1.5}(OH)_6$ hydroxychalcogenides, etc, are being developed for water-splitting applications [101–104]. The unique properties of the chalcogenides show promising performance in water-splitting. There are some drawbacks to TMDs, such as poor stability, conductivity, and activity, limiting the initial scale of applications. The synthesis of stable and active chalcogenides remains a big challenge [104–107].

2.4 Summary

In summary, the superior properties of the chalcogenides, such as layered regulated structure, atomic-scale arrangement, electronic transition, ideal electrical transport, etc, make them fascinating materials for a diverse field of applications. Highly active sites with more stable configurations of chalcogenides can be synthesized via various approaches. Additionally, the nano-engineering of materials (hybrid/composites) opens the way for the development of novel functional catalytic materials. However, achieving controlled architectures of chalcogenides with different components remains challenging. Recently, a huge volume of research has been devoted towards microstructure regulation (chalcogenides) on the atomic scale for energy and environmental remedial applications. In order to produce more stable device topologies, high active site chalcogenides with increased atomic efficiency should be developed. This chapter summarizes the impact of 2D chalcogenides in energy (conversion, storage, and conservation) and environmental applications with a few examples. Chalcogenides exhibit outstanding performance in energy and environmental applications but real-time applications require a sustainable environment, thus requiring future research focus on the long-term stability and durability of the chalcogenides.

Acknowledgement

The authors A A and R V M gratefully acknowledge FONDECYT Post-doctoral Project No. 3200076, Universidad Adolfo Ibanez, Chile, Santiago, for financial assistance. N T thanks his supervisor Dr N Sridewi for constant support and encouragement in his research activities. R C acknowledges the University Grant Commission (UGC) India for a fellowship.

References

[1] Whittingham M S 1978 Chemistry of intercalation compounds: metal guests in chalcogenide hosts *Prog. Solid State Chem.* **12** 41–99

[2] Bouroushian M 2010 *Electrochemistry of Metal Chalcogenides* (Berlin: Springer)

[3] Choi W, Choudhary N, Han G H, Park J, Akinwande D and Lee Y H 2017 Recent development of two-dimensional transition metal dichalcogenides and their applications *Mater. Today* **20** 116–30

[4] Marseglia E A 1983 Transition metal dichalcogenides and their intercalates *Int. Rev. Phys. Chem.* **3** 177–216

[5] Xia F, Wang H, Xiao D, Dubey M and Ramasubramaniam A 2014 Two-dimensional material nanophotonics *Nat. Photonics* **8** 899–907

[6] Greenwood N N and Earnshaw A 1997 *Chemistry of the Elements* (Amsterdam: Elsevier) pp 1–18

[7] Manna L, Scher E C and Alivisatos A P 2000 Synthesis of soluble and processable rod-, arrow-, teardrop-, and tetrapod-shaped CdSe nanocrystals *J. Am. Chem. Soc.* **122** 12700–6

[8] Gao M-R, Xu Y-F, Jiang J and Yu S-H 2013 Nanostructured metal chalcogenides: synthesis, modification, and applications in energy conversion and storage devices *Chem. Soc. Rev.* **42** 2986

[9] He Z, Yu S-H, Zhou X, Li X and Qu J 2006 Magnetic-field-induced phase-selective synthesis of ferrosulfide microrods by a hydrothermal process: microstructure control and magnetic properties *Adv. Funct. Mater.* **16** 1105–11

[10] Yao W-T, Yu S-H and Wu Q-S 2007 From mesostructured wurtzite ZnS-nanowire/amine nanocomposites to ZnS nanowires exhibiting quantum size effects: a mild-solution chemistry approach *Adv. Funct. Mater.* **17** 623–31

[11] Du W, Zhu J, Li S and Qian X 2008 Ultrathin β-In_2S_3 nanobelts: shape-controlled synthesis and optical and photocatalytic properties *Cryst. Growth Des.* **8** 2130–6

[12] Hwang H, Kim H and Cho J 2011 MoS_2 nanoplates consisting of disordered graphene-like layers for high rate lithium battery anode materials *Nano Lett.* **11** 4826–30
[13] Wu C, Yu S-H and Antonietti M 2006 Complex concaved cuboctahedrons of copper sulfide crystals with highly geometrical symmetry created by a solution process *Chem. Mater.* **18** 3599–601
[14] Yang S-L, Yao H-B, Gao M-R and Yu S-H 2009 Monodisperse cubic pyrite NiS_2 dodecahedrons and microspheres synthesized by a solvothermal process in a mixed solvent: thermal stability and magnetic properties *CrystEngComm.* **11** 1383
[15] Gao M-R, Lin Z-Y, Jiang J, Yao H-B, Lu Y-M, Gao Q, Yao W-T and Yu S-H 2011 Selective synthesis of Fe_7Se_8 polyhedra with exposed high-index facets and Fe_7Se_8 nanorods by a solvothermal process in a binary solution and their collective intrinsic properties *Chem. —Eur. J.* **17** 5068–75
[16] Wang X and Li Y 2006 Solution-based synthetic strategies for 1-D nanostructures *Inorg. Chem.* **45** 7522–34
[17] Lai C-H, Huang K-W, Cheng J-H, Lee C-Y, Lee W-F, Huang C-T, Hwang B-J and Chen L-J 2009 Oriented growth of large-scale nickel sulfide nanowire arrays via a general solution route for lithium-ion battery cathode applications *J. Mater. Chem.* **19** 7277
[18] Nayak P K and Cahen D 2014 Real world efficiency limits: the Shockley–Queisser model as a starting point *RSC Energy and Environment Series* (Cambridge: Royal Society of Chemistry) ch 17 pp 547–66
[19] Matthews P D, Akhtar M, Malik M A, Revaprasadu N and O'Brien P 2016 Synthetic routes to iron chalcogenide nanoparticles and thin films *Dalt. Trans.* **45** 18803–12
[20] Shukla S *et al* 2016 Origin of photocarrier losses in iron pyrite (FeS_2) nanocubes *ACS Nano* **10** 4431–40
[21] Omata T, Nose K and Otsuka-Yao-Matsuo S 2009 Size dependent optical band gap of ternary I–III–VI_2 semiconductor nanocrystals *J. Appl. Phys.* **105** 073106
[22] Le Donne A, Trifiletti V and Binetti S 2019 New earth-abundant thin film solar cells based on chalcogenides *Front. Chem.* **7** 297
[23] Santoni A, Biccari F, Malerba C, Valentini M, Chierchia R and Mittiga A 2013 Valence band offset at the CdS/Cu_2ZnSnS_4 interface probed by x-ray photoelectron spectroscopy *J. Phys. D: Appl. Phys.* **46** 175101
[24] Kumar M, Dubey A, Adhikari N, Venkatesan S and Qiao Q 2015 Strategic review of secondary phases, defects and defect-complexes in kesterite CZTS–Se solar cells *Energy Environ. Sci.* **8** 3134–59
[25] Arulraj A, Ramesh M, Subramanian B and Senguttuvan G 2018 *In-situ* temperature and thickness control grown 2D-MoS_2 via pulsed laser ablation for photovoltaic devices *Sol. Energy* **174** 286–95
[26] Arulraj A and Ramesh M 2018 Rational strategies for large-area perovskite solar cells *Rational Design of Solar Cells for Efficient Solar Energy Conversion* (Hoboken, NJ: Wiley) pp 307–37
[27] Arulraj A, Subramanian B, Ramesh M and Senguttuvan G 2019 Effect of active sites in pulsed laser deposited bimetallic $NiMoS_2$ thin films for solar energy conversion *Mater. Lett.* **241** 132–5
[28] Yang Z, Wang M, Shukla S, Zhu Y, Deng J, Ge H, Wang X and Xiong Q 2015 Developing seedless growth of ZnO micro/nanowire arrays towards ZnO/FeS_2/CuI P–I–N photodiode application *Sci Rep.* **5** 11377

[29] Wang Y-C, Wang D-Y, Jiang Y-T, Chen H-A, Chen C-C, Ho K-C, Chou H-L and Chen C-W 2013 FeS_2 nanocrystal ink as a catalytic electrode for dye-sensitized solar cells *Angew. Chemie Int. Ed.* **52** 6694–8

[30] Wang D-Y, Jiang Y-T, Lin C-C, Li S-S, Wang Y-T, Chen C-C and Chen C-W 2012 Solution-processable pyrite FeS_2 nanocrystals for the fabrication of heterojunction photodiodes with visible to NIR photodetection *Adv. Mater.* **24** 3415–20

[31] Li Y, Wang H, Zhang H, Liu P, Wang Y, Fang W, Yang H, Li Y and Zhao H 2014 A {0001} faceted single crystal NiS nanosheet electrocatalyst for dye-sensitised solar cells: sulfur-vacancy induced electrocatalytic activity *Chem. Commun.* **50** 5569

[32] Li F, Wang J, Zheng L, Zhao Y, Huang N, Sun P, Fang L, Wang L and Sun X 2018 *In situ* preparation of NiS_2/CoS_2 composite electrocatalytic materials on conductive glass substrates with electronic modulation for high-performance counter electrodes of dye-sensitized solar cells *J. Power Sources.* **384** 1–9

[33] Su A-L, Lu M-N, Chang C-Y, Wei T-C and Lin J-Y 2016 Scalable fabrication of efficient $NiCo_2S_4$ counter electrodes for dye-sensitized solar cells using a facile solution approach *Electrochim. Acta* **222** 1410–6

[34] Karthick R, Arulraj A, Ramesh M and Selvaraj M 2018 Free-standing graphene/NiMoS paper as cathode for quasi-solid state dye-sensitized solar cells *J. Colloid Interface Sci.* **530** 179–88

[35] Xiao J, Wan L, Yang S, Xiao F and Wang S 2014 Design hierarchical electrodes with highly conductive $NiCo_2S_4$ nanotube arrays grown on carbon fiber paper for high-performance pseudocapacitors *Nano Lett.* **14** 831–8

[36] Arulraj A and Nishanthi S T 2020 2D materials-based flexible supercapacitors for high energy storage devices *Nanostructured, Functional, and Flexible Materials for Energy Conversion and Storage Systems* (Amsterdam: Elsevier) pp 417–36

[37] Arulraj A, Ilayaraja N, Rajeshkumar V and Ramesh M 2019 Direct synthesis of cubic shaped Ag_2S on Ni mesh as binder-free electrodes for energy storage applications *Sci Rep.* **9** 10108

[38] Abu Dakka Y, Balamurugan J, Balaji R, Kim N H and Lee J H 2020 Advanced $Cu_{0.5}Co_{0.5}Se_2$ nanosheets and MXene electrodes for high-performance asymmetric supercapacitors *Chem. Eng. J.* **385** 123455

[39] Nguyen T T, Balamurugan J, Aravindan V, Kim N H and Lee J H 2019 Boosting the energy density of flexible solid-state supercapacitors via both ternary NiV_2Se_4 and $NiFe_2Se_4$ nanosheet arrays *Chem. Mater.* **31** 4490–504

[40] Balamurugan J, Nguyen T T, Aravindan V, Kim N H, Lee S H and Lee J H 2019 All ternary metal selenide nanostructures for high energy flexible charge storage devices *Nano Energy* **65** 103999

[41] Chebrolu V T, Balakrishnan B, Chinnadurai D and Kim H 2020 Selective growth of Zn–Co–Se nanostructures on various conductive substrates for asymmetric flexible hybrid supercapacitor with enhanced performance *Adv. Mater. Technol.* **5** 1900873

[42] Xia C, Jiang Q, Zhao C, Beaujuge P M and Alshareef H N 2016 Asymmetric supercapacitors with metal-like ternary selenides and porous graphene electrodes *Nano Energy.* **24** 78–86

[43] Yoo J J *et al* 2011 Ultrathin planar graphene supercapacitors *Nano Lett.* **11** 1423–7

[44] Feng N, Meng R, Zu L, Feng Y, Peng C, Huang J, Liu G, Chen B and Yang J 2019 A polymer-direct-intercalation strategy for MoS_2/carbon-derived heteroaerogels with ultrahigh pseudocapacitance *Nat. Commun.* **10** 1372

[45] Sellam A, Jenjeti R N and Sampath S 2018 Ultrahigh-rate supercapacitors based on 2-dimensional, 1T $MoS_{2x}Se_{2(1-x)}$ for AC line-filtering applications *J. Phys. Chem.* C **122** 14186–94
[46] Kanade C, Arbuj S, Kanade K, Kim K S, Yeom G Y, Kim T and Kale B 2018 Hierarchical nanostructures of nitrogen-doped molybdenum sulphide for supercapacitors *RSC Adv.* **8** 39749–55
[47] Pan U N, Sharma V, Kshetri T, Singh T I, Paudel D R, Kim N H and Lee J H 2020 Freestanding 1T-$Mn_xMo_{1-x}S_{2-y}Se_y$ and $MoFe_2S_{4-z}Se_z$ ultrathin nanosheet-structured electrodes for highly efficient flexible solid-state asymmetric supercapacitors *Small* **16** 2001691
[48] Holmberg K, Andersson P and Erdemir A 2012 Global energy consumption due to friction in passenger cars *Tribol. Int.* **47** 221–34
[49] Holmberg K, Andersson P, Nylund N-O, Mäkelä K and Erdemir A 2014 Global energy consumption due to friction in trucks and buses *Tribol. Int.* **78** 94–114
[50] Zheng J, Zhou Z, Zhang J, Li H and Yu H 2003 On the friction and wear behaviour of human tooth enamel and dentin *Wear* **255** 967–74
[51] Carvalho O, Buciumeanu M, Madeira S, Soares D, Silva F S and Miranda G 2015 Optimization of AlSi–CNTs functionally graded material composites for engine piston rings *Mater. Des.* **80** 163–73
[52] Demirci M T and Düzcükoğlu H 2014 Wear behaviors of polytetrafluoroethylene and glass fiber reinforced polyamide 66 journal bearings *Mater. Des.* **57** 560–7
[53] Xiao H, Kim S, He X, Zhou D, Li C and Liang H 2014 Friction pair evaluation of cartilage–diamond for partial joint repair *Carbon NY* **80** 551–9
[54] Vattikuti S V P and Byon C 2015 Synthesis and characterization of molybdenum disulfide nanoflowers and nanosheets: nanotribology *J. Nanomater.* **2015** 1–11
[55] Kalin M, Kogovšek J and Remškar M 2012 Mechanisms and improvements in the friction and wear behavior using MoS_2 nanotubes as potential oil additives *Wear* **280–81** 36–45
[56] Xiao H and Liu S 2017 2D nanomaterials as lubricant additive: a review *Mater. Des.* **135** 319–32
[57] Ratoi M, Niste V B, Walker J and Zekonyte J 2013 Mechanism of action of WS_2 lubricant nanoadditives in high-pressure contacts *Tribol. Lett.* **52** 81–91
[58] Lahouij I, Vacher B, Martin J-M and Dassenoy F 2012 IF-MoS_2 based lubricants: Influence of size, shape and crystal structure *Wear* **296** 558–67
[59] Kogovšek J and Kalin M 2014 Various MoS_2-, WS_2- and C-based micro- and nanoparticles in boundary lubrication *Tribol. Lett.* **53** 585–97
[60] Rabaso P, Ville F, Dassenoy F, Diaby M, Afanasiev P, Cavoret J, Vacher B and Le Mogne T 2014 Boundary lubrication: influence of the size and structure of inorganic fullerene-like MoS_2 nanoparticles on friction and wear reduction *Wear* **320** 161–78
[61] Shi X, Song S, Zhai W, Wang M, Xu Z, Yao J, Qamarud Din A and Zhang Q 2014 Tribological behavior of Ni_3Al matrix self-lubricating composites containing WS_2, Ag and hBN tested from room temperature to 800 °C *Mater. Des.* **55** 75–84
[62] Wang F, Shifa T A, Zhan X, Huang Y, Liu K, Cheng Z, Jiang C and He J 2015 Recent advances in transition-metal dichalcogenide based nanomaterials for water splitting *Nanoscale.* **7** 19764–88
[63] Peng W, Li Y, Zhang F, Zhang G and Fan X 2017 Roles of two-dimensional transition metal dichalcogenides as cocatalysts in photocatalytic hydrogen evolution and environmental remediation *Ind. Eng. Chem. Res.* **56** 4611–26

[64] Nagarajan L, Sampath kumar P, Arulraj A, Senguttuvan G and Kumaraguru K 2020 Solar assisted reduced graphene oxide as adsorbent for carbon dioxide and its kinetic studies *Physica* E **116** 113739

[65] Chen Y, Sun H and Peng W 2017 2D Transition metal dichalcogenides and graphene-based ternary composites for photocatalytic hydrogen evolution and pollutants degradation *Nanomaterials* **7** 62

[66] Hiragond C B, Kim H, Lee J, Sorcar S, Erkey C and In S-I 2020 Electrochemical CO_2 reduction to CO catalyzed by 2D nanostructures *Catalysts* **10** 98

[67] Gao D, Zhou H, Wang J, Miao S, Yang F, Wang G, Wang J and Bao X 2015 Size-dependent electrocatalytic reduction of CO_2 over Pd nanoparticles *J. Am. Chem. Soc.* **137** 4288–91

[68] Mistry H, Reske R, Zeng Z, Zhao Z-J, Greeley J, Strasser P and Cuenya B R 2014 Exceptional size-dependent activity enhancement in the electroreduction of CO_2 over Au nanoparticles *J. Am. Chem. Soc.* **136** 16473–6

[69] Reske R, Mistry H, Behafarid F, Roldan Cuenya B and Strasser P 2014 Particle size effects in the catalytic electroreduction of CO_2 on Cu nanoparticles *J. Am. Chem. Soc.* **136** 6978–86

[70] Zhu W, Michalsky R, Metin Ö, Lv H, Guo S, Wright C J, Sun X, Peterson A A and Sun S 2013 Monodisperse Au nanoparticles for selective electrocatalytic reduction of CO_2 to CO *J. Am. Chem. Soc.* **135** 16833–6

[71] Lee G-J, Hou Y-H, Huang H-T, Wang W, Lyu C and Wu J J 2020 Microwave-assisted solvothermal synthesis of chalcogenide composite photocatalyst and its photocatalytic CO_2 reduction activity under simulated solar light *Catalysts* **10** 789

[72] Gerschel P, Battistella B, Siegmund D, Ray K and Apfel U-P 2020 Electrochemical CO_2 reduction—the effect of chalcogenide exchange in Ni–isocyclam complexes *Organometallics* **39** 1497–510

[73] Collison D, Helliwell M, Jones V M, Mabbs F E, McInnes E J L, Riedi P C, Smith G M, Pritchard R G and Cross W I 1998 Single and double quantum transitions in the multi-frequency continuous wave electron paramagnetic resonance (cwEPR) of three six co-ordinate nickel(II) complexes: $[Ni(EtL)_2(Me_5dien)]$ and $[Ni(5\text{-methylpyrazole})_6]X_2$, $X = (ClO_4)^-$ or $(BF_4)^-$. The single crystal x-ray structure at room temperature of $[Ni(5\text{-methylpyrazole})_6](ClO_4)_2$ *J. Chem. Soc. Faraday Trans.* **94** 3019–25

[74] Krzystek J, Yeagle G J, Park J-H, Britt R D, Meisel M W, Brunel L-C and Telser J 2003 High-frequency and -field EPR spectroscopy of tris(2,4-pentanedionato)manganese(III): investigation of solid-state versus solution Jahn–Teller effects *Inorg. Chem.* **42** 4610–8

[75] Asadi M *et al* 2014 Robust carbon dioxide reduction on molybdenum disulphide edges *Nat. Commun.* **5** 4470

[76] Asadi M *et al* 2016 Nanostructured transition metal dichalcogenide electrocatalysts for CO_2 reduction in ionic liquid *Science* **353** 467–70

[77] Cadena C *et al* 2004 Why is CO_2So soluble in imidazolium-based ionic liquids? *J. Am. Chem. Soc.* **126** 5300–8

[78] Kuhl K P, Cave E R, Abram D N and Jaramillo T F 2012 New insights into the electrochemical reduction of carbon dioxide on metallic copper surfaces *Energy Environ. Sci.* **5** 7050

[79] Kuhl K P, Hatsukade T, Cave E R, Abram D N, Kibsgaard J and Jaramillo T F 2014 Electrocatalytic conversion of carbon dioxide to methane and methanol on transition metal surfaces *J. Am. Chem. Soc.* **136** 14107–13

[80] Calle-Vallejo F and Koper M T M 2013 Theoretical considerations on the electroreduction of CO to C_2 species on Cu(100) electrodes *Angew. Chemie Int. Ed.* **52** 7282–5
[81] Cheng J, Hu P, Ellis P, French S, Kelly G and Lok C M 2008 Brønsted–Evans–Polanyi relation of multistep reactions and volcano curve in heterogeneous catalysis *J. Phys. Chem.* C **112** 1308–11
[82] Evans M G and Polanyi M 1938 Inertia and driving force of chemical reactions *Trans. Faraday Soc.* **34** 11
[83] Yang D, Zhu Q, Chen C, Liu H, Liu Z, Zhao Z, Zhang X, Liu S and Han B 2019 Selective electroreduction of carbon dioxide to methanol on copper selenide nanocatalysts *Nat. Commun.* **10** 677
[84] Montoya J H, Peterson A A and Nørskov J K 2013 Insights into C–C coupling in CO_2 electroreduction on copper electrodes *ChemCatChem* **5** 737–42
[85] Varley J B and Nørskov J K 2013 First-principles calculations of Fischer–Tropsch processes catalyzed by nitrogenase enzymes *ChemCatChem* **5** 732–6
[86] Calle-Vallejo F and Koper M T M 2013 Theoretical considerations on the electroreduction of CO to C_2 species on Cu(100) electrodes *Angew. Chem.* **125** 7423–6
[87] Chan K, Tsai C, Hansen H A and Nørskov J K 2014 Molybdenum sulfides and selenides as possible electrocatalysts for CO_2 reduction *ChemCatChem* **6** 1899–905
[88] Yang D, Zhu Q, Sun X, Chen C, Guo W, Yang G and Han B 2020 Electrosynthesis of a defective indium selenide with 3D structure on a substrate for tunable CO_2 electroreduction to syngas *Angew. Chemie.* **132** 2374–9
[89] Zhou Y *et al* 2018 Dopant-induced electron localization drives CO_2 reduction to C_2 hydrocarbons *Nat. Chem.* **10** 974–80
[90] Abe R 2010 Recent progress on photocatalytic and photoelectrochemical water splitting under visible light irradiation *J. Photochem. Photobiol.* C **11** 179–209
[91] Caravaca A, Jones W, Hardacre C and Bowker M 2016 H_2 production by the photocatalytic reforming of cellulose and raw biomass using Ni, Pd, Pt and Au on titania *Proc. R. Soc.* A **472** 0054
[92] Zhao W, Wang S, Feng C, Wu H, Zhang L and Zhang J 2018 Novel cobalt-doped $Ni_{0.85}$ Se chalcogenides ($Co_xNi_{0.85-x}Se$) as high active and stable electrocatalysts for hydrogen evolution reaction in electrolysis water splitting *ACS Appl. Mater. Interfaces* **10** 40491–9
[93] Zeng M and Li Y 2015 Recent advances in heterogeneous electrocatalysts for the hydrogen evolution reaction *J. Mater. Chem.* A **3** 14942–62
[94] Zhou W, Hou D, Sang Y, Yao S, Zhou J, Li G, Li L, Liu H and Chen S 2014 MoO_2 nanobelts@nitrogen self-doped MoS_2 nanosheets as effective electrocatalysts for hydrogen evolution reaction *J. Mater. Chem.* A **2** 11358
[95] Shi Y and Zhang B 2016 Recent advances in transition metal phosphide nanomaterials: synthesis and applications in hydrogen evolution reaction *Chem. Soc. Rev.* **45** 1529–41
[96] Zhou M, Xiao K, Jiang X, Huang H, Lin Z, Yao J and Wu Y 2016 Visible-light-responsive chalcogenide photocatalyst Ba_2ZnSe_3: crystal and electronic structure, thermal, optical, and photocatalytic activity *Inorg. Chem.* **55** 12783–90
[97] Zhou P, Wu J, Yu W, Zhao G, Fang G and Cao S 2014 Vectorial doping-promoting charge transfer in anatase TiO_2 {001} surface *Appl. Surf. Sci.* **319** 167–72
[98] Reshak A H 2018 Active photocatalytic water splitting solar-to-hydrogen energy conversion: chalcogenide photocatalyst Ba_2ZnSe_3 under visible irradiation *Appl. Catal.* B **221** 17–26

[99] da Silva R, Barbosa R, Mançano R R, Durães N, Pontes R B, Miwa R H, Fazzio A and Padilha J E 2019 Metal chalcogenides Janus monolayers for efficient hydrogen generation by photocatalytic water splitting *ACS Appl. Nano Mater.* **2** 890–7
[100] Fan H, Huang J, Chen G, Chen W, Zhang R, Chu S, Wang X, Li C and Ostrikov K K 2019 Hollow Ni–V–Mo chalcogenide nanopetals as bifunctional electrocatalyst for overall water splitting *ACS Sustain. Chem. Eng.* **7** 1622–32
[101] Chava R K, Do J Y and Kang M 2018 Smart hybridization of Au coupled CdS nanorods with few layered MoS_2 nanosheets for high performance photocatalytic hydrogen evolution reaction *ACS Sustain. Chem. Eng.* **6** 6445–57
[102] Dan M, Zhang Q, Yu S, Prakash A, Lin Y and Zhou Y 2017 Noble-metal-free MnS/In_2S_3 composite as highly efficient visible light driven photocatalyst for H_2 production from H_2S *Appl. Catal.* B **217** 530–9
[103] Kim J, Jun M, Choi S, Jo J and Lee K 2019 Reactive nanotemplates for synthesis of highly efficient electrocatalysts: beyond simple morphology transfer *Nanoscale.* **11** 20392–410
[104] Wang B, Tang C, Wang H, Li B, Cui X and Zhang Q 2018 Anion-regulated hydroxysulfide monoliths as OER/ORR/HER electrocatalysts and their applications in self-powered electrochemical water splitting *Small Methods.* **2** 1800055
[105] Gao M-R, Liang J-X, Zheng Y-R, Xu Y-F, Jiang J, Gao Q, Li J and Yu S-H 2015 An efficient molybdenum disulfide/cobalt diselenide hybrid catalyst for electrochemical hydrogen generation *Nat. Commun.* **6** 5982
[106] Perkins F K, Friedman A L, Cobas E, Campbell P M, Jernigan G G and Jonker B T 2013 Chemical vapor sensing with monolayer MoS_2 *Nano Lett.* **13** 668–73
[107] Ahmadi M, Zabihi O, Jeon S, Yoonessi M, Dasari A, Ramakrishna S and Naebe M 2020 2D transition metal dichalcogenide nanomaterials: advances, opportunities, and challenges in multi-functional polymer nanocomposites *J. Mater. Chem.* A **8** 845–83

Chapter 3

Two-dimensional nanolayers for wearable supercapacitors

Neetu Talreja, Divya Chauhan and R V Mangalaraja

Two-dimensional (2D) nanolayered materials are gaining the interest of researchers after instructive research into graphene that makes innovative next-generation 2D materials. Researchers continue to pay attention to 2D materials because of their astonishing characteristics, mainly their large electron-active sites, high conductivity, specificity, and relatively good cyclic stability, thereby making them potential candidates for electrode materials and energy storage devices. Moreover, 2D nanolayered materials can be incorporated with other nanomaterials and polymers to produce numerous 2D composite materials, such as transition metal hydroxide (TMH), transition metal oxide (TMO), transition metal sulfide (TMS), and transition metal phosphide (TMP) based composites. Currently, these 2D nanolayered materials are fragile and easily experience structural damage during their use in wearable or skin-based devices. In this context, flexible 2D electrode materials might resolve such issues associated with commercially used electrode materials. Recently, the demand for flexible/wearable electrode materials has been increasing continuously and researchers have developed next-generation flexible supercapacitors. This chapter focuses on the numerous 2D nanolayer based materials, 2D nanolayer based composites, and flexible or wearable 2D nanolayer based energy storage devices. The challenges and future prospects of 2D nanolayer based energy storage devices will also be discussed.

3.1 Introduction

Currently, the requirements for energy in daily life are increasing continuously with the increase in the human population as well as industries. Numerous nanomaterials, including carbon-based materials, metal oxides, metal hydroxides, carbon nanotubes (CNTs), graphene, activated carbon, carbon nanofibers (CNFs), and

doi:10.1088/978-0-7503-5143-0ch3

conducting polymers, have been used for the production of electrodes for sensor, battery, and supercapacitor applications and energy storage devices. Such electrode materials for energy devices do not completely fulfill the requirements from the human or industrial perspectives due to their low cyclic stability, low performance, low energy density, and relatively low capacitance (Ashfaq *et al* 2018, Afreen *et al* 2018, Talreja *et al* 2020, Saha *et al* 2018, Chen and Dai 2013). In this perspective, researchers continue to focus on newer materials that more easily fulfill the social requirements for devices such as electric vehicles and stationary batteries.

Noteworthy investigation has been completed in the search for two-dimensional (2D) materials after the initial synthesis of graphene from graphite. The 2D nanolayered materials exhibit various characteristics, such as physical, chemical, optical, and electronic properties, and are applied in numerous areas such as drug delivery systems, agriculture, sensors, and energy devices. Numerous 2D nanolayered structures such as transition metal dichalcogenides (TMDs) (e.g. MoS_2, $MoSe_2$, WSe_2, WS_2, TiS_2, and TaS_2), hexagonal-boron nitride (h-BN), graphitic carbon-nitride (g-C_3N_4), layered double hydroxides (LDHs), layered metal oxides (LMOs), graphene, MXenes, metal nitrides, and black phosphorus (BP) have been used extensively in battery and supercapacitor applications (Kumar *et al* 2018, Zhang 2015, Tan *et al* 2017, Hu *et al* 2019, Carvalho *et al* 2020). These 2D nanolayered materials may be appropriate for electric double-layer capacitance (EDLC) because of their large surface area, and reactive edges and basal planes. The reactive edges or functional sites could be produced easily by using doping and chemical modification processes. 2D nanolayered materials act with intercalation pseudocapacitance due to their ions' intercalation ability. Moreover, 2D nanolayers have the potential ability to store energy in the 2D channels in between nanosheets using ion adsorption mechanisms that permit rapid expansion, intercalation of water molecules, and contraction of multi-layered 2D materials. The 2D nanolayered materials exhibit elevated volumetric capacitance because of their high packing density, sub-nanometer thickness, high mechanical strength, and high flexibility, that are necessary for supercapacitors/micro-supercapacitors. Additionally, electrical conductivity is one of the foremost characteristics required to store energy. Interestingly, some of the 2D nanolayered materials inherently possess high electrical conductivity that can be used easily in the development of electrodes without using electrically conducting additives. Usually, the conductivity of materials depends on the structures of the 2D nanolayer, which in turn depends on the crystal structure and many atomic layers (Da *et al* 2019, Ke and Wang 2016, Ogata *et al* 2017, Mendoza-Sánchez and Gogotsi 2016, Yuan *et al* 2020, Tao *et al* 2017, Sun *et al* 2018). However, researchers are still focusing on improving the conductivity of 2D nanolayers for high performing supercapacitors.

Numerous processes have been used to alter or tune the conductivity of 2D nanolayered materials, including surface functionalization by chemicals, metals, and non-metals. Despite the tremendous success of 2D nanolayered materials in energy storage applications, there remain some issues. One of the foremost challenges in developing high-performance 2D nanolayered materials for energy storage devices involves the restacking of nanolayers or sheets that prevents the complete utilization of the surface area and blocks access to electrolytes (Li *et al* 2020b, Li and Wang 2020, Khan *et al* 2019, Gui *et al* 2019). Such issues associated with 2D nanolayered materials are easily resolved by tailoring the morphology or structure of 2D nanolayered materials. Different synthesis processes have been developed to produce different shapes of nanosheets or nanolayers other than flat. The different shapes (such as curved or crumpled) of nanolayered structures generate porosity upon deposition and the current collectors' movement of ions is improved. The 2D nanolayered materials might be synthesized straight onto the current collectors in an upright direction, leaving appropriate channels for ion transport. In this context, numerous amalgamation processes have been applied to graphene that increase the porosity on the nanolayer or nanosheets. These pores might be beneficial for the movement of ions across the nanolayer or nanosheets. The improved movement of ions might lead to an improved performance for the supercapacitor. Currently, numerous materials including 2D nanolayer based electrode materials are used efficiently in various applications, including wearable devices (Mendoza-Sánchez and Gogotsi 2016, El Rouby 2015, Chen *et al* 2017, Kang *et al* 2019, Ishihara *et al* 2016, Li and Östling 2013). However, the fragile nature of these electrode materials means they easily develop structural deformities during operation, particularly in wearable devices. In this context, flexible and wearable electrode materials might resolve such issues associated with fragile electrode materials. Therefore, flexible electrode materials might be the next-generation materials for energy devices.

The demand for flexible materials is increasing with increasing technological advancements in daily life and population growth. A flexible energy storage device has unique characteristics, such as high flexibility, a higher charging–discharging rate, and economic viability that make it an ideal candidate for energy storage devices, in particular for a wearable device. Moreover, flexible material based supercapacitors have higher conductivity, exceptional stability in different environmental condition, and thermal stability with insignificant degradation under repetitive use in wearable devices. In this context, supercapacitors are well-known energy storage devices which have numerous advantages over batteries, such as high power density, higher capacitive retention, a fast charge–discharge rate, and long cyclic life (Lao *et al* 2018, Kanahashi *et al* 2020, Zhang *et al* 2009, Wang and Liu 2019, Carvalho *et al* 2020). Figure 3.1 shows an illustration of different 2D nanolayered materials.

3.2 2D nanolayers and their properties

Nanosheets/nanolayers are composed of a few or several layers of materials bonded together by weak van der Waals forces and a very strong in-plane bond. Usually, a single layer of 2D materials is ideal but a set of ten layers generally occurs. These materials are gaining interest continuously due to their remarkable characteristics, such as electronic and optical properties, and are highly promising materials for advanced technological applications. 2D nanolayer materials can be semiconductors, metals, insulators, or superconductors. Graphene is an extensively used 2D material that continues to gain the interest of researchers, mainly in energy storage devices, because of its unique optical and electronic characteristics. In addition to

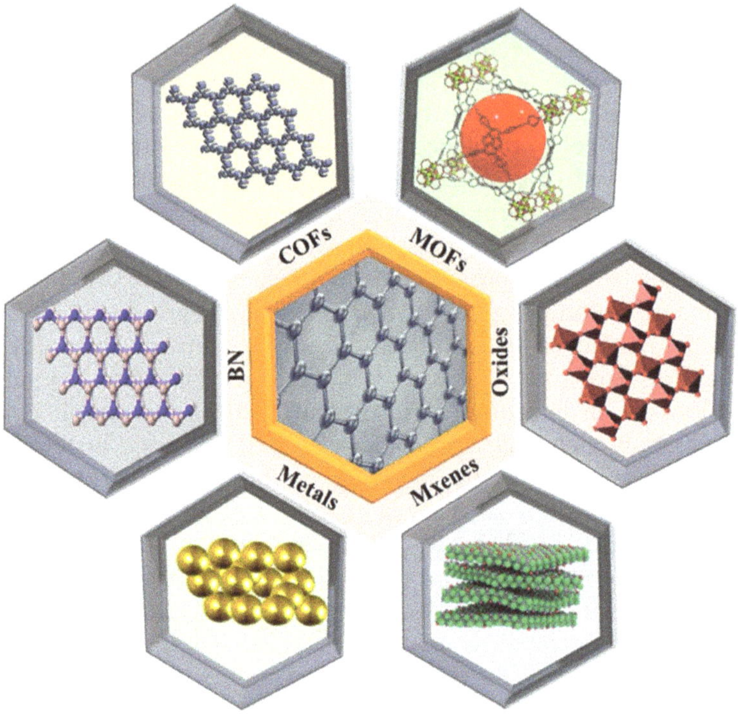

Figure 3.1. Illustration of different 2D nanolayered materials.

graphene, other newer 2D nanolayers have been discovered, such as TMDs, BP, C_3N_4, BN, clay minerals, etc. These 2D nanolayers are promising for energy storage devices due to their various advantageous and tunable properties: (i) excellent conductivity that increases the adsorption and desorption of the charges; (ii) high electrochemical activity that might be advantageous to the redox reaction process to improve pseudocapacitance; (iii) a high surface area providing large active sites that enhance electrochemical activity as well as adsorption/desorption of the charges; (iv) single-crystal structures the enhance fast charge transfer due to no grain boundaries; and (v) unsaturated coordination and dangling bonds at the active site of the edge that might be beneficial for stabilizing reaction intermediates (Zhao *et al* 2020, Liu *et al* 2019, Mannix *et al* 2017, Bromley *et al* 1972, Jana and Rao 2016, Brent *et al* 2017). Figure 3.2 shows a 2D nanolayered and its applications.

3.3 2D nanolayer based electrodes for supercapacitor applications

The continuously increasing demand for energy coupled with dwindling fossil fuel reserves is an undeniable threat to humanity; therefore improving sustainable electrochemical energy conversion and storage devices is necessary for society. In this perspective, 2D nanolayer based electrodes for supercapacitors are being considered as prospective energy storage devices that offer cleaner and eco-friendly energy. Usually, the properties of energy storage devices depend on the electrode

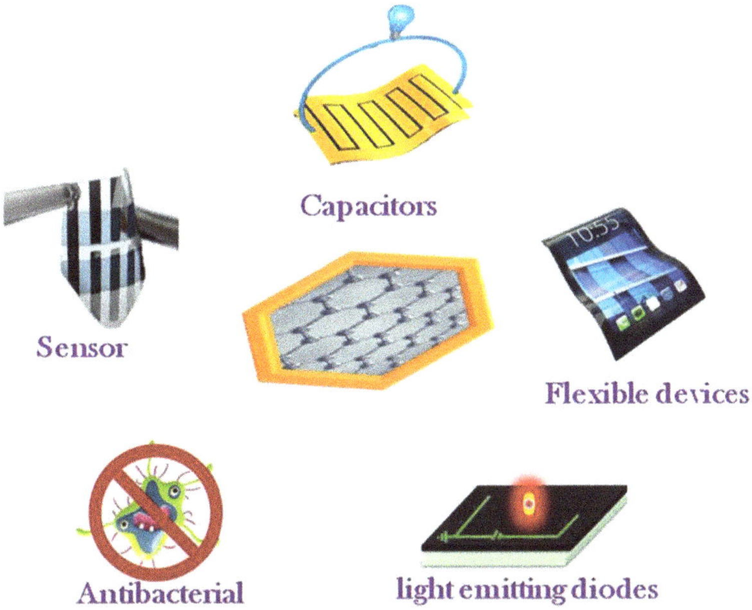

Figure 3.2. Graphical representation of a 2D nanolayer and its applications.

materials. The 2D nanolayered materials are an exceptional candidates for electrode materials because of their extraordinary physical, chemical, and optical properties, high surface area, and high conductivity (Liu and Peng 2017, Panda *et al* 2020, Palaniselvam and Baek 2015). Here, we focus on some of the recently developed 2D nanolayers.

3.3.1 Graphene based electrode materials for supercapacitors

Currently, 2D nanolayered materials including graphene are gaining interest from researchers due to their potential use in the development of high electrochemical performance supercapacitors. The large surface area and large ranges of pore size distribution in graphene provide astonishing potential that makes them the ultimate alternative to carbon-based nanomaterials in supercapacitor applications. Graphene is one of the active materials that takes part in the charge storage process, where ions might be stored electrostatically on the electrode. However, the surface of graphene is not completely available due to the agglomeration process. In other words, the agglomerated structure might restrict direct access to the charge storage surface leading to ionic resistance within the electrodes. Numerous studies have focused on graphene and graphene based composites for supercapacitor applications. For example, Chen *et al* synthesized a graphene WS_2 nanosheet (G–WS_2) composite. The G–WS_2 nanosheet composite was synthesized using an ice template approach that significantly increased the performance of the device. The synthesized G–WS_2 nanosheet composite consists of an irregular and craggy surface that might make electrolyte dissemination possible and augment the wettability. The data suggest that the nanosheet composite also contains the 1T WS_2 phase and covalent bonds at the G–WS_2 interface which improve the pseudocapacity over the composite surface. The fabricated device demonstrates high specific capacitance, with a rate ability of (79.9%) and exceptional cyclic stability (102.5%) due to the fast proton insertion within the WS_2 nanoflakes. The literature supports the applicability of composite nanolayers to enhance the supercapacitive performance by including the pseudocapacitive mechanism (Chen *et al* 2019a).

Non-metallic doping is another prospective approach to increase the faradaic activity over the graphene surface. Non-metallic doping, such as N, B, S, and O, is able to considerably alter the electronic structure. Therefore it might allow a superior interfacial capacitance value. On the other hand, surface functional groups might be affected in some critical way, such as wettability, surface charge, conductivity, and adsorption. Karaman *et al* synthesized nitrogen-doped graphene (N–graphene) by using a sol–gel process and observed the effect of N doping content on the supercapacitor performance. The data suggest that the prepared N–graphene demonstrates high volumetric capacitance (118 F cm^{-3}), coulombic efficiency (>99%), retention capacitance, and high stability (Karaman *et al* 2020). Another study by Lv *et al* synthesized a S-doped graphene manganese oxide composite (SG–MnO_2) electrode for supercapacitor applications. The data suggest that the S doping in the SG–MnO_2 based electrode materials causes defects that lead to a large surface area, and offer additional secure sites for MnO_2, thereby more space and areas to encourage the movement of ions in the electrolyte, and subsequently high electrochemical properties. The observed specific capacitance for the S–graphene–MnO_2 composite was (1311 F g^{-1}) with high retention (81.2%) after 5000 cycles showing the suitability of the material for energy applications (Lv *et al* 2020). Figure 3.3 shows an illustration of SG–MnO_x based electrode material.

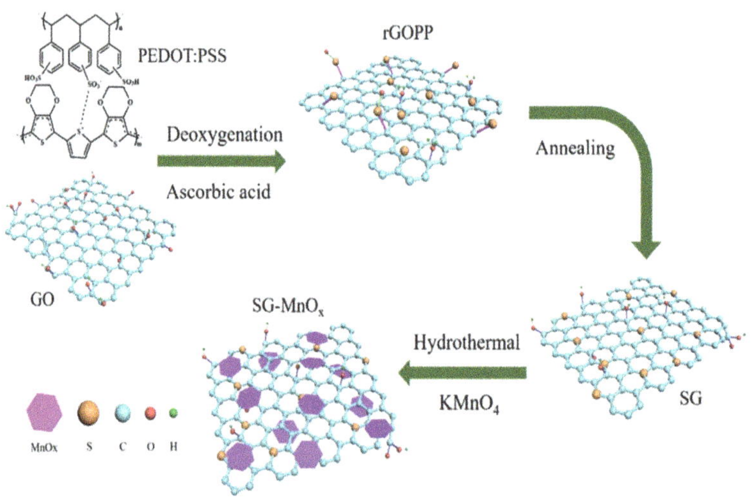

Figure 3.3. Graphical illustration of SG–MnO_x based electrode material. (Reproduced with permission from (Lv *et al* 2020). Copyright 2019 Elsevier.)

Polymer–graphene based composites were fabricated to utilize the benefits of graphene nanosheets and polymers, which provide flexibility, tunable resistance over a wide range, low cost, large conductivity, light weight, and exceptional electrochemical performance with a high surface area and mechanical strength. Several polymers, such as PVA, PAN, polypyrrole, and PVDF, etc, have been used for providing flexibility and to control resistance. George *et al* synthesized a polyaniline (PANI)–rGO composite based electrode for a supercapacitor. According to the study, the PANI was coated on the rGO using the *in situ* polymerization method which enhances the conductivity of the electrode to deliver high specific capacitance (600 F g^{-1}) using a three-electrode systems. Another study by Wang *et al* produced a graphene paper–polypyrrole nanocomposite as an electrode for flexible supercapacitor applications. Polypyrrole was coated on graphene paper using an electrodeposition technique. The data suggest that the prepared PANI–rGO composite based devices can be used effectively in supercapacitor applications because of their high areal capacitance, current density, extraordinary energy density, power density, and high cycling stability (George *et al* 2020).

3.3.2 Transition metal dichalcogenides (TMDs)

TMDs are a group of 2D-layered metallic materials that have high conductivity. Usually, numerous phases of 2D TMDs layers are semiconductors with poor conductivity or high resistance compared to that of the bulk forms. The single/few layered materials have considerably altered characteristics due to their charge carriers and changes in their electronic structure. Usually, in 2D TMDs there are electrochemical reactions on both surfaces which might be faster than traditional reactions. The unique characteristics of 2D TMD layered materials, such as their exceptional mechanical, chemical, and electrical properties, makes them promising candidates for electrochemical technologies. Numerous studies focus on TMD applicability as electrodes for supercapacitor applications. For example Zhou *et al* synthesized MoS_2 nanosheet arrays on carbon cloth using the hydrothermal method. The direct growth of MoS_2 on carbon cloth without the use of binders improves the electrochemical performance by facilitating electron transfer. A measured areal capacitance of MoS_2 over carbon cloth of 2236.6 mF cm^{-2} at 10 mA cm^{-2} was observed, with a cycle stability of 86.1% after 2000 cycles (Zhou *et al* 2019). Another study of TMDs as electrodes was performed by Meyer *et al* utilizing VS_2 over a CNT support using the hydrothermal method. The CNT support structure has an interconnected microporous structure and exhibits a specific capacitance value of 33 F g^{-1} at a current density of 0.1 A g^{-1} with a lowest impedance to electron transfer of 0.32 Ω, which was due to the fast ion transfer from the interconnected microporous structure (Meyer *et al* 2020). Li *et al* report the growth of $MoSe_2$ nanosheet over multi-walled CNTs (MW-CNTs) using a hydrothermal process, applied as an asymmetric supercapacitor electrode. The supercapacitor performance of the electrode materials was determined under asymmetric conditions, such as using a $MoSe_2$–MWCNT anode and an activated carbon cathode. The data suggest that the produced asymmetric supercapacitor has excellent performance with high specific capacitance and electric density as well as stability (Li *et al* 2020a).

The aforementioned study suggested that TMDs demonstrate high conductivity, surface area, and capacitive properties. Therefore, they became the center of attention as an efficient electrode material for electrochemical energy storage devices. The disadvantage of TMDs is their low interplanar electronic conductivity which sometimes limits their performance. To overcome this drawback numerous 2D–2D heterojunctions have been formed on the basis of the structure and alignment of bands, which exhibited better performance compared to their individual materials, attributed to electron–hole separation and ease of charge transfer. Wang *et al* synthesized a 2D–2D heterojunction with SnS_2–MoS_2 layers to improve electrode performance in supercapacitor applications. The data suggest that the prepared SnS_2–MoS_2 heterojunctions have high capacitance and stability. The exceptional performance of the SnS_2–MoS_2 heterojunction is due to the prevention of agglomeration and stacking in the electrochemical process (Wang *et al* 2020a). Another study by Li *et al* synthesized a $MoSe_2$–MoS_2 heterostructure. The data suggested that the $MoSe_2$–MoS_2 heterostructure interface provides more active sites for electrochemical reactions and facilitates charge transport because of the porous space within the intercalated nanosheet arrays, thereby producing a synergistic effect due to amalgamation of MoS_2–$MoSe_2$. The data clearly indicate that the $MoSe_2$–MoS_2 heterostructure shows outstanding specific capacitance with high stability (Li *et al* 2019).

3.3.3 MXene

MXene is one of the newly discovered 2D-layered materials that have captured global attention due to its exceptional properties, such as surface chemistry, conductivity, and 2D structure. MXene was produced from the MAX phase, where A is an IIIA- or IVA-group element, M represents a transition metal, and X is C or N. More than ten new MXenes have been developed. A study reports the applicability of MXene as an electrode for supercapacitor applications. In addition to its high conductivity, MXene can remain as a free-standing film to use in flexible devices. Wang *et al* synthesized a few-layer $Ti_3C_2T_x$ MXene from multi-layered $Ti_3C_2T_x$ using a flash-freezing-assisted delamination process. The interlayer spacing was increased due to the entrapment of water molecules within the multilayers, which enhanced the capacitance performance to 255 F g^{-1} compared to multilayer $Ti_3C_2T_x$ due to the large interlayer spacing which allows the easy migration of electrolyte ions (Wang *et al* 2020c). Figure 3.4 shows a schematic representation of the synthesis of the MXene layer.

Another study by He *et al* synthesized vanadium carbide (V_2CT_x) layers for a supercapacitor electrode. V_2CT_x was synthesized from the V_2AlC MAX phase by NaF + HCl etching. The fabricated V_2CT_xMXene-based electrodes exhibit a specific capacitance value of 181.1 F g^{-1}, and a volumetric capacitance of 317.8 F cm^{-3}, with a retention capacity of 89.1% over 5000 cycles (He *et al* 2020).

MXene and its composites have also been found to enhance capacitance performance. MXene provides excellent properties as an electrode material, however, its use in a small voltage window using an aqueous electrolyte limits its applicability. To overcome such issues, various synthesis processes are being used to obtain a large voltage window to

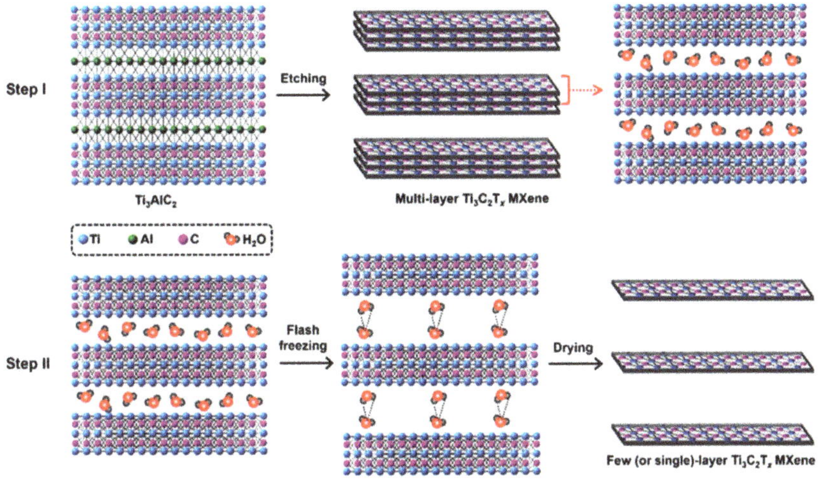

Figure 3.4. A schematic representation of the synthesis of the MXene layer. (Reproduced with permission from (Wang *et al* 2020c). Copyright 2020 Science Press and Dalian Institute of Chemical Physics, Chinese Academy of Sciences. Published by Elsevier.)

increase the energy and power density as they depends directly on the voltage window. Dall'Agnese *et al* synthesized a CNT/MXene composite to offer the potential of 2D MXene over large voltage windows which subsequently increased the power and energy density of the synthesized composite. The synthesized composite was tested under organic conditions and demonstrated specific capacitance and volumetric capacitance values of 85 F g^{-1} and 245 F cm^{-3}, respectively, at 2 mV s^{-1}, with a high rate capability and good cyclability. The CNTs served as additives and increased ion accessibility (Dall'Agnese *et al* 2016). Another study by Wang *et al* synthesized an LDH–MXene–LDH composite using a co-precipitation process and tested it as an electrode for supercapacitor applications. The data suggest that the prepared LDH–MXene–LDH composite has excellent capacitive performance with high energy and power density. Moreover, the MXene improved the conductivity of the composite whereas the LDH (Ni–Mg) enhanced the retention rate of the composite or provided more active sites, and subsequently high pseudocapacitance (Wang *et al* 2020b). Wang *et al* synthesized an rGO–MXene hybrid electrode with a fibrous shape synthesized from two different layered structures. The rGO–MXene hybrid electrode was synthesized using a hydrothermal process. The data suggest that the prepared rGO–MXene hybrid electrode materials showed high capacitive performance and specific capacitance. The specific capacitance and stability increased with an increasing number of MXene layers. Therefore MXene composites not only provide large energy and power density, but also provide the improvement of working in a large voltage window with organic and ionic electrolytes (Wang *et al* 2020c). Wu *et al* synthesized a Ti_3C_2MXene–polyaniline composite as an electrode using a chemical grafting method. The synthesized composite was then deposited over an FTO–glass substrate by a two-step electrochemical reaction. The interlaminar deposition of the polymer prevents the restacking of Ti_3C_2 layers and due to the conductivity of the polymer it shows excellent pseudocapacitance behavior that might be improve the surface area and interlaminar spacing of the Ti_3C_2 MXenes. The synthesized composites demonstrate an area capacitance of 228 mF cm^{-2} at a scan rate of 5 mV s^{-1} with 85% capacitance retention over 1000 cycles (Wu *et al* 2019).

3.3.4 Other 2D nanolayers

There are also several other 2D nanosheets which have shown suitability for use as electrode materials in supercapacitor applications. They can be used as an individual sheet or in the form of composites, for example transition metal oxides (MnO_2, Fe_3O_4, Co_3O_4, NiO, etc), and they offer outstanding properties such as high redox behavior, large surface area with a large pore size distribution which helps in electric double-layer capacitance, and high mechanical and chemical property. They are thus used extensively in various end applications, such as environmental remediation processes, drug delivery, antibiotic materials, medicines, sensors, agriculture, energy devices, batteries, and supercapacitors. TMOs can be modified easily with composites to form hybrid 2D TMOs, which have also been developed to produce advanced supercapacitor electrodes with high redox activity and improved cycle life (Yu *et al* 2015, Wang *et al* 2015, Kumar *et al* 2018, Omar *et al* 2019a, 2019b, Irsad *et al* 2020, Afreen *et al* 2020, Chauhan *et al* 2020).

LDH is a class of metal and clay composites that is considered as suitable for electrodes because of its high redox activity, environmental friendliness, and effective consumption of transition metal atoms. Moreover, its facile exfoliation process into single layer nanosheets and easy surface modification makes it an exceptional candidate for electrode materials (Li *et al* 2017b, Laipan *et al* 2020, Yan *et al* 2018, Wang *et al* 2018).

Another class of 2D materials is the phosphorous family named phosphorene, which is considered to be equivalent to graphene. Usually, it is a single atomic layer of black phosphorus that exists in different forms such as BP, blue phosphorous, and purple phosphorous. These have good electrical conductivity, thereby making them superior candidates for supercapacitor applications (Xu *et al* 2019, Korotcenkov 2019, Sorkin *et al* 2017, Zhang *et al* 2016).

3.4 Nanolayer based flexible wearable devices

Technological innovation, including the discovery of inexpensive, flexible, wearable, and sustainable new materials, is increasing continuously. The emergence of 2D materials has the potential to offer a wider material platform in the development of nano/atomic-level applications, such as water treatment processes, drug delivery, sensors, agriculture, and energy devices. 2D-material based devices with high-performance and high-quality optical encoders have enormous industrial application potentials. Currently, most 2D-material based devices are inflexible, and thus easily experience structural damage when used in wearable devices. Therefore the energy problem associated with wearable devices is one of the great challenges associated with them. Such issues associated with wearable devices might be resolved by maintenance-free flexible batteries or supercapacitor materials and wearable flexible devices (Chee *et al* 2016).

Flexible supercapacitors are of enormous interest due to their high flexibility, exceptional power density, lightweight materials, and simple handling. Usually, flexible supercapacitors consist of thin-film electrode materials that have high flexibility in planar form without any mechanical support. The strong adherence of the nanolayered materials on a flexible substrate allows mechanically flexibility. Free-standing materials or fibers could be used directly as electrodes without the incorporation of binders that offer high volumetric and gravimetric performance. Numerous nanolayer based flexible devices and materials have been developed with high supercapacitor performance. For example, Mendoza-Sanchez *et al* prepared a composite of graphene and manganese oxide nanosheet. This composite was synthesized through a single-step liquid-phase co-exfoliation method and used in a supercapacitor application. The data suggest that the prepared graphene and manganese oxide nanosheet composite had high capacitance with high surface area (Mendoza-Sánchez *et al* 2015). Choudhary *et al* synthesized WO_3–WS_2-nanowires on W foil for flexible nanowire supercapacitors. Figure 3.5 shows a schematic representation of the fabrication process of the WO_3–WS_2-nanowires on W foils for flexible nanowire supercapacitors along with XRD images and Raman spectra. The data suggest the high electrochemical performance of the prepared WO_3–WS_2-nanowire based flexible supercapacitors with high energy density (Choudhary *et al* 2016). Hao *et al* synthesized liquid exfoliated black phosphorus (BP) for flexible supercapacitor applications. The data suggest that the exfoliated BP has a high supercapacitor performance with high flexibility. The incorporation of graphene improves the conductivity and the supercapacitor performance (Hao *et al* 2016).

Wang *et al* synthesized CNTs–MoS_2 graphene based composite materials by using the hydrothermal process followed by a vacuum filtration and reduction process. The data suggest that the prepared CNTs–MoS_2 graphene based composite materials have high volumetric capacitance with high energy density. Moreover, the higher flexibility of the prepared CNTs–MoS_2 graphene based composite materials makes them a superior candidate for flexible supercapacitor applications (Wang *et al* 2017).

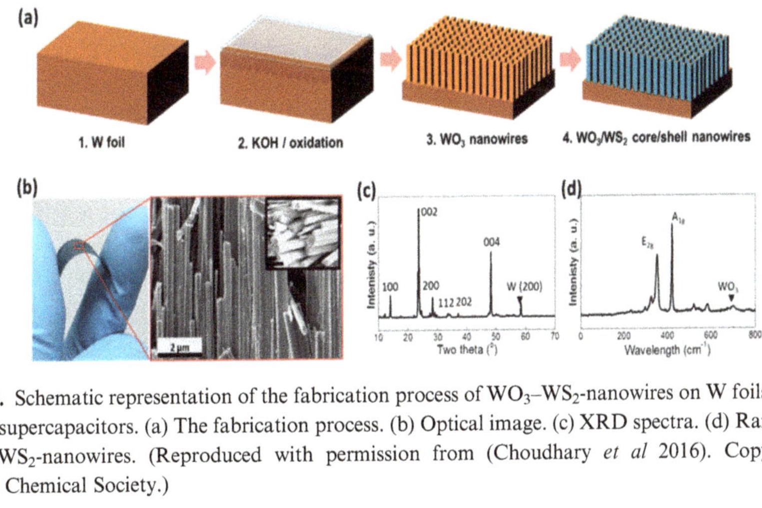

Figure 3.5. Schematic representation of the fabrication process of WO_3–WS_2-nanowires on W foils for flexible nanowire supercapacitors. (a) The fabrication process. (b) Optical image. (c) XRD spectra. (d) Raman spectra of WO_3–WS_2-nanowires. (Reproduced with permission from (Choudhary *et al* 2016). Copyright 2016 American Chemical Society.)

Liu *et al* synthesized binder-free WS_2 nanosheet based electrode materials for flexible asymmetric supercapacitors. The data suggest that the WS_2 nanosheet improved the crystallinity, with a high energy density of 0.97 mWh cm^{-3}, high stability, and flexibility (Liu *et al* 2017). Li *et al* synthesized a flexible graphene–MoS_2 composite film for supercapacitor applications. The data suggest that the prepared graphene–MoS_2 composite film, with its pseudocapacitive layer of MoS_2 and compact structure of the graphene film, developed extraordinary flexibility and stretchability of the composite, making it a superior candidate for flexible supercapacitor devices (Li *et al* 2017a). Another study by Singh *et al* fabricated a flexible supercapacitor using exfoliated graphene. The electrochemical exfoliation technique was used for the separation of nanosheets from graphite. The exfoliated graphene was then coated over A4 sized paper to prepare the graphene paper electrode. The fabricated graphene paper based flexible electrode was then measured using PVA/KOH electrolyte and exhibited an areal capacitance of 30 μF cm^{-2} using a gel electrolyte (Singh and Charu Tripathi 2018). Liang *et al* synthesized flexible WS_2–poly(3,4-ethylene dioxythiophene):poly(styrene sulfonate) (WS_2–PEDOT–PSS) based film as a miniature flexible supercapacitor electrode material. The synthesized WS_2–PEDOT–PSS based flexible electrode material show high electrical conductivity, high specific capacitance, and high volumetric capacitance, and produced a stable miniature device (Liang *et al* 2018). Zhou *et al* synthesized highly deformable and conductive free-standing pseudocapacitive electrode materials ($Ti_3C_2T_x$–MnO_2–NWs) through solution processing. The data suggest that the prepared Ti_3C_2Tx–MnO_2–NWs paper based electrode has remarkable aerial capacitance (205 mF cm^{-2}) and outstanding volumetric capacitance. Moreover, the Ti_3C_2Tx–MnO_2–NWs paper based electrode shows extraordinary flexibility and electrochemical stability (Zhou *et al* 2018). Fan *et al* synthesized a flexible nanoporous MXene film based electrode by incorporating $Fe(OH)_3$ nanoparticles with a diameter range of 3–5 nm. The data suggest that highly interconnected pores facilitate ion transfer efficiency without compromising density. The prepared nanoporous MXene film shows exceptional volumetric performance with high flexibility (Fan *et al* 2018). Lamberti *et al* synthesized MoS_2–rGO on Ti mesh as an electrode material for supercapacitors. The data suggest that the prepared MoS_2–rGO can be used effectively in supercapacitor applications and wearable devices (Lamberti 2018). Guo *et al* synthesized $V_2O_5{\cdot}nH_2O$–CNT based flexible electrode materials by compacting a self-standing layer for high performance supercapacitors. The data suggest that the prepared $V_2O_5{\cdot}nH_2O$–CNT based flexible electrode materials have high flexibility, stability, and capacitive performance (Guo *et al* 2019). Wei *et al* synthesized a self-assembled MoS_2 nanolayer based nanoflower using one-pot synthesis. The data suggest that the assembled MoS_2 nanolayer based nanoflowers have excellent electrochemical capacitance and flexibility (Wei *et al* 2019). Wang *et al* synthesized MoS_2 nanolayer-like H-MoS_2, MoS_2-CTAB, and pure multi-layered 1T-MoS_2 by using magneto-hydrothermal synthesis for solid-state flexible supercapacitors. The data suggest that the higher layer spacing of the MoS_2 nanolayer enhances the specific capacitance. Moreover, the hydrophilicity as well as conductivity of the MoS_2 layer also increased with the intercalation and leads to good supercapacitors performance

(Wang *et al* 2019b). Chen *et al* synthesized Au/graphene composite electrodes for flexible devices. The designed composite showed excellent transmittance, high conductivity, and high flexibility (Chen *et al* 2019b). Wang *et al* synthesized MXene encapsulated with graphene using a plasma exfoliation technique. The data suggest that the prepared MXene encapsulated with graphene had two times higher capacitance compared with that of MXene, high mechanical stability, and an exceptional charge–discharge rate (Wang *et al* 2019a). Kumar *et al* synthesized rGO–WO_3–WS_2 nanowire based electrodes using electrochemical deposition and chemical vapor deposition processes for supercapacitor applications. The data suggest that the structural integrity of these electrodes offers high electrochemical, stability, and capacitance performance (Kumar *et al* 2020). Chen *et al* synthesized rGO–MoS_2–poly(3,4-ethylene dioxythiophene) (rGO–MoS_2–PEEP) for flexible supercapacitor applications. The data suggest that the prepared flexible rGO–MoS_2–PEEP electrode materials have high electrochemical properties. Moreover, the conductivity of the composite is enhanced by the incorporation of rGO within the materials. Additionally, the MoS_2 and PEEP improve the electronic and ionic transport, which enhances areal capacitance, thereby providing high-performance flexible supercapacitors (Chen *et al* 2020). Hamra *et al* exfoliated graphene nanoplates using a microwave exfoliation technique. The advantage of the microwave exfoliation technique is to minimize restacking and also to increase the conductivity of the exfoliated layers. The exfoliated layers were subsequently used as electrodes for supercapacatiors by mixing graphene oxide, microwave exfoliate graphene oxide, and polypropylene (GO-M-GO-PP). The prepared GO–M–GO–PP based electrode materials have high specific capacitance (137.2 F g^{-1}) and high cycling stability with 1000 charge–discharge cycles. The prepared GO–M–GO–PP based electrode shows the high flexibility up to an 180° angle for up to 50 repetitive bends. The data suggest that the prepared GO–M–GO–PP based electrode materials might be suitable for supercapacitor applications (Hamra *et al* 2020).

MXene and polymer composites are also being developed to achieve flexibility for next-generation smart wearable devices. Some conducting polymers also help to improve the conductivity and enhance applicability within the 2D layers to provide flexible support. For example, Ling *et al* synthesized $Ti_3C_2T_x$–polymer composites (polydiallyldimethylammoniumchloride or polyvinyl alcohol). The $Ti_3C_2T_x$–PVA composite film has high conductivity. The synthesized composite exhibits an outstanding capacitance performance of ~530 F cm^{-3} for MXene–PVA-KOH at 2 mV s^{-1}, demonstrating the applicability of polymers in flexible devices (Ling *et al* 2014). Ling *et al* synthesized flexible MXene–PVA composite electrode materials for supercapacitor applications. The data suggest that the prepared MXene–PVA and MXene have high electrical conductivity of 2.4×10^4 and 2.4×10^5 S m^{-1}, respectively. Moreover, the high tensile strength of the MXene–PVA compares well with PVA and MXene, thereby allowing high flexibility of the composite. The prepared MXene–PVA composite has a high capacitive performance. The higher conductivity and flexibility make it a superior candidate for supercapacitor applications (Ling *et al* 2014). Qin *et al* synthesized flexible micro-supercapacitors using polymer MXene composite film. The synthesized MXene based composites show high areal capacitance and outstanding cycling stability. The prepared polymer MXene composite film shows tremendous electrochemical properties (Qin *et al* 2019). Based on the aforementioned studies, the data suggest that 2D nanolayers such as MoS_2, rGO, WO_3, WS_2, and MXene are extraordinary candidates for flexible supercapacitor applications. Moreover, the high interspacing between the two layers also improves the hydrophilicity and conductivity, thereby enabling high supercapacitor performance. The incorporation of binders might increase the flexibility or mechanical strength of the composite, thereby allowing ease-of-use in wearable devices. Moreover, 2D nanolayered materials can be used as flexible electrodes for supercapacitor applications. Electrical double-layer capacitors are provided by high surface area carbon-based materials such as graphene and its composites, while transition metal oxides provide pseudocapacitors through a faradaic activity which increases the overall device performance. The design and fabrication of flexible devices requires polymer supports. However, there are also some 2D–2D composites that can provide flexibility together with high conductivity. Table 3.1 summarizes the different flexible electrode materials for supercapacitor applications. The data suggest that flexible 2D materials have high capacitance and energy density and can be used to make superior supercapacitors.

Table 3.1. Different flexible electrode materials for supercapacitor applications.

S. No.	Flexible electrode materials	Capacitance performance ($F\ C^{-1}\ m^{-2}$)	Energy density ($mWh\ cm^{-3}$)	Stability (cycles)	Reference
1.	MXene–PVA	$530\ F\ C^{-1}\ m^{-3}$		100 000	Ling *et al* (2014)
2.	WO_3–WS_2-nanowires	0.074	60	30 000	Choudhary *et al* (2016)
3	BP with graphene	13.75		30 000	Hao *et al* (2016)
4.	WS_2 nanosheet	0.93	0.97	10 000	Liu *et al* (2017)
5.	CNTs–MoS_2–graphene	0.129	3.2	10 000	Wang *et al* (2017)
6.	Graphene–MoS_2 composite film	0.70	1.72	300	Li *et al* (2017a)
7.	WS_2–PEDOT–PSS	0.086		5000	Liang *et al* (2018)
8.	Ti_3C_2Tx–MnO_2–NWs	0.205	0.057	10 000	Zhou *et al* (2018)
9.	Nanoporous MXene	$1142\ F\ C^{-1}\ m^{-3}$	6300		Fan *et al* (2018)
10.	MOS_2 intercalated rGO	0.17			Lamberti (2018)
11.	$V_2O_5{\cdot}nH_2O$–CNTs	521	0.0174	1000	Guo *et al* (2019)
12.	MoS_2 nanolayer based nanoflower	$1120\ F\ C^{-1}\ m^{-2}$	0.073	2000	Wei *et al* (2019)
13.	1T-MoS_2	$320\ F\ g^{-1}$	0.0431	30 000	Wang *et al* (2019b)
14.	rGO–WO_3–WS_2-nanowires	0.055	0.0105	10 000	Kumar *et al* (2020)
15.	rGO–MoS_2–PEEP	0.241	0.0014	5000	Chen *et al* (2020)

3.5 Conclusion and future prospects

In this chapter we have reviewed recent progress in the synthesis of 2D nanolayer electrode materials for supercapacitors. The 2D nanolayers exhibits high surface area, exceptional mechanical stability, and pseudocapacitive behavior, which make them ideal candidates for supercapacitor applications. Researchers are still focusing on flexible, wearable, and high-performance electrode materials for supercapacitor applications. In this context, the development of newer 2D nanolayered materials with high redox reactions and high pseudocapacitive performance might be a necessary or important step towards flexible supercapacitors with high performance. Moreover, the selection of electrode materials and the design or electrodes are very important steps towards high-performance flexible supercapacitors. The correct selection of the substrate and electrolytes is also necessary for higher flexibility and high performance of the electrode materials. In conclusion, the 2D nanolayered materials might provide the next-generation flexible supercapacitors with high energy density as well as high electrochemical performance.

Acknowledgment

The author (Neetu Talreja) acknowledges financial support by CONICYT through FONDECYT project No. 3190515.

References

Afreen S, Omar R A, Talreja N, Chauhan D and Ashfaq M 2018 Carbon-based nanostructured materials for energy and environmental remediation applications *Approaches in Bioremediation: The New Era of Environmental Microbiology and Nanobiotechnology* ed R Prasad and E Aranda (Cham: Springer)

Afreen S, Talreja N, Chauhan D and Ashfaq M 2020 Polymer/metal/carbon-based hybrid materials for the detection of heavy metal ions ed K A Abd-Elsalam *Multifunctional Hybrid Nanomaterials for Sustainable Agri-Food and Ecosystems* (Amsterdam: Elsevier) ch 15

Ashfaq M, Verma N and Khan S 2018 Novel polymeric composite grafted with metal nanoparticle-dispersed CNFs as a chemiresistive non-destructive fruit sensor material *Mater. Chem. Phys.* **217** 216–27

Brent J R, Savjani N and O'Brien P 2017 Synthetic approaches to two-dimensional transition metal dichalcogenide nanosheets *Prog. Mater Sci.* **89** 411–78

Bromley R A, Murray R B and Yoffe A D 1972 The band structures of some transition metal dichalcogenides. III. Group VIA: trigonal prism materials *J. Phys.* C **5** 759–78

Carvalho B R *et al* 2020 Nonlinear dark-field imaging of one-dimensional defects in monolayer dichalcogenides *Nano Lett.* **20** 284–91

Chauhan D, Afreen S, Talreja N and Ashfaq M 2020 Multifunctional copper polymer-based nanocomposite for environmental and agricultural applications *Multifunctional Hybrid Nanomaterials for Sustainable Agri-Food and Ecosystems* ed K A Abd-Elsalam (Amsterdam: Elsevier) ch 8

Chee W K, Lim H N, Zainal Z, Huang N M, Harrison I and Andou Y 2016 Flexible graphene-based supercapacitors: a review *J. Phys. Chem.* C **120** 4153–72

Chen P-Y, Liu M, Wang Z, Hurt R H and Wong I Y 2017 From flatland to spaceland: higher dimensional patterning with two-dimensional materials *Adv. Mater.* **29** 1605096

Chen T and Dai L 2013 Carbon nanomaterials for high-performance supercapacitors *Mater. Today* **16** 272–80

Chen W, Yu X, Zhao Z, Ji S and Feng L 2019a Hierarchical architecture of coupling graphene and 2D WS2 for high-performance supercapacitor *Electrochim. Acta* **298** 313–20

Chen Y, Bai J, Yang D, Sun P and Li X 2020 Excellent performance of flexible supercapacitor based on the ternary composites of reduced graphene oxide/molybdenum disulfide/poly (3,4-ethylenedioxythiophene) *Electrochim. Acta* **330** 135205

Chen Y, Fu X-Y, Yue Y-Y, Zhang N, Feng J and Sun H-B 2019b Flexible and transparent supercapacitor based on ultrathin Au/graphene composite electrodes *Appl. Surf. Sci.* **467–8** 104–11

Choudhary N, Li C, Chung H-S, Moore J, Thomas J and Jung Y 2016 High-performance one-body core/shell nanowire supercapacitor enabled by conformal growth of capacitive 2D WS_2 layers *ACS Nano* **10** 10726–35

Da Y, Liu J, Zhou L, Zhu X, Chen X and Fu L 2019 Engineering 2D architectures toward high-performance micro-supercapacitors *Adv. Mater.* **31** 1802793

Dall'Agnese Y, Rozier P, Taberna P-L, Gogotsi Y and Simon P 2016 Capacitance of two-dimensional titanium carbide (MXene) and MXene/carbon nanotube composites in organic electrolytes *J. Power Sources* **306** 510–5

El Rouby W M A 2015 Crumpled graphene: preparation and applications *RSC Adv.* **5** 66767–96

Fan Z, Wang Y, Xie Z, Xu X, Yuan Y, Cheng Z and Liu Y 2018 A nanoporous MXene film enables flexible supercapacitors with high energy storage *Nanoscale* **10** 9642–52

George D, James G, Manjally J K, Thomas J, Shine Kuriakose M, Aryamol K S, Benny A and George S C 2020 Graphene/PANI electrode material characterization for supercapacitor application *Mater. Today: Proc.* **24** 1734–41

Gui R, Jin H, Sun Y, Jiang X and Sun Z 2019 Two-dimensional group-VA nanomaterials beyond black phosphorus: synthetic methods, properties, functional nanostructures and applications *J. Mater. Chem.* A **7** 25712–71

Guo K, Li Y, Li C, Yu N and Li H 2019 Compact self-standing layered film assembled by $V_2O_5{\cdot}nH_2O$/CNTs 2D/1D composites for high volumetric capacitance flexible supercapacitors *Science China Materials* **62** 936–46

Hamra A A B, Lim H N, Huang N M, Gowthaman N S K, Nakajima H and Rahman M M 2020 Microwave exfoliated graphene-based materials for flexible solid-state supercapacitor *J. Mol. Struct.* **1220** 128710

Hao C *et al* 2016 Flexible all-solid-state supercapacitors based on liquid-exfoliated black-phosphorus nanoflakes *Adv. Mater.* **28** 3194–201

He H, Xia Q, Wang B, Wang L, Hu Q and Zhou A 2020 Two-dimensional vanadium carbide (V_2CT_x) MXene as supercapacitor electrode in seawater electrolyte *Chin. Chem. Lett.* **31** 984–7

Hu T, Mei X, Wang Y, Weng X, Liang R and Wei M 2019 Two-dimensional nanomaterials: fascinating materials in biomedical field *Sci. Bull.* **64** 1707–27

Irsad T N, Chauhan D, Rodríguez C, Mera A C and Ashfaq M 2020 Nanocarriers: an emerging tool for micronutrient delivery in plants *Plant Micronutrients: Deficiency and Toxicity Management* ed T Aftab and K R Hakeem (Cham: Springer)

Ishihara H, Chen Y-C, De Marco N, Lin O, Huang C-M, Limsakoune V, Chou Y-C, Yang Y and Tung V 2016 Electrohydrodynamic-assisted assembly of hierarchically structured, 3D crumpled nanostructures for efficient solar conversions *Sci. Rep.* **6** 38701

Jana M K and Rao C N R 2016 Two-dimensional inorganic analogues of graphene: transition metal dichalcogenides *Philos. Trans.* A **374** 20150318

Kanahashi K, Pu J and Takenobu T 2020 2D Materials for large-area flexible thermoelectric devices *Adv. Energy Mater.* **10** 1902842

Kang J, Lim T, Jeong M H and Suk J W 2019 Graphene papers with tailored pore structures fabricated from crumpled graphene spheres *Nanomaterials* **9** 815

Karaman C, Bayram E, Karaman O and Aktaş Z 2020 Preparation of high surface area nitrogen doped graphene for the assessment of morphologic properties and nitrogen content impacts on supercapacitors *J. Electroanal. Chem.* **868** 114197

Ke Q and Wang J 2016 Graphene-based materials for supercapacitor electrodes—a review *J. Materiomics* **2** 37–54

Khan K, Tareen A K, Aslam M, Zhang Y, Wang R, Ouyang Z, Gou Z and Zhang H 2019 Recent advances in two-dimensional materials and their nanocomposites in sustainable energy conversion applications *Nanoscale* **11** 21622–78

Korotcenkov G 2019 Black phosphorus-new nanostructured material for humidity sensors: achievements and limitations *Sensors* **19** 1010

Kumar K S, Choudhary N, Jung Y and Thomas J 2018 Recent advances in two-dimensional nanomaterials for supercapacitor electrode applications *ACS Energy Lett.* **3** 482–95

Kumar K S, Choudhary N, Pandey D, Hurtado L, Chung H-S, Tetard L, Jung Y and Thomas J 2020 High-performance flexible asymmetric supercapacitor based on rGO anode and WO_3/WS_2 core/shell nanowire cathode *Nanotechnology* **31** 435405

Laipan M, Yu J, Zhu R, Zhu J, Smith A T, He H, O'Hare D and Sun L 2020 Functionalized layered double hydroxides for innovative applications *Mat. Horiz.* **7** 715–45

Lamberti A 2018 Flexible supercapacitor electrodes based on MoS_2-intercalated rGO membranes on Ti mesh *Mater. Sci. Semicond. Process.* **73** 106–10

Lao J, Xie H, Shi Z, Li G, Li B, Hu G-H, Yang Q and Xiong C 2018 Flexible regenerated cellulose/boron nitride nanosheet high-temperature dielectric nanocomposite films with high energy density and breakdown strength *ACS Sustain. Chem. Eng.* **6** 7151–8

Li C, Zhou Y, Huo P and Wang X 2020a Fabricated high performance ultrathin $MoSe_2$ nanosheets grow on MWCNT hybrid materials for asymmetric supercapacitors *J. Alloys Compd.* **826** 154175

Li J and Östling M 2013 Prevention of graphene restacking for performance boost of supercapacitors—a review *Crystals* **3** 163–90

Li N, Lv T, Yao Y, Li H, Liu K and Chen T 2017a Compact graphene/MoS_2 composite films for highly flexible and stretchable all-solid-state supercapacitors *J. Mater. Chem.* A **5** 3267–73

Li S, Zang W, Liu X, Pennycook S J, Kou Z, Yang C, Guan C and Wang J 2019 Heterojunction engineering of $MoSe_2/MoS_2$ with electronic modulation towards synergetic hydrogen evolution reaction and supercapacitance performance *Chem. Eng. J.* **359** 1419–26

Li X, Du D, Zhang Y, Xing W, Xue Q and Yan Z 2017b Layered double hydroxides toward high-performance supercapacitors *J. Mater. Chem.* A **5** 15460–85

Li X and Wang J 2020 One-dimensional and two-dimensional synergized nanostructures for high-performing energy storage and conversion *InfoMat* **2** 3–32

Li Z, Gadipelli S, Li H, Howard C A, Brett D J L, Shearing P R, Guo Z, Parkin I P and Li F 2020b Tuning the interlayer spacing of graphene laminate films for efficient pore utilization towards compact capacitive energy storage *Nat. Energy* **5** 160–8

Liang A *et al* 2018 Robust flexible WS_2/PEDOT:PSS film for use in high-performance miniature supercapacitors *J. Electroanal. Chem.* **824** 136–46

Ling Z, Ren C E, Zhao M-Q, Yang J, Giammarco J M, Qiu J, Barsoum M W and Gogotsi Y 2014 Flexible and conductive MXene films and nanocomposites with high capacitance *Proc. Natl Acad. Sci.* **111** 16676–81

Liu S, Zeng Y, Zhang M, Xie S, Tong Y, Cheng F and Lu X 2017 Binder-free WS_2 nanosheets with enhanced crystallinity as a stable negative electrode for flexible asymmetric supercapacitors *J. Mater. Chem.* A **5** 21460–6

Liu Y, Huang Y and Duan X 2019 Van der Waals integration before and beyond two-dimensional materials *Nature* **567** 323–33

Liu Y and Peng X 2017 Recent advances of supercapacitors based on two-dimensional materials *Appl. Mater. Today* **8** 104–15

Lv Q, Hao H, Ge M and Li W 2020 S-doped graphene/mixed-valent manganese oxides composite electrode with superior performance for supercapacitors *J. Alloys Compd.* **819** 152970

Mannix A J, Kiraly B, Hersam M C and Guisinger N P 2017 Synthesis and chemistry of elemental 2D materials *Nat. Rev. Chem.* **1** 0014

Mendoza-Sánchez B, Coelho J, Pokle A and Nicolosi V 2015 A 2D graphene-manganese oxide nanosheet hybrid synthesized by a single step liquid-phase co-exfoliation method for supercapacitor applications *Electrochim. Acta* **174** 696–705

Mendoza-Sánchez B and Gogotsi Y 2016 Synthesis of two-dimensional materials for capacitive energy storage *Adv. Mater.* **28** 6104–35

Meyer E, Bede A, Mutukwa D, Taziwa R and Zingwe N 2020 Optimization, and analysis of carbon supported VS_2 nanocomposites as potential electrodes in supercapacitors *J. Energy Storage* **27** 101074

Ogata C, Kurogi R, Awaya K, Hatakeyama K, Taniguchi T, Koinuma M and Matsumoto Y 2017 All-graphene oxide flexible solid-state supercapacitors with enhanced electrochemical performance *ACS Appl. Mater. Interfaces* **9** 26151–60

Omar R A, Afreen S, Talreja N, Chauhan D and Ashfaq M 2019a Impact of nanomaterials in plant systems *Plant Nanobionics: Advances in the Understanding of Nanomaterials Research and Applications* vol 1 ed R Prasad (Cham: Springer)

Omar R A, Afreen S, Talreja N, Chauhan D, Ashfaq M and Srituravanich W 2019b Impact of nanomaterials on the microbial system *Microbial Nanobionics: State-of-the-Art* vol 1 ed R Prasad (Cham: Springer)

Palaniselvam T and Baek J-B 2015 Graphene based 2D-materials for supercapacitors *2D Mater.* **2** 032002

Panda P K, Grigoriev A, Mishra Y K and Ahuja R 2020 Progress in supercapacitors: roles of two dimensional nanotubular materials *Nanoscale Adv.* **2** 70–108

Qin L, Tao Q, Liu X, Fahlman M, Halim J, Persson P O Å, Rosen J and Zhang F 2019 Polymer–MXene composite films formed by MXene-facilitated electrochemical polymerization for flexible solid-state microsupercapacitors *Nano Energy* **60** 734–42

Saha S, Samanta P, Murmu N C and Kuila T 2018 A review on the heterostructure nanomaterials for supercapacitor application *J. Energy Storage* **17** 181–202

Singh R and Tripathi C 2018 Electrochemical exfoliation of graphite into graphene for flexible supercapacitor application *Mater. Today: Proc.* **5** 1125–30

Sorkin V, Cai Y, Ong Z, Zhang G and Zhang Y W 2017 Recent advances in the study of phosphorene and its nanostructures *Crit. Rev. Solid State Mater. Sci.* **42** 1–82

Sun Z, Talreja N, Tao H, Texter J, Muhler M, Strunk J and Chen J 2018 Catalysis of carbon dioxide photoreduction on nanosheets: fundamentals and challenges *Angew. Chem. Int. Ed.* **57** 7610–27

Talreja N, Jung S, Yen L T H and Kim T 2020 Phenol-formaldehyde-resin-based activated carbons with controlled pore size distribution for high-performance supercapacitors *Chem. Eng. J.* **379** 122332

Tan C *et al* 2017 Recent advances in ultrathin two-dimensional nanomaterials *Chem. Rev.* **117** 6225–331

Tao H, Gao Y, Talreja N, Guo F, Texter J, Yan C and Sun Z 2017 Two-dimensional nanosheets for electrocatalysis in energy generation and conversion *J. Mater. Chem.* A **5** 7257–84

Wang B, Hu R, Zhang J, Huang Z, Qiao H, Gong L and Qi X 2020a 2D/2D SnS_2/MoS_2 layered heterojunction for enhanced supercapacitor performance *J. Am. Ceram. Soc.* **103** 1088–96

Wang J and Liu B 2019 Electronic and optoelectronic applications of solution-processed two-dimensional materials *Sci. Technol. Adv. Mater.* **20** 992–1009

Wang K, Zheng B, Mackinder M, Baule N, Qiao H, Jin H, Schuelke T and Fan Q H 2019a Graphene wrapped MXene via plasma exfoliation for all-solid-state flexible supercapacitors *Energy Storage Mater.* **20** 299–306

Wang S, Zhu J, Shao Y, Li W, Wu Y, Zhang L and Hao X 2017 Three-dimensional MoS_2@CNT/RGO network composites for high-performance flexible supercapacitors *Chem.—Eur. J.* **23** 3438–46

Wang W, Jiang D, Chen X, Xie K, Jiang Y and Wang Y 2020b A sandwich-like nano-micro LDH–MXene–LDH for high-performance supercapacitors *Appl. Surf. Sci.* **515** 145982

Wang X *et al* 2019b Unveiling highly ambient-stable multilayered 1T-MoS_2 towards all-solid-state flexible supercapacitors *J. Mater. Chem.* A **7** 19152–60

Wang X, Dong L, Liu W, Huang Y, Pu X, Wang J, Kang F, Li J and Xu C 2020c Few-layer $Ti_3C_2T_x$ MXene delaminated via flash freezing for high-rate electrochemical capacitive energy storage *J. Energy Chem.* **48** 233–40

Wang Y, Guo J, Wang T, Shao J, Wang D and Yang Y-W 2015 Mesoporous transition metal oxides for supercapacitors *Nanomaterials* **5** 1667–89

Wang Y, Yan D, El Hankari S, Zou Y and Wang S 2018 Recent progress on layered double hydroxides and their derivatives for electrocatalytic water splitting *Adv. Sci.* **5** 1800064

Wei S, Zhou R and Wang G 2019 Enhanced electrochemical performance of self-assembled nanoflowers of MoS_2 nanosheets as supercapacitor electrode materials *ACS Omega* **4** 15780–8

Wu W, Niu D, Zhu J, Gao Y, Wei D, Liu X, Wang F, Wang L and Yang L 2019 Organ-like Ti_3C_2 mxenes/polyaniline composites by chemical grafting as high-performance supercapacitors *J. Electroanal. Chem.* **847** 113203

Xu Y, Shi Z, Shi X, Zhang K and Zhang H 2019 Recent progress in black phosphorus and black-phosphorus-analogue materials: properties, synthesis and applications *Nanoscale* **11** 14491–527

Yan A-L, Wang X-C and Cheng J-P 2018 Research progress of NiMn layered double hydroxides for supercapacitors: a review *Nanomaterials* **8** 747

Yu Z, Tetard L, Zhai L and Thomas J 2015 Supercapacitor electrode materials: nanostructures from 0 to 3 dimensions *Energy Environ. Sci.* **8** 702–30

Yuan S, Pang S-Y and Hao J 2020 2D transition metal dichalcogenides, carbides, nitrides, and their applications in supercapacitors and electrocatalytic hydrogen evolution reaction *Appl. Phys. Rev.* **7** 021304

Zhang H 2015 Ultrathin two-dimensional nanomaterials *ACS Nano* **9** 9451–69

Zhang J L *et al* 2016 Epitaxial growth of single layer blue phosphorus: a new phase of two-dimensional phosphorus *Nano Lett.* **16** 4903–8

Zhang S, Lanty G, Lauret J-S, Deleporte E, Audebert P and Galmiche L 2009 Synthesis and optical properties of novel organic–inorganic hybrid nanolayer structure semiconductors *Acta Mater.* **57** 3301–9

Zhao S, Dong B, Wang H, Wang H, Zhang Y, Han Z V and Zhang H 2020 In-plane anisotropic electronics based on low-symmetry 2D materials: progress and prospects *Nanoscale Adv.* **2** 109–39

Zhou C, Wang J, Yan X, Yuan X, Wang D, Zhu Y and Cheng X 2019 Vertical MoS_2 nanosheets arrays on carbon cloth as binder-free and flexible electrode for high-performance all-solid-state symmetric supercapacitor *Ceram. Int.* **45** 21534–43

Zhou J, Yu J, Shi L, Wang Z, Liu H, Yang B, Li C, Zhu C and Xu J 2018 A conductive and highly deformable all-pseudocapacitive composite paper as supercapacitor electrode with improved areal and volumetric capacitance *Small* **14** 1803786

Chapter 4

Micro-mesoporous carbon-based nanostructured materials for flexible supercapacitors

Mohammad Ashfaq, Divya Chauhan and R V Mangalaraja

Carbon-based materials have been utilized for advanced energy applications, including lithium ions batteries, fuel cells, and supercapacitors, due to their exceptional properties, such as large pore-size distribution (PSD), ease of surface functionalization, high chemical reactivity, and mechanical strength. Moreover, their low cost makes them suitable for the small scale industrial development of energy storage systems, such as alkali-ion batteries, fuel cells, and supercapacitors with high performance. Additionally, the micro- and mesoporosity of carbon-based materials have the potential to tune the properties of Li-ion batteries and supercapacitors. Therefore, carbon-based materials have the potential for next-generation, smart, flexible, and wearable material based electronic devices. This chapter will focus on providing a summary of the current understanding of carbon -based porous materials and their composites. The chapter mainly focuses on recent research on carbon-based materials applicable in advanced energy applications. The fundamental properties and advantages/disadvantages of these micro/mesoporous carbon-based materials will also be considered in detail. Some possible future research directions will also be discussed.

4.1 Introduction

The continuous need for cost-effective and socio-economic development has increased the demand for smart miniature portable electronic devices and the electronics needed for newer and smarter applications. Smart miniature portable electronic devices and electronics appliances require exceptional materials characteristics, such as high performance ability, flexibility, and high energy storage ability, thereby providing an easily developed self-powered, wearable, and deformable energy supply (Iqbal *et al* 2019, Zhu *et al* 2014, Liu *et al* 2019). Therefore, the

doi:10.1088/978-0-7503-5143-0ch4

rapidly developing smart miniature portable electronic devices and electronics need to have large capacitance performance to minimize the size and volume of these advanced energy applications. These devices, including smart miniature portable electronic devices, alkali-ion batteries, and supercapacitors, have captured the attention of researchers. Alkali-ion batteries with high rate performance, large energy density, and long cycle retention are the most widely accepted technologies for the development of various industrial needs, such as electric vehicles and smart wearable and portable electronic devices. However, supercapacitors bridge the gap between batteries and capacitors and are documented as being the most suitable the majority of high power energy devices, including smart electronics, hybrid electric automobiles, digital infrastructure, and renewable energy systems, due to their outstanding properties, such as elevated power density, long life cycle, quick charge–discharge rate, and improved performance. The most important part of all energy systems is the electrode material. In this context, a perfect electrode material or at least newer electrode materials should be developed and designed. Furthermore, these newly developed materials must be designed to be environmentally friendly, flexible, wearable, cost-effective, and high-performing (Gao *et al* 2019, Park *et al* 2013, Chee *et al* 2016).

Numerous carbon-based flexible supercapacitor (FSC) electrode materials, such as activated carbon, biomass derived carbon, carbon nanotubes (CNTs), carbon nanofibers (CNFs), carbon-aerogel, carbon beads, and carbon-xerogel based composites have shown potential due to their higher charging–discharging rate, higher power density, higher deformer compatibility, and lightweight materials compared to traditional materials. However, increasing the energy density without addressing the power density of the FSC is one of the greatest challenges and might be degrade various applications including in industrial contexts. To combat this greatest challenge in the field of electronics, researchers are focusing continuously on newer materials or modified existing materials based on their porosity, such as the meso- and microporosity of various materials including carbon (Li *et al* 2018, Wang *et al* 2020, Zhao *et al* 2015, De *et al* 2015).

Recently, micro- and mesoporous materials (such as micro- and mesoporous carbon, metal oxides, and composites) have gained significant attention from researchers due to their tunable porosity that directly affects the specific surface area (SSA) of the materials and ultimately the performance of the advanced energy application. Mesoporous carbon (pores ranging from 2 nm to 50 nm, obtained from numerous carbon-based materials), due to its higher SA, large pore-size distribution (PSD), unusual or disordered pore shapes, optimum electroconductivity, and chemical stability that enhance the effectiveness of catalysts, has a wide ranges of applicability in energy conversion and energy storage materials for use as different electrode materials. With the help of such extraordinary materials, we can easily achieve better performing energy storage materials (Xin and Song 2015, Eftekhari and Fan 2017, Pérez–Mayoral *et al* 2019). Figure 4.1 shows a graphical representation of the pore-size distribution of micro-, meso-, and macroporous carbon-based materials.

There are various ways to improve the energy storage ability and performance of supercapacitors for energy applications. (i) Disordered mesoporous materials, in

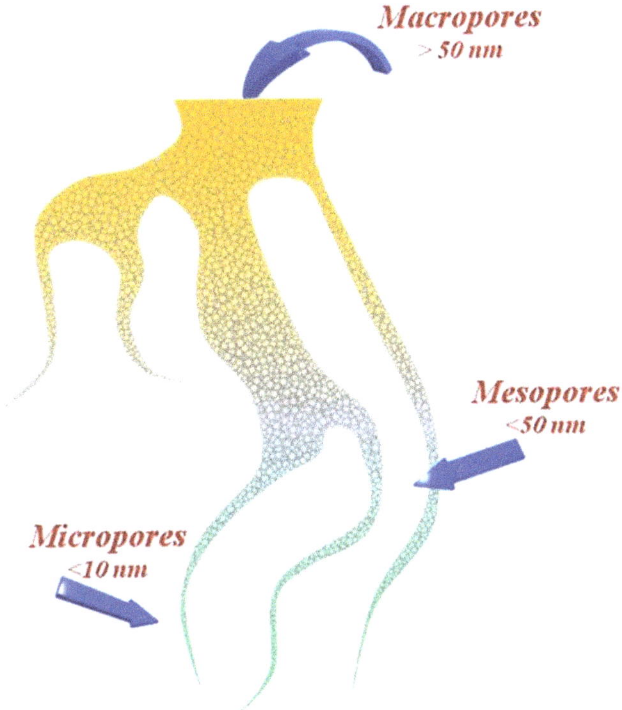

Figure 4.1. Graphical representation of the pore-size distribution of micro-, meso-, and macroporous carbon-based materials.

particular carbon-based materials, have been synthesized by using different processes such as metal catalysis, carbonization of organic hydrogels, carbonization of polymers or polymeric composites, and synthesis of various nanomaterials using a template process. (ii) Ordered or uniform mesoporous carbon-based materials with different pore sizes or porous structures have been synthesized using various template processes including silica, such as SBA-15, MCM, and noncommercial mesoporous silica-based materials, with subsequent conversion of the structure by removing the template through an etching process. (ii) Ordered/uniform mesoporous carbon contains graphitic pores that can be synthesized using a soft carbon template such as a block co-polymer and amphiphilic surfactants. (iv) Highly ordered graphite can be synthesized using a high temperature, which might be beneficial for high energy storage as well as supercapacitor performance. (v) Designing a silica template to synthesize hierarchically ordered mesoporous carbon materials. These processes are considered as beneficial for the synthesis of mesoporous carbon materials. However, some processes require a large amount of inorganic or organic dopants. There are four basic requirements for mesoporous carbon atoms: (i) a precursor-component self-assembly tendency; (ii) pore-forming and carbon components; (iii) high temperature; and (iv) crosslinking ability and high mechanical strength of the carbon. Moreover, synthesis conditions, such as temperature, crosslinking agents, the mechanical strength of the carbon, and pressure, affect the porous texture of the materials (Ghosh *et al* 2019, Tan *et al* 2017, Xin and Song 2015, Eftekhari and Fan 2017, De *et al* 2015, Pérez–Mayoral *et al* 2019). Therefore, strategic approaches to the synthesis of mesoporous carbon might be beneficial to optimize a uniform pore size and texture.

4.2 Synthesis of micro- and mesoporous carbon-based materials

4.2.1 Carbonization and activation process

The carbonization and activation process is one of the most important processes to create a highly porous structure within the materials with a large PSD. Generally, there are two different types of activation process, chemical activation processes that utilize a mixture of carbon-based materials and chemical compounds heated at high temperature, and physical activation processes which involve carbonization of a precursor in a partially gasified atmosphere (carbon dioxide or steam) in a temperature range of 600 °C–900 °C. The high temperature treatment of a mixture of KOH with materials creates micropores/mesopores within carbon-based materials. A high number of micropores/mesopores within the carbon-based materials leads to a high SSA and micropore volume of the materials. However, the average mesoporous texture might be decreased due to the alkali metal adhering at the surface. Moreover, an alkali metal that is mixed into carbonaceous materials can be removed easily through a simple washing procedure (Choi and Ryoo 2007). Usually, the surface functional group is not incorporated within porous carbon materials during the synthesis process. Functional groups, such as phenol, ketone, ether, and carboxylic acid, adhere easily to the surface of carbon-based materials with the help of hydrogen bonding, electrostatic attraction, and covalent bonding. Moreover, the oxidation process might be improved through the wettability of the pores with *y* polar solvents that enhance the fraction of microporosity as well as the SSA of the materials, in particular carbon-based materials (Cheng and Teng 2003). Usually, the oxidation process depends on the potency of the oxidants that consist of oxidizing gases (such as oxygen, air, nitric oxide, and ozone) and oxidizing solutions (such as nitric acid, hypochlorite, dichromates, chlorates, persulfates, and peroxides, etc) (Otake and Jenkins 1993, Vinke *et al* 1994, Pradhan and Sandle 1999, Chen *et al* 2003, Negara *et al* 2020). It is important to mention here that the conditions for the oxidation process must be selected extremely carefully to avoid unnecessary corrosion and structural collapse of the carbon skeleton. A mild oxidant such as hypochlorite has the potential to change the structural texture, and the surface structural texture might influence the composition of the surface groups or the meso-, macro-, and microporosity of the carbon-based materials. It is important to mention here that all surface oxides are not stable. The surface oxide might be irrevocably detached by the reduction in KOH by using a cathodic potential sweep (Vinke *et al* 1994, Cheng and Teng 2003). On the other hand, these wobbly surface oxygen functional groups might tend to detach from the activated surface of carbon in the form of CO and CO_2, which can be examined using a temperature-programmed desorption process. The oxygenated surface group might be eliminated

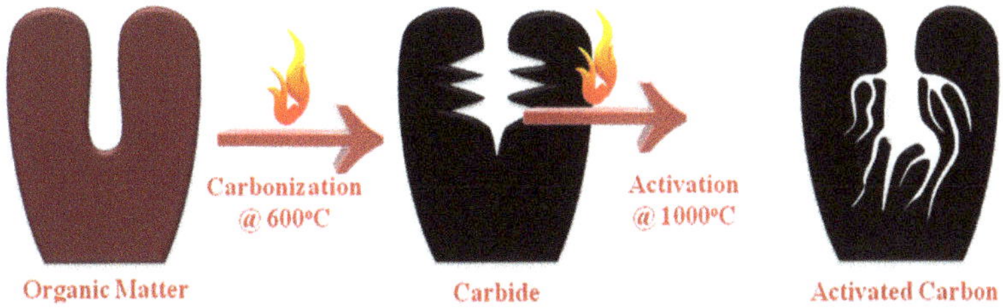

Figure 4.2. An illustration of the carbonization of micro-, meso-, and macroporous carbon.

by heat treatments that create additional porosity within the materials. Other aspects are activation using CO_2 for the creation of mesopores over the surface of carbon-based materials at a high temperature (~1000 °C). The mesoporous structure was found to be highly stable at ~700 °C (Yan *et al* 2008). Numerous studies suggest that the carbonization and activation process is advantageous as it creates meso-, and microporosity of the carbon-based materials easily, which further improves energy storage capacity and supercapacitor performance. Figure 4.2 shows an illustration of the carbonization of micro-, meso-, and macroporous carbon.

For example, Zhang *et al* synthesized one-dimensional carbon nano-chains (CNCs) from onion for supercapacitor applications. The CNCs were activated by using KOH to create micro/mesoporosity of the samples. The micro/mesoporosity of the samples increased the SSA as well as optimized the PSD results of the samples, and thus high supercapacitor performance was observed around 112.7 F at 50 mV s^{-1} for 6 M KOH due to improving the SA and micro-mesoporosity of the materials (Zhang *et al* 2015). Figure 4.3 shows TEM images of activated CNC materials. Upon increasing the temperature the defects in the outer layer of the CNCs increased, whereas at 900 °C defects increased with exposure time. The data suggest that the structural changes might be influenced by various factors, either higher temperature or longer exposure at a certain temperature, that create micro-, meso-, and macroporosity of the materials, thereby improving the energy storage performance. Prakesh *et al* synthesized porous carbon using pigeon-pea stalk powder as a carbon precursor. Initially the pigeon-pea stalk powder was mixed with KOH, which acted as an activating agent, and was activated at 800 °C to achieve a total pore volume of ~0.1774 cm^3 g^{-1}. The highest pore volume of the carbon-based materials was observed due to the KOH activation process (Om Prakash *et al* 2020). Cheng *et al* synthesize sulfonated porous microspheres using $ZnCl_2$ as an activating agent (Chang *et al* 2015). Qian *et al* synthesized porous carbon-based materials using a sol–gel assisted chemical activation process. The data suggest that the high SA and pore volume of the composite due to the utilization of H_2SO_4 as an activating agent (Qian *et al* 2019). Another study by Zou *et al* synthesized highly porous carbon utilizing a combination of template and chemical activation approaches. The data suggest that the prepared porous carbon exhibited high SA and pore volume. Boron was used as the template and H_3PO_4 had a dual role as (i) an activating agent and (ii) a reactant that helped to remove embedded boron

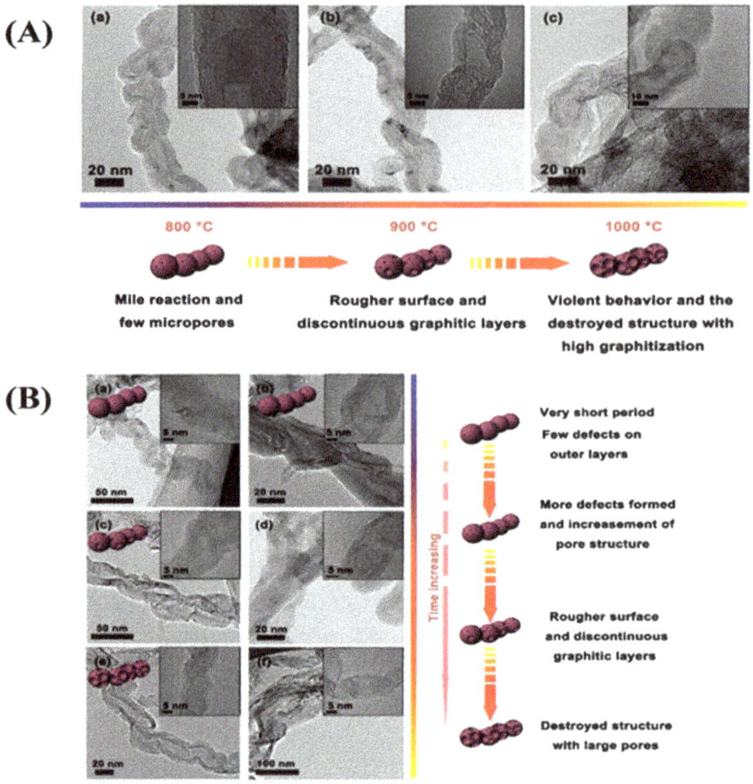

Figure 4.3. TEM images of activated CNCs (a) with a different temperatures (800 °C to 1000 °C) and (b) at 900 °C for 5 h to 30 h exposures. (Reproduced with permission from (Zhang *et al* 2015). Copyright 2015 American Chemical Society.)

during carbonization in the form of BPO_4, which is involved in the generation of excessive pore texture on the carbon surface and also expanded the diameter of the pores (Zou *et al* 2017).

4.2.2 Sulfonation process

Carbon-based materials can be functionalized by using sulfonic groups, which exhibit potential ability as well as being eco-friendly catalysts. Generally, surface functionalization of carbon-based materials eradicates the requirement for liquid acids in various catalytic reactions that might be reprocessed several times. For example, Budarin *et al* synthesized mesoporous materials using the pyrolysis of starch and incorporated sulfonated groups. The data suggest that the incorporation of the sulfonated group (starborn) provides a highly active, renewable, and environmentally friendly catalyst. The hydrophobicity and hydrophilicity are tuned easily with the help of the starborn sulfonated group that might be beneficial for various end applications (Budarin *et al* 2007). Wang *et al* synthesized ordered mesoporous silica-based carbon materials treated with sulfonic acid by the reduction of diazonium and 4-benzene-diazoniumsulfonate salt using hypophosphorous acid (Wang *et al* 2007). Another study focused on ordered mesoporous carbon and functionalization of the sulfonate group through the incorporation of sulfuric acid fumes with carbon within the autoclave (Xing *et al* 2007). In general, carbon-based materials functionalized with sulphonated groups lead to hydrophobicity or hydrophilicity of the materials, which easily improves the performance of various applications, including energy storage.

4.2.3 Halogenation process

The surface functionalization of the materials transforms the physical properties of carbon-based materials significantly. Fluorination is used to develop the hydrophobicity or mesoporosity of carbon-based materials. Usually, fluorine reacts with the hydrogen in C–H and unsaturated bonds of the carbon materials. The mesoporosity of carbon-based materials is tuned easily with the varying of the fluorine and carbon ratio. The carbon-based materials functionalized with fluorine have a superhydrophobic surface that exhibits a high-water contact angle with a low adsorption process. Carbon foam is one example of a fluorine functionalized surface. The superhydrophobic surface might be beneficial for various end applications, including energy storage (Mukhopadhyay *et al* 2004). In general, the surface functionalization of carbon-based materials with various functional groups, such as fluorination and sulfonations, tunes the meso/microporosity easily. The meso/microporosity of the materials could play a crucial role in the performance of advanced energy applications and supercapacitor applications.

4.3 Graphene: a revolution in energy storage

Graphene was the first two-dimensional (2D) material, and was first exfoliated by Andre Geim and Kostya Novoselov in 2004. Graphene is a one layer thick graphitic form of carbon and has unique properties, such as excellent mechanical strength, extremely high stiffness, elevated thermal conductivity (5000 W m^{-1} K^{-1}), high SSA (2630 m^2 g^{-1}), and promising electronic transportation that might be utilized in various end applications. In addition to its several advantages, there are also numerous disadvantages, such as its zero band gaps and restacking properties. Graphene is mainly found in three different forms, which are graphene oxide (GO), reduced graphene oxide (rGO), and functionalized reduced graphene oxide (fr-GO). All these forms of graphene have resemblance with each other or that of graphene. All forms of graphene possess excellent physical and chemical properties. Due to these outstanding characteristics graphene is one of the most promising materials in the group of 2D materials. Accordingly, to fulfill industrial and research requirements, research in the field of graphene and its composites has increased in the past decades. In this regard, there are several literature reports on the applicability of graphene in energy storage applications. Various forms of graphene and their composites have been used so far in numerous applications such as agriculture, sensors, water treatment processes, energy, electrodes for supercapacitors, lithium-ion batteries, and fuel cells (Afreen *et al* 2018, 2020, Chauhan *et al* 2020, Ashfaq *et al* 2019, Omar *et al* 2019).

4.3.1 Graphene and graphene composites in supercapacitors

Graphene and its composites have several advantages in energy storage over other carbon-based materials, due to their high SSA, large pore tuning and size distribution, and low resistance path with short ion-diffusion, which increases the capacitive performance of electrodes in several energy applications. For example, Wang *et al* synthesized porous graphene using partial combustion methods resulting in a controlled pore size of 3.5 nm and an SA of 1815 m^2 g^{-1}. The graphene synthesized using this method has high conductivity and hence can be applied as the electrode material for supercapacitor applications as the electrolyte can diffuse easily through the controlled pores of 3.5 nm (Wang *et al* 2019). Another study by Chen *et al* synthesized doped graphene for supercapacitor applications. Nitrogen and sulfur co-doping were applied over porous graphene-based aerogels. The material was synthesized using a hydrothermal process and exhibited a hierarchical structure with a maximum pore size of 2.5 nm and a PSD of 2–100 nm. The prepared material exhibited a high energy density of 100.7 Wh kg^{-1} (Chen *et al* 2018).

GO and graphene composites also demonstrate excellent performance in energy storage due to their wide PSD and ability to tune the pore size from the micro- to meso-level. For example, Wu *et al* synthesized a graphene–Ni_3S_2 composite for an electrode in a supercapacitor application. This synthesized nanocomposite structure has a high SA and a pore size of 4 nm to make the structure mesoporous for the perfect diffusion of electrolyte ions to exhibit a specific capacitance of 741 F g^{-1} over the synthesized composite (Wu *et al* 2018). Graphene oxide and its composites are also known as excellent electrode materials in energy storage due to their ease of functionalization and pore tuning through minor chemical and physical modification of their properties. For example, Singh *et al* synthesized nitrogen-doped graphene oxide as a flexible solid-state supercapacitor, in which nitrogen-doped porous graphene oxide exhibited a specific capacitance of 230 F g^{-1} at 1 A g^{-1}, which was due to the porous nature of the nitrogen-doped graphene oxide, as with a larger SA the movement of electrolyte ions is faster, facilitating electric double layer capacitance over the electrode surface (Singh *et al* 2017). Figure 4.4 shows a schematic representation of the synthesis of nitrogen-doped porous graphene. The figure shows the step-wise synthesis of nitrogen-doped rGO by H_2O_2 treatment followed by hydrolysis. Graphene composites also demonstrate the property of flexibility, which is another advantage of using graphene-based composites in energy storage to miniaturize devices with the combination of porous texture to facilitate electron transport. For example, Guo *et al* synthesized a free-standing porous MnO_2–graphene composite using a template method. Owing to the porous structure of graphene, fast ion transport inside the porous electrode surface results in a large capacitance value (Guo *et al* 2015). In general, the mesoporosity of carbon-based materials might be beneficial for various applications, mainly energy storage applications.

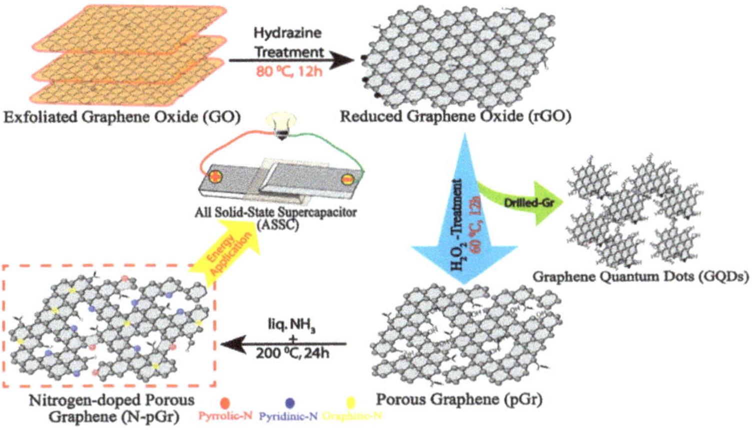

Figure 4.4. Schematic representation of the synthesis of nitrogen-doped porous graphene. (Reproduced with permission from (Singh *et al* 2017). Copyright 2017 Elsevier.)

4.4 Other carbon-based materials in supercapacitors

Carbon-based materials (such as carbon nanofibers (CNFs) and carbon nanotubes (CNTs)) are some of the most versatile materials due to their exceptional characteristics, such as a high SA, meso/micro/macroporosity, non-toxicity, and ease of functionalize, and thus is used extensively in various end applications, including drug delivery systems and devices, wound healing, agricultural applications, antibiotic materials, water remediation, sensors, and energy (Ashfaq *et al* 2013, 2014, 2016, 2017a, 2017b, 2018, Singh *et al* 2013, Kumar *et al* 2018, Bhadauriya *et al* 2018, Talreja *et al* 2014, 2016, 2020, Khare *et al* 2013, Kumar and Talreja 2019).

Carbon-based advanced energy applications, in particular supercapacitors, are a recent focus among advanced energy applications. Compared to batteries, supercapacitors have a high power density, large cycling stability, better Columbia efficiency, and a short charge–discharge time. Carbon-based materials, mainly CNFs, CNTs, aerogels, graphene, and mesoporous carbon electrodes, have received widespread attention and are increasing in popularity as one of the most advanced materials for energy storage applications. Here we discuss in detail different forms of activated carbon for energy storage applications.

4.4.1 CNTs and CNFs

CNTs and CNFs are one-dimensional (1D) forms of carbon-based materials. CNTs, in the form of single-walled or multi-walled or in the form of metallic or nonmetallic composites, are highly suitable as electrodes for supercapacitors due to their high conductivity, light weight, and large SA. CNTs are considered as excellent materials for energy applications. Several works have been published on CNTs to demonstrate the feasibility of CNTs as supercapacitor electrodes. For example, Wang *et al* synthesized mesoporous CNFs for supercapacitor applications. The mesoporous CNFs were synthesized using phenolic resol as a monomer by self-assembly of triblock copolymers. The data suggest that the prepared mesoporous CNFs have a high SA (1424 m^2 g^{-1}) with high capacitance (152 F g^{-1}). The high capacitance ability of the prepared CNFs was observed due to the mesoporous opening of the outer surface and controlled pore size (Wang *et al* 2009). Li *et al* synthesized mesoporous CNFs and mesoporous ZnO nanotubes using self-template assembly for supercapacitor application The data suggest that the mesoporous carbon has an excellent SA that leads to good performance of the supercapacitor (Li *et al* 2011). Agyemang *et al* synthesized a CNT–CNF composite in polyacrylonitrile solution using a chemical polymerization process. The synthesized composite shows a specific capacitance of 1119 F g^{-1} at 1 A g^{-1} with excellent cycling stability and 98% retention up to more than 2000 cycles. The author also prepared a CNF–polymer composite but the reported capacitance was 278 F g^{-1}, which was far less than for the CNT–CNF–polymer composite. The combined synergetic effects of all the materials are the major reason for the high capacitance, with CNT incorporation increasing SA while the pseudocapacitive property of PANI increases the electrical conductivity of the fabricated composite. CNFs contain many edge sites on the outer wall to help to improve high capacitance (Agyemang *et al* 2018). Lee *et al* synthesized mesoporous MnO_2–CNTs and nanodot MoO_3–CNTs using a co-precipitation process. The data suggest that the mesoporous MnO_2–CNTs have a higher SA and conductivity compared to those of the nanodot MoO_3–CNTs, thereby a higher capacitance performance (Lee *et al* 2018). Another study by Kim *et al* synthesized an olygoaniline–CNF composite as an electrode for supercapacitor applications. Due to the porosity of the activated CNFs, mesopores are created over the surface of the CNFs and help to produce a high capacitance value (Liu *et al* 2017a). Jalil *et al* synthesized PANI–MnO_2–CNFs using an electrospinning process for supercapacitor applications. The data suggest that the prepared PANI–MnO_2–CNFs have excellent supercapacitor performance as well as high electrical conductivity (Jalil *et al* 2020). The aforementioned literature suggests that CNTs and CNFs are exceptional candidates for advanced energy applications. Moreover, the high SA, high conductivity, and mesoporous structure of CNTs and CNFs improve capacitance performance.

4.4.2 Carbon aerogels/composites

Carbon-based aerogels/composites are the most interesting porous form of carbon with an interconnected 3D network. Due to several properties, such as large SSA, interconnected pores, high conductivity, and well-regulated structure, they meet the demands of advanced energy applications. Carbon-based aerogels can be synthesized mainly by carbonization/activation of biowaste or resorcinol/formaldehyde or melamine/formaldehyde, which exhibits high SSA with a wide PSD and specific capacitance. However, their practical application still needs attention due to their low mechanical strength and complicated synthesis process. Recently, several biomaterials, such as lignin, cellulose, and starch, have attracted widespread interest due to their biodegradability and renewability (Wang *et al* 2020, Li *et al* 2019, Ji *et al* 2020). For example, Yu *et al* synthesized phenol-formaldehyde resin carbon aerogel using ion catalyzed synthesis for Li/Na batteries. The data suggest that the prepared Fe catalysis microporous carbon aerogel has high conductivity due to the free-standing graphene nanosheets within the graphite flakes. The higher SA and high capacitance performance are due to the mesoporosity and graphitization (Yu *et al* 2016). Xu *et al* synthesized powdery carbon aerogels from Pluronic P123. The powdery carbon aerogel exhibits an interconnected 3D microporous network which allows efficient pore surface utilization and a fast response for electron transfer. The specific capacitance reported for the powdery hydrogel was as 384 F g^{-1} at a low current density of 10 mA g^{-1} which is due to the well formed porous network of the electrode material (Xu *et al* 2017). Another study by Gao *et al* synthesized N-doped carbon aerogels for supercapacitors. The data suggest that the prepared N-doped carbon aerogels have high SA, high charge density, and quick movements of ions, and thereby high capacitance performance. This proves the applicability of biomass-derived aerogels for micro-supercapacitor applications due to their flexibility, stability, and low synthesis cost (Gao *et al* 2020).

Similarly, numerous carbon composites have been synthesized and used effectively in advanced energy applications. For example, Lv *et al* synthesized ordered mesoporous carbon materials by using a soft template method utilizing the process of self-assembly. The prepared mesoporous carbon FDU-15 material was activated using KOH. The data suggest that the KOH activation enhances the SA and tuned the pore size of the carbon-based materials. Moreover, a high KOH/carbon ratio developed a microporous texture within the carbon, whereas a low KOH/carbon ratio led to mesoporosity. The optimized micro-mesoporous texture might provide high SA (1410 m^2 g^{-1}), a large pore volume (0.73 cm^3 g^{-1}), and high supercapacitor performance (200 F g^{-1}) (Lv *et al* 2012). Tang *et al* synthesized Zn-methylimidazole based nitrogen-doped graphitic carbon shells (Zn–GCS) for supercapacitors. The data suggest that the prepared Zn–GCS obtained graphitic carbon due to catalytic graphitization of Co. Interestingly, the mesoporous structure was obtained after the removal of Co. The graphitization and mesoporous structure of the Zn–GCS improved the conductivity and SA, leading to good performance of supercapacitors (Tang *et al* 2015). Talreja *et al* synthesized metal incorporated phenol-formaldehyde resin activated carbon beads (M–CMP) for supercapacitor applications. The prepared metal of the M–CMP was leached out to produce uniform pores within the beads (M–CMP–S). The prepared M–CMP–S had mesopores over a range of 2–50 nm by varying the metal ions. The mesoporous M–CMP–S had excellent supercapacitance ability with high energy and power density (Talreja *et al* 2020).

In general, carbon-based materials such as aerogels and composites have been used effectively in advanced energy applications. The lightweight materials such as carbon-aerogels are superior electrode materials due to their network-like structure, mesoporosity, and high SSA. The conductivity of the carbon aerogels could be tuned by graphitization and incorporation of catalysis. Moreover, high performance capacitance might be improved with the help of mesopores or adjusting the size of the pores and the conductivity of the materials.

4.5 Flexible carbon-based material energy storage devices

Currently, most supercapacitors and batteries are based on inflexible and easily broken electrodes that are unsuitable for advanced flexible energy applications due to structural failure or deformation during operation. For utilization in advanced energy applications, flexible or wearable electrode materials require high electronic and ionic conductivity during operation. Therefore, there is a need to develop flexible and unbreakable electrode materials for advanced wearable energy applications. In this context, researchers are attracted to the lightweight, flexible carbon-based high performance energy storage devices, including supercapacitors and batteries. The flexible advanced energy devices have the potential to provide power sources for artificial skin, portable electronics, and wearable devices. The development of such flexible or wearable electrode materials with high mechanical strength as well as high performance is one of the greatest current challenges (Shao *et al* 2015, Zhong *et al* 2014, Weng *et al* 2016, Liu *et al* 2017b, Yang and Gao 2019, Huang *et al* 2019). Numerous efforts have been made in the design and fabrication of flexible carbon-based electrode materials for the development of batteries and supercapacitors, that are based mainly on carbon-based materials, and polymeric binders on flexible substrates such as fabric, paper, and plastics (Lee *et al* 2013, Stoppa and Chiolerio 2014, Khan *et al* 2016). For example, Cheng *et al* synthesized graphene–MNO_2–CNT based composite materials for supercapacitor applications. The data suggest that the prepared graphene–MNO_2–CNT based composite materials have exceptional supercapacitor performance (372 F g^{-1}) with high mechanical strength (Cheng *et al* 2012). Chen *et al* synthesized CNT polymer-based hydrogels for electrode materials for flexible batteries. For this poly(3,4-ethylene dioxythiophene) and poly(4-styrene sulfonate) (PEDOT:PSS) are mixed with CNTs to form a 3D interconnection hydrogel network. Ammonium persulfate acts as a gelation agent while the CNTs help to induce mechanical robustness and conductivity and also provide an ideal scaffold for the growth of interconnecting channels. These CNT–conducting polymer hydrogels perform very well as electrodes due to their fast ion transport and a large tolerance for strain during volumetric changes (Chen *et al* 2014). Zhao *et al* synthesized Ni–Mn layered double hydroxide microcrystal doped CNTs (Ni–Mn–LDH–CNTs) by an *in situ* growth process for supercapacitors. The data suggest that the prepared three-dimensional (3D) structures of the Ni–Mn–LDH–CNTs might be beneficial for high performance supercapacitors. Moreover, the mesoporous structures enhanced the SA of the Ni–Mn–LDH–CNT materials leading to better supercapacitor performance (Zhao *et al* 2014). Another study focused on free-standing flexible MXene–CNTs based paper by using alternating layers of MXene and CNTs using a simple filtration process. The data suggest that the prepared flexible MXene–CNT based paper has a high capacitance performance (Zhao *et al* 2015). Miao *et al* synthesized PANI–CNF based composites by an electrospinning process. The deposition of PANI nanoflakes on CNFs has advantages for supercapacitor performance as increasing the loading of PANI on CNFs increased the supercapacitor performance (Miao *et al* 2016). Li *et al* manufactured micro/mesoporous CNFs entrenched with well-ordered carbon (M-CNFs) using an electrospinning process for a flexible supercapacitor. The data suggest that the high capacitance performance of the prepared M-CNFs was observed due to mesoporosity and graphitization. The higher mesoporous structure and graphitization of the M-CNFs enhances the SA and conductivity of the M-CNF materials. Interestingly, M-CNFs have high flexibility, bending up to a 135 °C angle (Li *et al* 2018). Xu *et al* synthesized activated carbon coated with CNTs as a core–shell nanocomposite for supercapacitor electrode material. A chemical activation process using KOH as an activating agent was used to create pores of 2 nm to 60 nm which easily facilitate the electrolyte ions from the pore of CNTs. The data suggest that the prepared mesoporous activated carbon coated CNT core–shell nanocomposites have excellent stability and capacitance (97 F g^{-1}). The polymeric composites with CNTs provide flexibility together with high mechanical strength and support, and thereby high performance (Xu *et al* 2019). Table 4.1 summarizes the different materials and their supercapacitor performance. The data suggest that flexible, lightweight mesoporous carbon has exceptional supercapacitor performance. Moreover, flexible and wearable electrode materials might be the next-generation materials for advance energy applications.

Table 4.1. Different carbon-based materials and their supercapacitor performance.

S. No.	Materials	Process	Supercapacitor performance (F g^{-1})	SSA (m^2 g^{-1})	Stability (cycle)	References
1.	Mesoporous CNFs	Self-assembly	152	1424	1000	Wang *et al* (2009)
2.	Mesoporous CNFs	Calcination	200	2071	1000	Li *et al* (2011)
3.	Carbon FDU-15	KOH activation	200	1410	Excellent stability	Lv *et al* (2012)
4.	Ni–Mn–LDH–CNTs	*In situ* growth process	2960	198	2000	Zhao *et al* (2014)
5.	CNC	KOH activation	112.7	470	Excellent stability	Zhang *et al* (2015)
6.	Zn–GCS	Pyrolysis	270	1499	10 000	Tang *et al* (2015)
7.	MXene–CNT based paper	Filtration process	390		10 000	Zhao *et al* (2015)
8.	PANI–CNFs	Electrospinning	407		6000	Miao *et al* (2016)
9.	Nitrogen/sulfur-doped graphene-based aerogels		203.2	110.9	3000	Chen *et al* (2018)
10.	Graphene/Ni_3S_2 composite	Template synthesis	741	125.6	5000	Wu *et al* (2018)

(*Continued*)

Table 4.1. (*Continued*)

S. No.	Materials	Process	Supercapacitor performance (F g^{-1})	SSA (m^2 g^{-1})	Stability (cycle)	References
11.	CNT/CNF composite		1119	22.2	2000	Agyemang *et al* (2018)
12.	M-CNFs	Electrospinning	272	533	3000	Li *et al* (2018)
13.	Activated carbon coated CNTs core–shell nanocomposite	KOH activation	97	289	10 000	Xu *et al* (2019)
14.	PANI–MnO_2–CNFs	Electrospinning	937.66		5000	Jalil *et al* (2020)

4.6 Conclusion and future prospects

The remarkable progress in the field of advanced energy applications has been achieved through the development of carbon-based materials such as graphene, CNTs, CNFs, aerogels, carbon composites, and flexible and wearable electrode materials. The porous (micro-, meso-, and macroporous) carbon materials with new textures have stimulated new opportunities to optimize their surface and electronic properties for various end applications. The incorporation of new surface functional groups, polymers, and surface coatings tunes the size of the pores easily in developing micro-, macro-, and mesoporous carbon-based materials. Moreover, surface oxidation and sonication of metals from carbon materials can also be used to tune the micro-, macro-, and mesoporosity of carbon-based materials. Numerous advanced methods are allowing the design of ordered structural materials to be controlled for use in high performance storage electrode materials. However, most materials produce inflexible and easily broken electrodes that are inappropriate for flexible advanced energy applications due to structural failure and/or deformation during operation. Therefore researchers continue to find novel approaches for the development of flexible and wearable carbon-based materials as electrodes that can be used easily in wearable devices without any structural failure. The flexible carbon-based materials might be next-generation materials for high-performing advanced energy applications. With the help of the approaches mentioned in this chapter we can easily modify the structure and pores of carbon-based materials, which could improve advanced energy applications or resolve the challenges of flexible and wearable advanced energy applications.

Acknowledgment

The authors acknowledge financial support by CONICYT through FONDECYT project No. 319581.

References

Afreen S, Omar R A, Talreja N, Chauhan D and Ashfaq M 2018 Carbon-based nanostructured materials for energy and environmental remediation applications *Approaches in Bioremediation: The New Era of Environmental Microbiology and Nanobiotechnology* ed R Prasad and E Aranda (Cham: Springer)

Afreen S, Talreja N, Chauhan D and Ashfaq M 2020 Polymer/metal/carbon-based hybrid materials for the detection of heavy metal ions *Multifunctional Hybrid Nanomaterials for Sustainable Agri-Food and Ecosystems* ed K A Abd-Elsalam (Amsterdam: Elsevier) ch 15

Agyemang F O, Tomboc G M, Kwofie S and Kim H 2018 Electrospun carbon nanofiber–carbon nanotubes coated polyaniline composites with improved electrochemical properties for supercapacitors *Electrochim. Acta* **259** 1110–9

Ashfaq M, Khan S and Verma N 2014 Synthesis of PVA–CAP-based biomaterial *in situ* dispersed with Cu nanoparticles and carbon micro-nanofibers for antibiotic drug delivery applications *Biochem. Eng. J.* **90** 79–89

Ashfaq M, Singh S, Sharma A and Verma N 2013 Cytotoxic evaluation of the hierarchical web of carbon micronanofibers *Ind. Eng. Chem. Res.* **52** 4672–82

Ashfaq M, Talreja N, Chuahan D and Srituravanich W 2019 Carbon nanostructure-based materials: a novel tool for detection of Alzheimer's disease *Biological, Diagnostic and Therapeutic Advances in Alzheimer's Disease: Non-Pharmacological Therapies for Alzheimer's Disease* ed G M Ashraf and A Alexiou (Singapore: Springer)

Ashfaq M, Verma N and Khan S 2016 Copper/zinc bimetal nanoparticles-dispersed carbon nanofibers: a novel potential antibiotic material *Mater. Sci. Eng.* C **59** 938–47

Ashfaq M, Verma N and Khan S 2017a Carbon nanofibers as a micronutrient carrier in plants: efficient translocation and controlled release of Cu nanoparticles *Environ. Sci. Nano* **4** 138–48

Ashfaq M, Verma N and Khan S 2017b Highly effective Cu/Zn–carbon micro/nanofiber–polymer nanocomposite-based wound dressing biomaterial against the *P. aeruginosa* multi- and extensively drug-resistant strains *Mater. Sci. Eng.* C **77** 630–41

Ashfaq M, Verma N and Khan S 2018 Novel polymeric composite grafted with metal nanoparticle-dispersed CNFs as a chemiresistive non-destructive fruit sensor material *Mater. Chem. Phys.* **217** 216–27

Bhadauriya P, Mamtani H, Ashfaq M, Raghav A, Teotia A K, Kumar A and Verma N 2018 Synthesis of yeast-immobilized and copper nanoparticle-dispersed carbon nanofiber-based diabetic wound dressing material: simultaneous control of glucose and bacterial infections *ACS Appl. Bio Mater.* **1** 246–58

Budarin V L, Clark J H, Luque R and Macquarrie D J 2007 Versatile mesoporous carbonaceous materials for acid catalysis *Chem. Commun.* **6** 634–6

Chang B, Guo Y, Yin H, Zhang S and Yang B 2015 Synthesis of sulfonated porous carbon nanospheres solid acid by a facile chemical activation route *J. Solid State Chem.* **221** 384–90

Chauhan D, Afreen S, Talreja N and Ashfaq M 2020 Multifunctional copper polymer-based nanocomposite for environmental and agricultural applications *Multifunctional Hybrid Nanomaterials for Sustainable Agri-Food and Ecosystems* ed K A Abd-Elsalam (Amsterdam: Elsevier) ch 8

Chee W K, Lim H N, Zainal Z, Huang N M, Harrison I and Andou Y 2016 Flexible graphene-based supercapacitors: a review *J. Phys. Chem.* C **120** 4153–72

Chen X, Farber M, Gao Y, Kulaots I, Suuberg E M and Hurt R H 2003 Mechanisms of surfactant adsorption on non-polar, air-oxidized and ozone-treated carbon surfaces *Carbon* **41** 1489–500

Chen Y, Liu Z, Sun L, Lu Z and Zhuo K 2018 Nitrogen and sulfur co-doped porous graphene aerogel as an efficient electrode material for high performance supercapacitor in ionic liquid electrolyte *J. Power Sources* **390** 215–23

Chen Z, To J W F, Wang C, Lu Z, Liu N, Chortos A, Pan L, Wei F, Cui Y and Bao Z 2014 A three-dimensionally interconnected carbon nanotube–conducting polymer hydrogel network for high-performance flexible battery electrodes *Adv. Energy Mater.* **4** 1400207

Cheng P-Z and Teng H 2003 Electrochemical responses from surface oxides present on HNO_3-treated carbons *Carbon* **41** 2057–63

Cheng Y, Lu S, Zhang H, Varanasi C V and Liu J 2012 Synergistic effects from graphene and carbon nanotubes enable flexible and robust electrodes for high-performance supercapacitors *Nano Lett.* **12** 4206–11

Choi M and Ryoo R 2007 Mesoporous carbons with KOH activated framework and their hydrogen adsorption *J. Mater. Chem.* **17** 4204–9

De S, Balu A M, van der Waal J C and Luque R 2015 Biomass-derived porous carbon materials: synthesis and catalytic applications *ChemCatChem* **7** 1608–29

Eftekhari A and Fan Z 2017 Ordered mesoporous carbon and its applications for electrochemical energy storage and conversion *Mater. Chem. Front.* **1** 1001–27

Gao W, Ota H, Kiriya D, Takei K and Javey A 2019 Flexible electronics toward wearable sensing *Acc. Chem. Res.* **52** 523–33

Gao Y, Zheng S, Fu H, Ma J, Xu X, Guan L, Wu H and Wu Z-S 2020 Three-dimensional nitrogen doped hierarchically porous carbon aerogels with ultrahigh SSA for high-performance supercapacitors and flexible micro-supercapacitors *Carbon* **168** 701–9

Ghosh S, Yong W D, Jin E M, Polaki S R, Jeong S M and Jun H 2019 Mesoporous carbon nanofiber engineered for improved supercapacitor performance *Korean J. Chem. Eng.* **36** 312–20

Guo W-H, Liu T-J, Jiang P and Zhang Z-J 2015 Free-standing porous manganese dioxide/graphene composite films for high performance supercapacitors *J. Colloid Interface Sci.* **437** 304–10

Huang S, Liu Y, Zhao Y, Ren Z and Guo C F 2019 Flexible electronics: stretchable electrodes and their future *Adv. Funct. Mater.* **29** 1805924

Iqbal S, Khatoon H, Hussain Pandit A and Ahmad S 2019 Recent development of carbon based materials for advanced energy application *Mater. Sci. Energy Technol.* **2** 417–28

Jalil N A, M Abdah M A A, Azman N H N and Sulaiman Y 2020 Polyaniline and manganese oxide decorated on carbon nanofibers as a superior electrode material for supercapacitor *J. Electroanal. Chem.* **867** 114188

Ji Y, Zou H, Zhang P, Ding H, Yang X, Li B, Wei J and Wei X 2020 Carbon aerogels with mutual support structures constructed by hybrid hydrogels: robust energy storage materials *Mater. Today Commun.* **25** 101444

Khan Y, Ostfeld A E, Lochner C M, Pierre A and Arias A C 2016 Monitoring of vital signs with flexible and wearable medical devices *Adv. Mater.* **28** 4373–95

Khare P, Talreja N, Deva D, Sharma A and Verma N 2013 Carbon nanofibers containing metal-doped porous carbon beads for environmental remediation applications *Chem. Eng. J.* **229** 72–81

Kumar D and Talreja N 2019 Nickel nanoparticles-doped rhodamine grafted carbon nanofibers as colorimetric probe: naked eye detection and highly sensitive measurement of aqueous Cr^{3+} and Pb^{2+} *Korean J. Chem. Eng.* **36** 126–35

Kumar R, Ashfaq M and Verma N 2018 Synthesis of novel PVA–starch formulation-supported Cu–Zn nanoparticle carrying carbon nanofibers as a nanofertilizer: controlled release of micronutrients *J. Mater. Sci.* **53** 7150–64

Lee T H, Pham D T, Sahoo R, Seok J, Luu T H T and Lee Y H 2018 High energy density and enhanced stability of asymmetric supercapacitors with mesoporous MnO_2@CNT and nanodot MoO_3@CNT free-standing films *Energy Storage Mater.* **12** 223–31

Lee Y H *et al* 2013 Wearable textile battery rechargeable by solar energy *Nano Lett.* **13** 5753–61

Li F, Xie L, Sun G, Kong Q, Su F, Cao Y, Wei J, Ahmad A, Guo X and Chen C-M 2019 Resorcinol-formaldehyde based carbon aerogel: preparation, structure and applications in advanced energy application *Microporous Mesoporous Mater.* **279** 293–315

Li W, Zhang F, Dou Y, Wu Z, Liu H, Qian X, Gu D, Xia Y, Tu B and Zhao D 2011 A self-template strategy for the synthesis of mesoporous carbon nanofibers as advanced super-capacitor electrodes *Adv. Energy Mater.* **1** 382–6

Li Y, Ou-Yang W, Xu X, Wang M, Hou S, Lu T, Yao Y and Pan L 2018 Micro-/mesoporous carbon nanofibers embedded with ordered carbon for flexible supercapacitors *Electrochim. Acta* **271** 591–8

Liu P, Han J-J, Jiang L-F, Li Z-Y and Cheng J-N 2017a Polyaniline/multi-walled carbon nanotubes composite with core–shell structures as a cathode material for rechargeable lithium-polymer cells *Appl. Surf. Sci.* **400** 446–52

Liu W-J, Jiang H and Yu H-Q 2019 Emerging applications of biochar-based materials for energy storage and conversion *Energy Environ. Sci.* **12** 1751–79

Liu Y, He K, Chen G, Leow W R and Chen X 2017b Nature-inspired structural materials for flexible electronic devices *Chem. Rev.* **117** 12893–941

Lv Y, Zhang F, Dou Y, Zhai Y, Wang J, Liu H, Xia Y, Tu B and Zhao D 2012 A comprehensive study on KOH activation of ordered mesoporous carbons and their supercapacitor application *J. Mater. Chem.* **22** 93–9

Miao F, Shao C, Li X, Lu N, Wang K, Zhang X and Liu Y 2016 Polyaniline-coated electrospun carbon nanofibers with high mass loading and enhanced capacitive performance as free-standing electrodes for flexible solid-state supercapacitors *Energy* **95** 233–41

Mukhopadhyay S M, Pulikollu R V and Roy A K 2004 Surface modification of a microcellular porous solid: carbon foam *Appl. Surf. Sci.* **225** 223–8

Negara D N K P, Nindhia T G T, Surata I W, Hidajat F and Sucipta M 2020 Textural characteristics of activated carbons derived from tabah bamboo manufactured by using H_3PO_4 chemical activation *Mater. Today Proc.* **22** 148–55

Om Prakash M, Raghavendra G, Ojha S and Panchal M 2021 Characterization of porous activated carbon prepared from arhar stalks by single step chemical activation method *Mater. Today Proc.* **39** 1476–81

Omar R A, Afreen S, Talreja N, Chauhan D and Ashfaq M 2019 Impact of nanomaterials in plant systems *Plant Nanobionics: Advances in the Understanding of Nanomaterials Research and Applications*vol 1 ed R Prasad (Cham: Springer)

Otake Y and Jenkins R G 1993 Characterization of oxygen-containing surface complexes created on a microporous carbon by air and nitric acid treatment *Carbon* **31** 109–21

Park S, Vosguerichian M and Bao Z 2013 A review of fabrication and applications of carbon nanotube film-based flexible electronics *Nanoscale* **5** 1727–52

Pérez-Mayoral E, Matos I, Bernardo M and Fonseca I M 2019 New and advanced porous carbon materials in fine chemical synthesis. Emerging precursors of porous carbons *Catalysts* **9** 133

Pradhan B K and Sandle N K 1999 Effect of different oxidizing agent treatments on the surface properties of activated carbons *Carbon* **37** 1323–32

Qian M, Wang Z, Li Z, Xu J, Sun P, Lin J, Lin T and Huang F 2019 Sol–gel assisted chemical activation for nitrogen doped porous carbon *Microporous Mesoporous Mater.* **286** 18–24

Shao Y, El-Kady M F, Wang L J, Zhang Q, Li Y, Wang H, Mousavi M F and Kaner R B 2015 Graphene-based materials for flexible supercapacitors *Chem. Soc. Rev.* **44** 3639–65

Singh S, Ashfaq M, Singh R K, Joshi H C, Srivastava A, Sharma A and Verma N 2013 Preparation of surfactant-mediated silver and copper nanoparticles dispersed in hierarchical carbon micro-nanofibers for antibacterial applications *New Biotechnol.* **30** 656–65

Singh S K, Dhavale V M, Boukherroub R, Kurungot S and Szunerits S 2017 N-doped porous reduced graphene oxide as an efficient electrode material for high performance flexible solid-state supercapacitor *Appl. Mater. Today* **8** 141–9

Stoppa M and Chiolerio A 2014 Wearable electronics and smart textiles: a critical review *Sensors* **14** 11957–92

Talreja N, Jung S, Yen L T H and Kim T 2020 Phenol-formaldehyde-resin-based activated carbons with controlled pore size distribution for high-performance supercapacitors *Chem. Eng. J.* **379** 122332

Talreja N, Kumar D and Verma N 2014 Removal of hexavalent chromium from water using Fe-grown carbon nanofibers containing porous carbon microbeads *J. Water Process Eng.* **3** 34–45

Talreja N, Verma N and Kumar D 2016 Carbon bead-supported ethylene diamine-functionalized carbon nanofibers: an efficient adsorbent for salicylic acid *CLEAN—Soil Air Water* **44** 1461–70

Tan J, Han Y, He L, Dong Y, Xu X, Liu D, Yan H, Yu Q, Huang C and Mai L 2017 *In situ* nitrogen-doped mesoporous carbon nanofibers as flexible freestanding electrodes for high-performance supercapacitors *J. Mater. Chem.* A **5** 23620–7

Tang J, Salunkhe R R, Liu J, Torad N L, Imura M, Furukawa S and Yamauchi Y 2015 Thermal conversion of core–shell metal–organic frameworks: a new method for selectively functionalized nanoporous hybrid carbon *J. Am. Chem. Soc.* **137** 1572–80

Vinke P, van der Eijk M, Verbree M, Voskamp A F and van Bekkum H 1994 Modification of the surfaces of a gas activated carbon and a chemically activated carbon with nitric acid, hypochlorite, and ammonia *Carbon* **32** 675–86

Wang C, Liang W, Yang Y, Liu F, Sun H, Zhu Z and Li A 2020 Biomass carbon aerogels based shape-stable phase change composites with high light-to-thermal efficiency for energy storage *Renew. Energy* **153** 182–92

Wang K, Wang Y, Wang Y, Hosono E and Zhou H 2009 Mesoporous carbon nanofibers for supercapacitor application *J. Phys. Chem.* C **113** 1093–7

Wang L, Tan H, Chen J, Zhang H, Li Z and Qiu H 2019 Porous graphene synthesized by partial combustion for high-performance supercapacitors *Mater. Lett.* **252** 345–8

Wang X, Liu R, Waje M M, Chen Z, Yan Y, Bozhilov K N and Feng P 2007 Sulfonated ordered mesoporous carbon as a stable and highly active protonic acid catalyst *Chem. Mater.* **19** 2395–7

Weng W, Chen P, He S, Sun X and Peng H 2016 Smart electronic textiles *Angew. Chem. Int. Ed.* **55** 6140–69

Wu P, Wang D, Ning J, Zhang J, Feng X, Dong J and Hao Y 2018 Novel 3D porous graphene/ Ni_3S_2 nanostructures for high-performance supercapacitor electrodes *J. Alloys Compd.* **731** 1063–8

Xin W and Song Y 2015 Mesoporous carbons: recent advances in synthesis and typical applications *RSC Adv.* **5** 83239–85

Xing R, Liu Y, Wang Y, Chen L, Wu H, Jiang Y, He M and Wu P 2007 Active solid acid catalysts prepared by sulfonation of carbonization-controlled mesoporous carbon materials *Microporous Mesoporous Mater.* **105** 41–8

Xu F, Xu J, Xu H, Lu Y, Yang H, Tang Z, Lu Z, Fu R and Wu D 2017 Fabrication of novel powdery carbon aerogels with high SAs for superior energy storage *Energy Storage Mater.* **7** 8–16

Xu J, Wang X, Zhou X, Yuan N, Ge S and Ding J 2019 Activated carbon coated CNT core–shell nanocomposite for supercapacitor electrode with excellent rate performance at low temperature *Electrochim. Acta* **301** 478–86

Yan Y, Wei J, Zhang F, Meng Y, Tu B and Zhao D 2008 The pore structure evolution and stability of mesoporous carbon FDU-15 under CO_2, O_2 or water vapor atmospheres *Microporous Mesoporous Mater.* **113** 305–14

Yang Y and Gao W 2019 Wearable and flexible electronics for continuous molecular monitoring *Chem. Soc. Rev.* **48** 1465–91

Yu Z-L *et al* 2016 Ion-catalyzed synthesis of microporous hard carbon embedded with expanded nanographite for enhanced lithium/sodium storage *J. Am. Chem. Soc.* **138** 14915–22

Zhang M, He C, Liu E, Zhu S, Shi C, Li J, Li Q and Zhao N 2015 Activated carbon nanochains with tailored micro-meso pore structures and their application for supercapacitors *J. Phys. Chem.* C **119** 21810–7

Zhao J, Chen J, Xu S, Shao M, Zhang Q, Wei F, Ma J, Wei M, Evans D G and Duan X 2014 Hierarchical NiMn layered double hydroxide/carbon nanotubes architecture with superb energy density for flexible supercapacitors *Adv. Funct. Mater.* **24** 2938–46

Zhao M-Q, Ren C E, Ling Z, Lukatskaya M R, Zhang C, Van Aken K L, Barsoum M W and Gogotsi Y 2015 Flexible MXene/carbon nanotube composite paper with high volumetric capacitance *Adv. Mater.* **27** 339–45

Zhong J, Zhang Y, Zhong Q, Hu Q, Hu B, Wang Z L and Zhou J 2014 Fiber-based generator for wearable electronics and mobile medication *ACS Nano* **8** 6273–80

Zhu J, Yang D, Yin Z, Yan Q and Zhang H 2014 Graphene and graphene-based materials for energy storage applications *Small* **10** 3480–98

Zou Z, Zhao J, Xue J, Huang R and Jiang C 2017 Highly porous carbon spheres prepared by boron-templating and reactive H_3PO_4 activation as electrode of supercapacitors *J. Electroanal. Chem.* **799** 187–93

IOP Publishing

Nanostructured Materials for Sustainable Energy and Environmental Remediation

R V Mangalaraja, R Udayabhaskar, P Sathishkumar and Joydeep Dutta

Chapter 5

Platinum based alloy nanostructure electrocatalyst for oxygen reduction reaction in polymer electrolyte membrane fuel cells

Moorthi Lokanathan, Bhalchandra Kakade and R V Mangalaraja

5.1 Introduction

Fuel cells with near-zero emissions and high energy conversion efficiencies have emerged as potential alternative energy conversion devices for stationary, mobile, and portable power applications [1–3]. They are the most compact electrochemical energy conversion devices which generate electrical energy from chemicals with water vapour as a by-product [4]. Thus low-temperature fuel cells, specifically proton exchange membrane fuel cells (PEMFCs), are considered environmentally sustainable and highly effective power sources because of their high efficiency, high energy density, simple operation, and rapid start-up/shut-down cycle under ambient conditions [5]. However, the efficient commercialization of fuel cells is limited by the sluggish kinetics of the oxygen reduction reaction (ORR) at the cathode of the PEMFCs. In addition, metal and non-metal catalysts (except for platinum) dissolve significantly at the cathode due to the formation of an acidic environment. Limitations of platinum are its higher cost and low availability. Hence it is crucial to promote electrocatalysts with improved catalytic activities and lower Pt content to achieve the true commercialization of fuel cells. Significant improvement in electrocatalytic activity can be accomplished by the dispersion of Pt based nano-structure electrocatalysts on high-surface-area support materials with the additional improvement of reduced noble metal utilization [7, 8]. This chapter presents an outline of platinum as monomer, binary, and ternary alloy nanostructures for cathode electrocatalysts in low-temperature fuel cells.

doi:10.1088/978-0-7503-5143-0ch5

5.2 Polymer electrolyte membrane fuel cells (PEMFCs)

5.2.1 The working principle of PEMFCs

In a PEMFC, the anode and cathode catalyst layers are separated by a polymer or proton electrolyte membrane, which together is called a membrane electrode assembly (MEA), and the polymer electrolyte membrane allows only protons to conduct through it.

A diagram of the usual PEMFC single cell is shown in figure 5.1. At the anode, hydrogen splits into a proton (H^+) and electron (e^-), called the hydrogen oxidation reaction (HOR), with an equilibrium potential of 0 V in the electrolyte at the platinum electrode surface [5–7], as shown in equation (5.1). The proton passes through a membrane (polyelectrolyte) due to its polymeric nature, while the electron passes through the external circuit to conduct electricity. At the cathode, oxygen is reduced to an anion with an equilibrium potential of 1.23 V, called the oxygen reduction reaction (ORR), as shown in equation (5.2).

$$\text{Anode: } H_2 \rightarrow 2H^+ + 2e^- \quad E^\circ = 0.00 \text{ V.} \tag{5.1}$$

$$\text{Cathode: } 1/2O_2 + 2H^+ + 2e^- \rightarrow H_2O \quad E^\circ = 1.23 \text{ V.} \tag{5.2}$$

$$\text{Overall: } H_2 + 1/2O_2 \rightarrow H_2O \quad E_{\text{Cell}} = 1.23 \text{ V.} \tag{5.3}$$

Further, the reduced oxygen fuses with the H^+ ions penetrating through the membrane and external circuit electron (e^-) and forms water (H_2O) as a by-product. Finally, an overall cell potential of 1.23 V is obtained due to the anodic HOR and cathodic ORR processes, as represented in equation (5.3).

5.2.2 Challenges of PEMFCs

Despite the zero carbon emission and higher theoretical efficiency of PEMFCs, there arise several concerning challenges, such as overpotentials because of reactant activation (due to the slow kinetics of ORR and HOR), electrical resistances (contacts and electrodes give resistance to electron conductance and the polymer electrolyte membrane gives resistance to proton conduction), and mass transport at elevated current densities (through the gas diffusion layer), as shown in figure 5.2. Notably, activation overpotentials are of vast importance for current studies aiming to advance the activity of the catalyst layer, particularly the cathodic part of the PEM fuel cell, where platinum plays a vital role in commercial fuel cells.

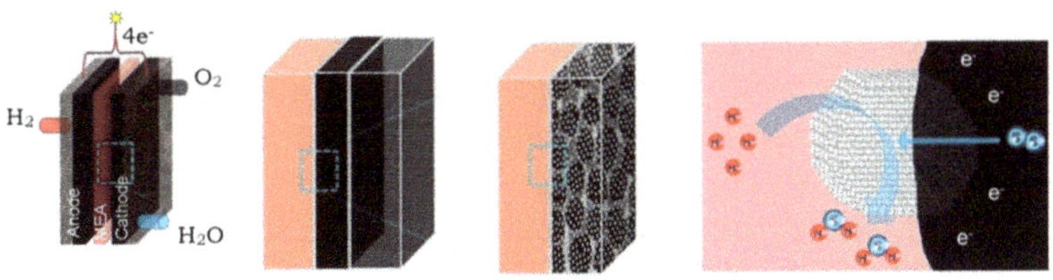

Figure 5.1. Schematic illustration of a typical single PEMFC.

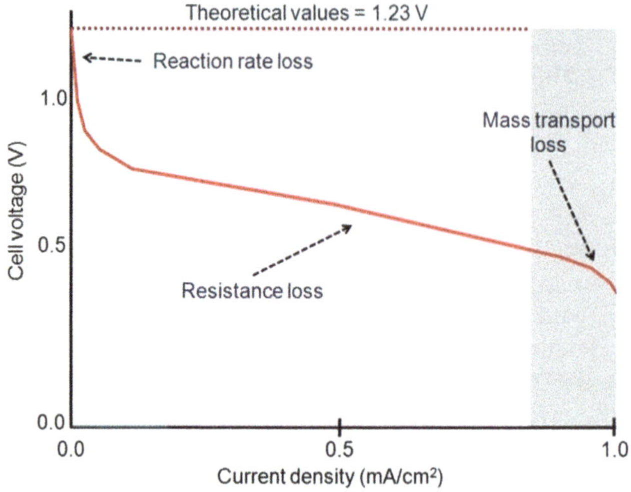

Figure 5.2. Three losses in a PEMFC due to various overpotentials.

5.2.3 ORR mechanisms on Pt and Pt based alloy surfaces

The mechanism for ORR on a Pt surface proposed by Toda *et al* [9] is shown in figure 5.3(a), and was somewhat adapted from that proposed by Yeager *et al* [10]. An increase in the donation of an electron from 2π of O_2 to the Pt surface is attained by an increase in 5d vacancies of the Pt surface by alloying it with transition metals (Ni, Fe, Co, etc), resulting in increasing O_2 adsorption, which leads to weakening of the O–O bond [11]. Upon back-donation from the Pt-5d orbitals to the adsorbed O_2-π^* orbitals an instantaneous O–O bond breaking takes place. Further, the back-donation process suppressed suddenly by the 5d vacancy of the Pt surface increases when it is alloyed with above 50% transition metal (beyond optimum content), resulting in the formation of stronger Pt–O bonds leading to a reduction of the reaction rate, as shown in figure 5.3(b). Therefore, alloying Pt with an optimal composition of transition metals could perhaps furnish an enhancement in the kinetics of the ORR and thus this is necessary to understand in the current status of the research in this area before committing to such a significant problem as reducing the activation over-potential in a typical fuel cell.

5.3 Electrochemical measurement and activity calculation

5.3.1 Preparation of the catalyst ink and a modified working electrode

The electrochemical measurements were accomplished in a typical three-electrode electrochemical cell system. The reversible hydrogen electrode (RHE) and a platinum (Pt) wire were used as the reference and counter electrode, respectively. The glassy carbon (GC) electrode was polished, followed by careful rinsing and ultrasonication for ultra-cleaning, using ultrapure water. The catalyst ink was drop-casted on the cleaned GC electrode utilizing a trivial modification of a previous report [12]. In brief, 2 mg catalyst was dissolved in 1 ml stock solution (containing

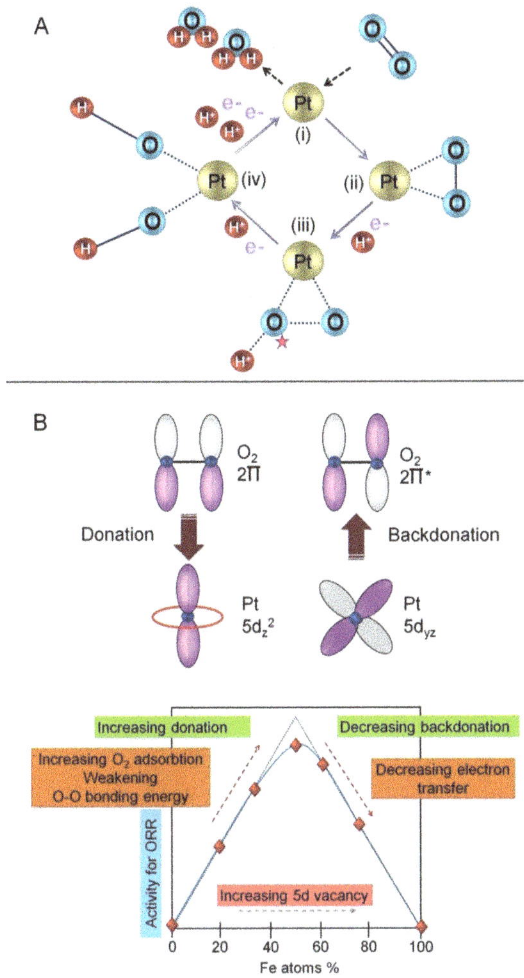

Figure 5.3. (a) ORR mechanisms on a Pt surface [9]. (b) Illustration of proposed ORR enhancement mechanism through alloying Pt with Fe-group metals [9].

200 μl IPA, 796 μl DI, water and 4 μl Nafion). Additionally, sonication was used to produce a uniform slurry followed by this catalyst ink being drop cast onto the GC electrode surface. The electrode was dried overnight and constant Pt loading was maintained for all catalysts for a better evaluation of activities. The rotating disc electrode (RDE) and rotating ring disc electrode (RRDE) and a standard three-electrode electrochemical cell are shown in figure 5.4.

5.3.2 Cyclic voltammetry (CV)

The cyclic voltammograms (CVs) were carried out in N_2 saturated 0.1 M $HClO_4$ solution to clean the electrode surface electrochemically until a constant CV curve was obtained. The electrochemical active surface area (ECSA) was determined

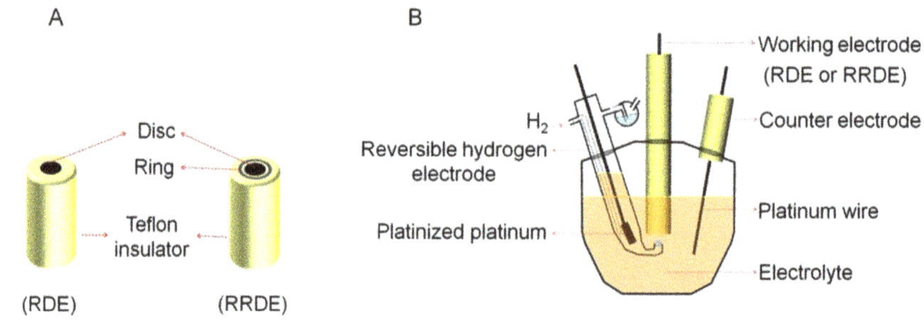

Figure 5.4. (a) Two different working electrodes (RDE and RRDE). (b) Schematic illustration of a three-electrode electrochemical cell.

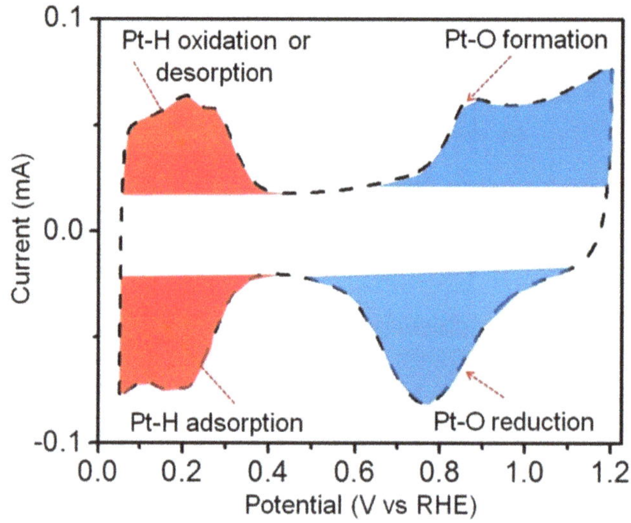

Figure 5.5. Typical cyclic voltommetry of Pt and Pt based alloys.

under similar conditions. The H-under potential deposition region was used to obtain the CSA of the catalyst. Figure 5.5 shows the typical CV of the carbon-supported Pt based alloy catalyst and the area under the Pt–H desorption/adsorption region utilized to determine the ECSA:

$$\mathrm{ECSA} = \frac{Q_{\mathrm{H}}}{Q_{\mathrm{Pt}} \times L_{\mathrm{Pt}}} \times 10^5\,(\mathrm{m^2 g_{Pt}^{-1}}), \tag{5.4}$$

where Q_{H} is the charge required for H desorption/adsorption, Q_{Pt} is the charge required for H desorption/adsorption in-unit surface area Pt polycrystalline (210 μC), and L_{Pt} is the metal loading (Pt) [13].

5.3.3 Rotating disc electrode (RDE) study

Anodic ORR polarization curves were recorded by rotating the working electrode (RDE) at 1600 rpm (to separate the diffusion contribution from the

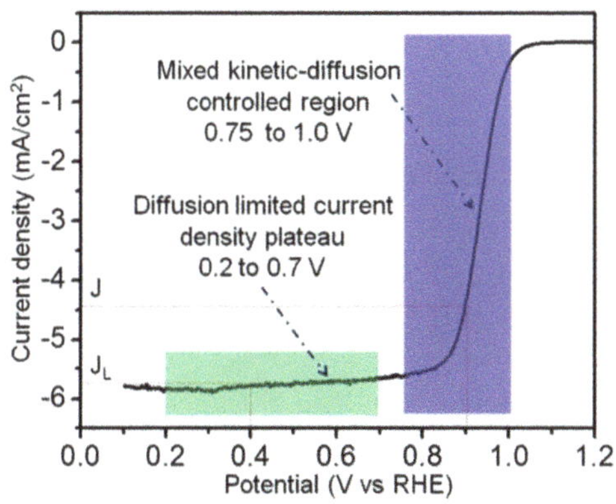

Figure 5.6. Typical ORR polarization curve of Pt and Pt based alloys.

kinetic current, the Koutecky–Levich approach could be followed) to deduce ORR activities in terms of specific activity and mass activity. Figure 5.6 exhibits the typical linear sweeping voltammogram (LSV) of carbon-supported Pt-based alloy catalyst. The kinetic current density (5.6) was extracted from the ORR polarization curves by the Koutecky–Levich relation (5.5):

$$\frac{1}{J} = \frac{1}{J_K} + \frac{1}{J_L}, \tag{5.5}$$

$$J_K = \frac{J_L \times J}{J_L - J} \ (\text{mA cm}^{-2}), \tag{5.6}$$

$$\text{Mass activity: } I_m = \frac{J_K}{L_{Pt}} \ (\text{mA mg}_{Pt}^{-1}), \tag{5.7}$$

$$\text{Specific activity: } I_s = \frac{J_K \times Q_{Pt}}{Q_H} \ (\mu\text{A cm}_{Pt}^{-2}), \tag{5.8}$$

where J is the current density at 0.9 V versus RHE, J_K is the kinetic current density, and J_L is the limiting diffusion current density. Further, the mass activity (I_m) is determined by normalizing J_K by mass loading, as shown in equation (5.7), and the specific activity (I_s) is determined by normalizing the J_K by the ECSA of the catalysts, as shown in equation (5.8) [13].

5.3.4 Rotating ring disc electrode (RRDE) study

In the case of rotating ring-disk electrode (RRDE) studies, the working electrodes were prepared by a similar method to that of the RDE studies. Additionally, the ring potential was kept at a certain constant value versus RHE. The amount of peroxide (H_2O_2) production and the number of electron transfers (n) were determined by

$$\%H_2O_2 = 200 \times \frac{I_r}{I_d N + I_r}, \tag{5.9}$$

$$n = 4 \times \frac{I_d N}{I_d N + I_r}, \tag{5.10}$$

where I_r and I_d are the ring and disk current, respectively, and N is the current collection efficiency of the Pt ring [14].

5.3.5 Durability or stability test

A very important parameter is the durability of the catalysts (Pt) used to determine superior and viable electrocatalysis. The durability test was carried out with a scan rate of 100 mV s^{-1} (using a triangular E–t wave) in N_2 or O_2 saturated 0.1 M $HClO_4$ with cycling between the potential window of 0.6 V to 1.0 V versus RHE.

5.4 A platinum based nanostructure electrocatalyst for ORR

As discussed previously, PEMFCs are efficient and environmentally friendly, making them the best option power sources for stationary and portable automotive applications [15]. However, high cost of platinum (Pt), the low durability of the catalysts and the support materials, and the sluggish kinetics of ORR (cathodic overpotential), are the most important limitations to the commercialization of such fuel cells [16, 17]. Other issues associated with hydrogen infrastructure, such as production, storage, and protection, are similarly significant areas to be considered in detail. Consequently, the reduction of Pt content (low or ultralow Pt content) without sacrifice the electrocatalytic activity and stability can be a binary perception to be esteemed for permanent PEMFC technology. As mentioned previously, alloying Pt with the group B metals (transition metals) should resolve the essential issues of the low Pt content approach. To ascertain an efficient electrocatalyst with sustained activity a number of approaches have been designed; a few of the significant examples, such as Pt monometallic, and Pt based bimetallic and trimetallic nanostructure are discussed here for comparison purposes.

5.4.1 Pt monometallic nanostructures

A methodical investigation of the Pt metal loading and enhanced catalyst interruption dependant oxygen reduction reaction (ORR) activities has been considered [18]. High ORR activity and superior durability have been attained through the embellishment of Pt nanoparticles (NPs) on single-walled carbon nanotubes (SWCNTs) [19]. Platinum nanocubes with 7 nm size show four-fold superior current density compared to polyhedrals with 3 nm size or truncated cubic Pt NPs with 5 nm size, demonstrating the influence of nanoparticle shape and size on the ORR performance [20]. A multi-octahedral (a quasi-octahedral shape) of Pt nanocrystals that consists of stepped surface atoms, and an abundant number of edges and corners demonstrate greater specific activity owing to highly exposed surfaces with {111} and {100} facets [21]. An ultrathin diameter with 1.3 ± 0.4 nm of unsupported

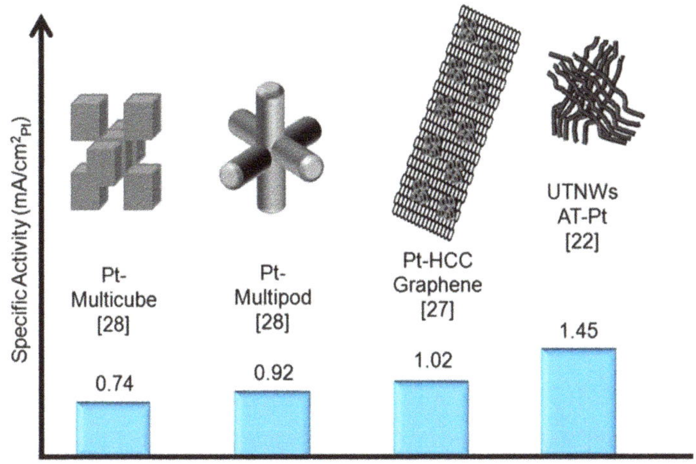

Figure 5.7. Comparative specific activities (I_s) of shape-dependent Pt monometallic electrocatalysts [22, 27, 28].

Pt nanowires was exfoliated in the course of acid washing that exhibited a specific activity of 1.45 mA cm_{Pt}^{-2} which was four times higher than for the Pt nanotube nanostructures and seven times higher than that of commercial Pt/C [22].

The influence of the surface steps of Pt nanoparticles on ORR activity has been studied through *in situ* x-ray absorption techniques [23]. For 2.2 nm size, in a range of particle sizes from 1 nm to 5 nm, the Pt nanoparticles with {111} facets demonstrated four times superior activity compared to the other sizes and this was also established by DFT calculations [24]. Additionally, Pt nanoparticles with 2.27 nm size evenly dispersed over tungsten oxide nanowires with a carbon paper support, called Pt/C–$W_{18}O_{49}$–NWs/C, was effectively synthesized through a chemical vapour deposition method and expressed superior electrocatalytic activity and excellent durability [25]. Pt nanowire arrays with diameters of 2–5 nm formed on sulphur-doped graphene, which act as nucleation sites, and concentrated surface coverage revealed two to three times superior ORR electrocatalytic activity [26]. A unique hybrid structure of highly concave cubic (HCC) Pt nanocrystals demonstrated approximately six times (20.847 A mg_{Pt}^{-1}) and seven times (21.02 mA cm^{-2}) higher mass and specific activity than commercial Pt/C, respectively, along with a surprising positive shift in the $E_{1/2}$ (0.967 V) value compared to that of commercial Pt/C [27]. (100) facets over Pt cubes, and Pt multicubes and -pods (a Pt branched structure) were prepared, among which the Pt multicubes displayed surprising ORR activity (superior current density and a positive shift in $E_{1/2}$) comparable to active Pt based catalysts with improved preservation of the activity after the durability test due to high-index facets at the intersection of two cubes [28]. The inter-atomic distance modification of Pt atoms diminished the adsorption tendency of oxygenated species to the Pt surface in the case of the highly crystalline nature of strained Pt films [29]. The overall ORR activities of Pt based monometallic nanostructures with different geometries of electrocatalysts are tabulated in table 5.1 and the topmost activities are shown in figure 5.7.

Table 5.1. Activity comparisons of Pt monometallic electrocatalysts.

		Activity @ 0.9 V versus RHE			
Pt monometallic nanostructure	Shape of nanostructure	ECSA ($m^2\,g_{Pt}^{-1}$)	I_m (mA mg_{Pt}^{-1})	I_s (mA cm_{Pt}^{-2})	Reference
Pt/CB	Particles			0.65 (0.8 V)	[18]
Pt/SWCNT	Particles	17.8			[19]
Pt multi-octahedral	Multi-octahedral	1.05 cm^2	0.28	0.8 (0.85 V)	[21]
Pt UTNWs	Ultrathin nanowires			0.52	[22]
Acid treated Pt UTNWs				1.45	
Surface step on Pt NPs	Particles			0.75 (0.85 V)	[23]
2.5 nm Pt NPs	Particles	80	0.12	0.09 (0.93 V)	[24]
Pt–$W_{18}O_{49}$–NW/carbon paper	Nanowires	63.5	0.83	1.3	[25]
Pt nanowire array in G and SG	Nanowire array	40		0.7 (0.85 V)	[26]
Pt nanocubes	Nanocubes			0.7 (0.85 V)	
Pt highly concave cubes–graphene	Highly concave cubes	7.3 cm^2	0.86	1.02	[27]
Pt multicubes	Multicubes		0.55	0.74	[28]
Pt multipods	Multipods		0.72	0.92	
Relaxed Pt film	Film			0.021	[29]
Strained Pt film	Film			0.168	

5.4.2 Pt based binary alloy nanostructures

As discussed previously, the electronic structure of Pt can be tuned by varying the position of the metal d states relative to the Fermi level, which could be attained by alloying Pt with 3d transition metals (polycrystalline form of Pt_3M alloy films, where M = Ni, Co, Fe, and Ti), as a result, the ORR activity could be boosted (figure 5.7) [30]. The d-band centre position in their electronic structure is highly optimized in the unique surface of $Pt_3Ni(111)$ compared to bare Pt with (111) facets and the former exhibited ten times higher electrocatalytic activity than the latter, as well as 90 times higher activity than a commercial Pt/C [31]. The balance between the adsorption energies of the reactive intermediates and surface coverage by a spectator (blocking) species acts as an indicator to estimate the highest catalytic activity for Pt_3Co through optimizing the donation and back-donation attained by alloying 3d transition metal with Pt exhibiting volcano-type behaviour (higher activity conditions) [32]. Peng *et al* reported that Pt nanoparticles on palladium nanoparticles support excellent ORR activity compared to pure Pt [33]. A value of 4.5 nm of Pt_3Co electrocatalyst demonstrated excellent specific activity and electrochemically active surface area (ECSA) compared to 3 nm to 9 nm of Pt_3Co nanoparticles, which strongly confirmed the influence of size on electrocatalytic activity [34]. Interestingly, extraordinary durability was displayed by core–shell Pd_9Au_1/Pt nanoparticles because of their shifting towards the positive side of the oxidation potential [35]. Mn–Pt binary alloy nanocubes demonstrated higher ORR activity than Pt/C nanoparticles, revealing that the nanocubes play a vital role in ORR activity [36]. An efficient investigation of structure-dependent electrocatalytic activity revealed ordered (face-centred tetragonal) fct-FePt and disordered (face-centred cubic) fcc-FePt binary alloys [37]. Pt bimetallic (Pt–M, M = Pd, Ni, Co, Fe) materials with nanocrystals of octahedral and cubic morphologies were prepared by the GRAILS method and their specific activities were 1.26 and 0.85 mA cm_{Pt}^{-2}, respectively [38]. Similar sized Pt_3Ni binary nanoctahedra and nanocube alloys comprising {111} and {100} facets, respectively, were also studied, where the nanoctahedra of Pt_3Ni alloy with {111} facets exhibits five times higher electrocatalytic activity than nanocubes with {100} facets [39]. Pt_3Ni binary alloy with a truncated octahedral geometry conveyed four times better ORR catalytic activity compared to Pt/C, as reported by Wu *et al* [40]. A size of 7 nm for FePt nanoparticles supported on graphene also demonstrated excellent ORR activity and enduring stability compared to Pt/C [41]. A Pt_3Co binary alloy nanostructure with truncated octahedra shape revealed surprising ORR activity and greater durability than Pt_3Co binary alloy nanostructure with nanocubes and commercial Pt/C [42]. The optimized shape and surface composition selectivity were revealed by 9.5 nm of octahedral PtNi nanoparticles which produced excellent ORR activity with ten times higher specific (~3.14 mA cm_{Pt}^{-2}) and mass (~1.45 A mg_{Pt}^{-1}) activity, respectively, than Pt/C, which are nearly the predicted theoretical values [43]. Pt–M binary alloys (M = Au, Ni, Pd) with icosahedral geometries showed extraordinary electrocatalytic activity with specific and mass activity of 1.83 mA cm_{Pt}^{-2} and 0.62 A mg_{Pt}^{-1}, respectively [44]. Kakade *et al* reported an interesting trend in the ORR activity of CoPt and PdPt

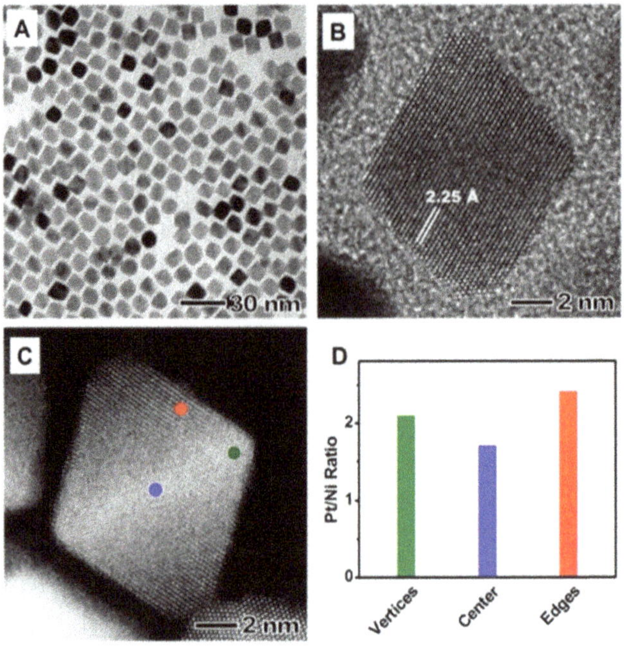

Figure 5.8. (a) Transmission electron microscopy and (b) High resolution transmission electron microscopy images of as-obtained Pt–Ni octahedra with 9 nm edge length synthesized in benzyl ether containing Olyelamine and olic acid at low concentrations. (c) High angle anular darkfield scanning transmission electron microscopy image and (d) energy dispersive spectroscopy spot analysis for the elemental distribution of Pt/Ni ratios at different positions of the as-obtained Pt–Ni octahedron. (Reproduced with permission from [46]. Copyright 2013 American Chemical Society.)

binary alloys [45]. Choi *et al* reported 51 times higher mass (3.3 A mg_{Pt}^{-1}) and specific activity compared to Pt/C at 0.9 V attained by identical 9 nm of Pt–Ni octahedra with {111} facets (figure 5.8) [46]. Pt and Ni atoms on Pt_xNi_{1-x} nano-octahedra were systematically investigated by Cui *et al*, where Ni atoms were favourably presented in their {111} facet region and Pt atoms were common at corners and edges [47].

Compared to Pt/C, 36 times and 22 times higher mass and specific activities were expressed by Pt_3Ni nanoframe electrocatalysts with 3D molecular accessibility [48]. A Pt shell on {111} facet rich Pd octahedrons had 28 times superior area-specific activity than A Pt shell on {100} facet rich Pd nanocubes, revealing that the ORR activity is boosted by particular facets of specific shape [49]. Three atomic layer thick ordered Pt_3Co binary intermetallic material demonstrated 300% and 200% specific and mass activity, respectively, compared to disordered Pt_3Co binary alloy nanoparticles [50]. Two times enhanced mass activity compared to the Department of Energy (DOE) US target for PEMFC was exhibited by platinum coated nickel nanowires [51]. The Pt–Ni octahedron expressed enhanced mass activity of 1.96 A mg_{Pt}^{-1} and specific activity of 3.99 mA cm_{Pt}^{-2} at 0.9 V versus RHE, respectively, because of the influence of CO in the synthesis of the same electrocatalyst reported by Zhang *et al* [52]. Platinum coated on cobalt nanowire core–shell electrocatalysts revealed specific and mass activity of

2.4 mA cm_{Pt}^{-2} and 793 mA mg_{Pt}^{-1}, respectively [53]. A highly stable and active octahedron PtNi binary alloy with {111} facets of a Pt-rich surface and Ni-rich sublayer was prepared by heating pristine octahedra of the PtNi binary alloy nanostructure [54]. Zhao *et al* reported that an ordered $CoPt_3$ intermetallic with a secondary solid solution (where Co acts as the solute) has been a significant factor for improved electrocatalytic activity compared to other stoichiometries [55]. Pt_xY on a mesoporous carbon support expressed three-fold enhancements in electrochemical activity (ORR) than Pt/C, where the $Pt_xY600h5$ catalyst demonstrated specific and mass activities of 1.570 mA cm_{Pt}^{-2} and 0.586 A mg_{Pt}^{-1}, respectively [56]. Escudero-Escribano *et al* reported that Pt_5M (M is a lanthanide series element) demonstrated three to six times superior ORR activity than Pt/C [57]. Sub-2 nm thick Rh-doped Pt nanowires on a carbon support (figure 5.9) exhibited approximately eight times and six times higher mass and specific activities, respectively, and revealed greater durability due to higher utilization of Pt atoms [58].

The ORR activity of hollow graphitic spheres with mesoporous shell supported Pt_xCo ($x = 5$, 3 and 1) binary alloys was investigated systematically, among which the composition of PtCo@HGS displayed high electrocatalytic activity and excellent durability compared to other compositions [59]. The relation between the ORR activities of their composition/surface structure were studied systematically, among which Pt_1Ni_1 octahedral nanoparticles exhibited seven and a half times and seven times superior area-specific and mass activities, respectively [60]. The high-index facets of segregated Pt skin associated with a zigzag geometry of Pt_3Fe nanowires boosted ORR activity with 2.11 A mg^{-1} and 4.34 mA cm^{-2} for mass and specific

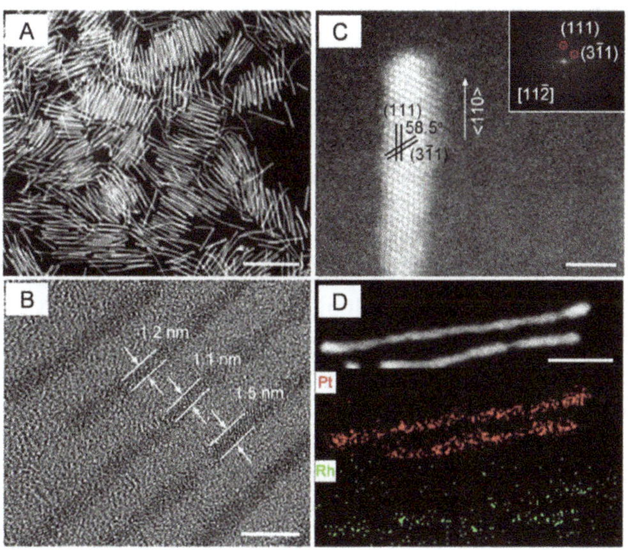

Figure 5.9. Structural and compositional characterizations of ultrathin Rh-doped Pt NWs. (a) HAADF-STEM image. Scale bar = 50 nm. (b) TEM image. Scale bar = 5 nm. (c) Atomic-resolution HAADF-STEM image. The inset shows the corresponding FFT pattern. Scale bar = 2 nm. (d) EDX mapping of Pt and Rh. Scale bar = 10 nm. (Reproduced with permission from [58]. Copyright 2017 American Chemical Society.)

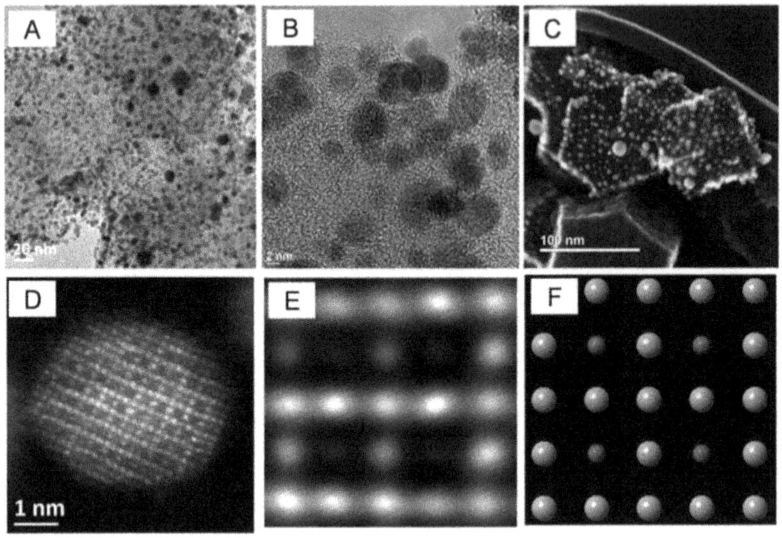

Figure 5.10. (a) TEM, (b) HRTEM, and (c) dark field STEM images of Pt/40Co-NC-900 catalyst. (d) Atomic-resolution HAADF-STEM image of Pt_3Co nanoparticle. (e) A zoom-in of the super lattice feature from (d). (f) The simulated HAADF-STEM images of L12 ordered Pt_3Co close along [100]. (Reproduced with permission from [63]. Copyright 2018 American Chemical Society.)

activity, respectively, and along with the distinctive geometry demonstrated surprising stability up to 50000 potential cycles [61]. An L_{10}-FePt/Pt core and 5 Å thick Pt shell ordered FePt core–shell (8 nm) binary alloy showed excellent activity and significant stability even up to 30 000 potential cycles [62]. Ordered Pt_3Co nanostructures on a carbon support (figure 5.10) were attained through Co-doped MOF embedded with Pt nanoparticles during optimized Co wt% and temperature [63].

Pristine $PtNi_{1.5}$ octahedra systematically evolved at elevated temperatures through *in situ* TEM images revealed the transformation of cuboctahedra or truncated octahedra with highly stable facets of PtNi nanostructure [64]. A nanoframe geometry that allows plentiful mass diffusion pathways and with a vast extent of high-indexed facets of Pt–Ni binary alloy multi-frames expressed a 30 times greater mass activity of 1.51 A mg_{Pt}^{-1} at 0.93 V than Pt/C because the unique frame promotes catalytically active surfaces [65]. The combined assembly of Ni_2P/GC with an ultralow Pt content consisting of accelerated charge and mass transport and superior conductivity furnished excess O_2 adsorption active sites exhibited 0.93 V in $E_{1/2}$ and 1.03 V positive shift in onset potentials with an outstanding mass activity of 565.3 mA mg_{Pt}^{-1} with excellent stability [66]. Qin *et al* reported that an ultrathin fcc-Pt shell on a hcp-PtBi core binary alloy revealed five times greater activity than Pt/C due to the intermetallic phase with core–shell architecture [67]. The overall electrocatalytic activity of Pt based binary alloys with different geometries/assemblies of catalysts for ORR applications have been discussed and a few representative activities are presented in figure 5.11.

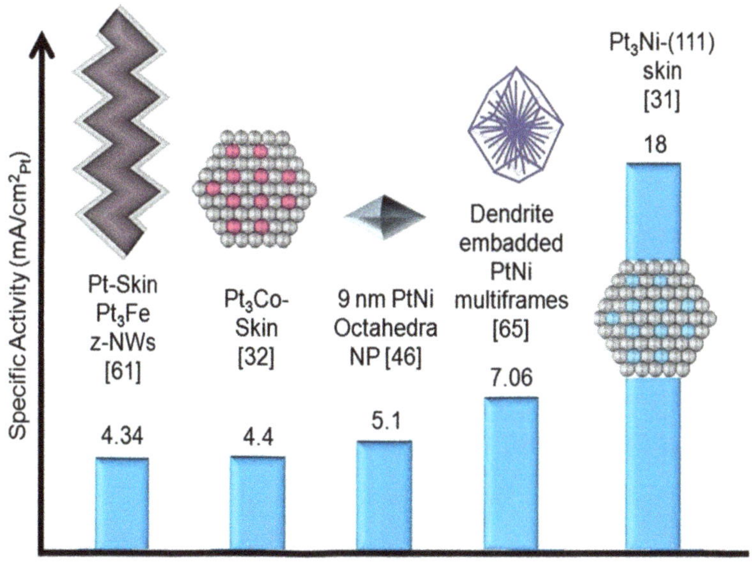

Figure 5.11. Comparative bar diagram of I_s for different geometry nanosystems of Pt based binary alloy electrocatalysts [31, 32, 46, 61, 65].

5.4.3 Pt based ternary alloy nanostructures

The extraordinary durability during 60 000 potential cycles was demonstrated by the tailored morphology and composition profile of multimetallic core–shell Au–$FePt_3$ nanoparticles reported by Wang *et al* [68]. The ORR activity order of systematically investigated PtNiFe ternary alloy nanocrystals is as follows: nanocubes < polyhedrons < nanowires = octahedrons [69]. The thin films surfaces of $Pt_3(MN)_1$ [M, N = Co, Ni, or Fe] polycrystalline ternary systems displayed four times higher ORR electrocatalytic activity than that of bare Pt single metallic electrocatalyst [70].

Ultralow Pt content with concomitant enhancement in the ORR activity and stability was attained through a Pt monolayer shell on a PdAu binary core. In particular, a higher stability even beyond 100 000 stability cycles between 0.6 and 1.0 V versus RHE was demonstrated by the $Pt_{ML}/Pd_9Au_1/C$ electrocatalyst. Further, it retains its activity and shows extraordinary stability under an elevated potential range of 0.6–1.4 V versus RHE with only mild Pd losses [71]. Loukrakpam *et al* successfully investigated how the ORR activity of bimetallic (PtIr and PtCo) and trimetallic (PtIrCo) systems is associated with the atomic arrangement through surface sensitive techniques, and the ternary nanoalloys demonstrated enhanced ORR activity because of their extra structural tunability compared to the binary nanoalloys [72]. Li *et al* found that particles size and ORR activity follows a volcano-like trend, where 13 nm sized Fe-doped Pt_3Ni octahedron expressed three times and eight times superior mass and specific activities, respectively, than Johnson Matthey Pt/C [73]. The fibrous membranes of Pt coated niobium-doped titania (Nb-TiO_2) with 10 at.% of Nb, synthesized by electrospinning

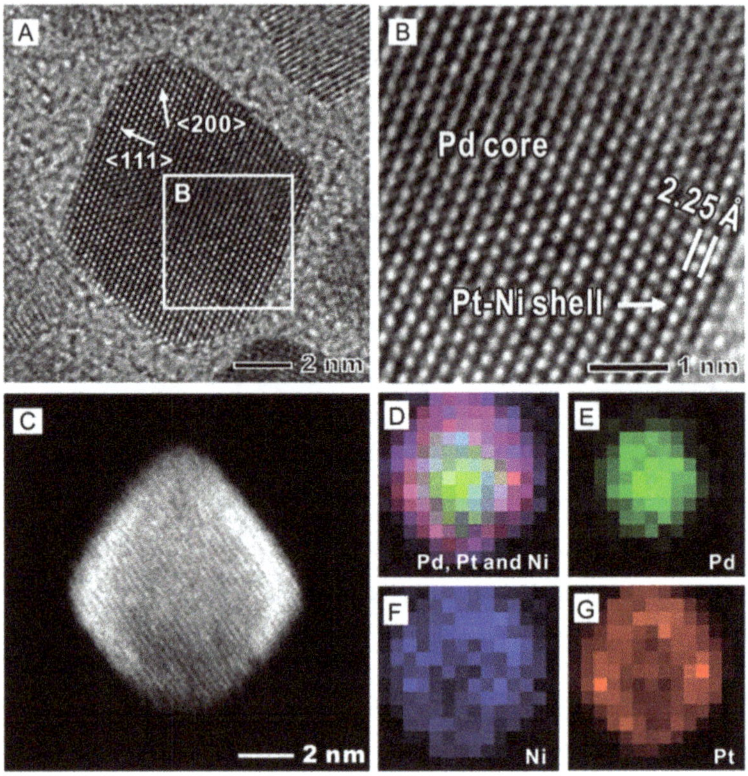

Figure 5.12. (a) High-resolution TEM image taken from a single Pd@Pt–Ni octahedron with an edge length of 8 nm. (b) High-resolution TEM image of the region marked by a box in (a). The lattice spacing of 2.25 Å for both the Pd core and the Pt–Ni shell confirms the epitaxial overgrowth of Pt–Ni on Pd. (c) HAADF-STEM image of a single Pd@Pt–Ni octahedron. The brighter shell on a relatively darker core indicates the difference in atomic number between Pt and Pd. (d)–(g) EDS mapping of elemental distributions for (d) Pd, Pt, and Ni, (e) Pt and Ni, (f) Pd and Ni, and (g) Pd and Pt. (Reproduced with permission from [76]. Copyright 2014 American Chemical Society.)

and atomic layer deposition of Pt, exhibited 20 times enhanced ORR electrocatalytic activity than Pt/C. Further, if the atomic layer of the Pt element is increased correspondingly the catalytic activity also increased [74]. Ordered intermetallic Pt_2FeCo/C and Pt6FeCo/C retains its ECSA and ORR activity even after 10 000 potential cycles under the membrane electrode assembly (MEA) testing reported by Tamaki *et al* [75]. An overall edge length of 8 nm and 6 nm of core–shell octahedral nanoparticles of carbon-supported Pd@Pt–Ni electrocatalyst (figure 5.12) exhibited 14 times and 12.5 times enhanced specific and mass activity, respectively [76]. The following trend in their activities has been successfully determined: Pt–Co/C < Pt/C < Pt–Sn/C < Pt–Co–Sn/C . Among these the higher activity of ternary alloy Pt–Co–Sn is due to its preparation using an impregnation method followed by reduction using a $NaBH_4$ agent as well as oxophilicity of the Pt–Co–Sn nanoparticles [77]. Pt/Ni atomic ratios from 0.6–1.8 of $Pd@Pt_{1.8}Ni$ core–shell octahedral nanocrystals exhibited five times enhanced

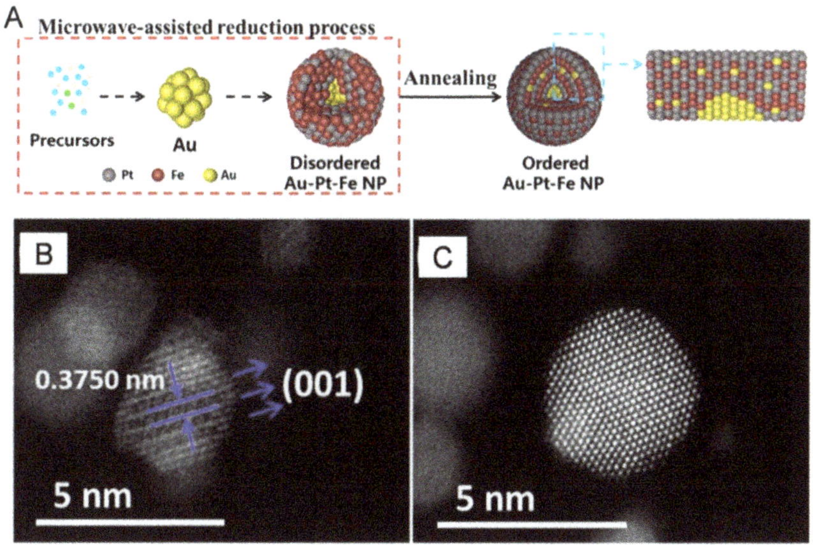

Figure 5.13. (a) Synthesis scheme of Au–PtFe/C–H and Au–PtFe3/C–H NPs; (b) enlarged and (c) atomic-resolution HAADF-STEM images of Au–PtFe/C–H NPs at different regions. (Reproduced with permission from [85]. Copyright 2018 American Chemical Society.)

ORR activity compared to Pt/C [78]. A Rh atom doped Pt–Ni octahedral ternary alloy nanoparticle catalyst expressed outstanding mass activity of 1.14 A mg_{Pt}^{-1} and the activity was retained even beyond 30 000 potential cycles [79]. Zhao *et al* reported that Pt–$Ni_{0.55}$–$Co_{0.1}$/C ternary alloy octahedra on carbon support demonstrated astonishing specific and mass activities of 5.05 mA cm^{-2} and 2.80 mA mg_{Pt}^{-1}, respectively, which are around 20.2 times and 14.7 times higher than for Pt/C [80]. The sub-nanometre thickness of the Pt–Pd shell on a nonporous core of Au acted as a supportless catalyst, revealing an initial mass activity of 1.14 A mg_{Pt}^{-1}. Surprisingly, this further improved to 1.471 A mg_{Pt}^{-1} even after 30 000 stability cycles and then maintained its activity up to 70 000 stability cycles without any losses [81]. Park *et al* reported that a binary alloy of PtCu dendrite surrounded by PtCuNi ternary alloy nanoframe displayed surprising electrocatalytic activity of 11 times and 16 times higher mass and specific activity than commercial Pt/C, respectively (figure 5.13) [82].

The existence of Mo oxides on the surface of Mo–PtNi/C octahedral NPs, which makes them highly tolerant of acidic conditions, and the (111) facet rich surface are highly favourable for ORR activity [83]. Four to five atomic layer thick mono-, binary- and trimetallic sub-nanometre Pt alloy nanowires displayed 32 times (4.20 A mg_{Pt}^{-1}) and 27 times (5.11 mA cm^{-2}) the mass and specific activities, respectively, of Pt/C [84]. The introduction of a trace amount of Au (0.2wt%) on PtFe binary alloy produced a Au–PtFe/C ternary alloy that demonstrated superior electrocatalytic activity compared to JM Pt/C [85]. The overall ORR activities of the Pt based ternary alloy nanostructures with different geometries and compositions of electrocatalysts are tabulated in table 5.2 and topmost activities are presented in figure 5.14.

Table 5.2. Activity comparison of Pt based trimetallic nanostructure electrocatalysts.

		Activity @ 0.9 V versus RHE			
Pt based ternary alloy nanostructure	Shape of nanostructure	ECSA ($m^2\ g_{Pt}^{-1}$)	I_m ($mA\ mg_{Pt}^{-1}$)	I_s ($mA\ cm_{Pt}^{-2}$)	Reference
Multimetallic Au–$FePt_3$ core–shell	Core–shell	35	550	1.5	[68]
PtFeNi nanocubes	Nanocubes		5.94	0.0074	[69]
PtFeNi polyhedrons	Polyhedrons		4.67	0.0061	
PtFeNi octahedrons	Octahedrons		4.02	0.0052	
PtFeNi nanowires	Nanowires		3.95	0.0048	
Pt_3FeCo skin	Skin			1.15	[70]
Pt_3FeCo skeleton	Skeleton			0.8	
Pt_3FeNi skin	Skin			1.5	
Pt_3FeNi skeleton	Skeleton			1	
Pt_3CoNi skin	Skin			1.75	
Pt_3CoNi skeleton	Skeleton			1.2	
Pt–ML–PdAu–C	Core–shell		110	0.38	[71]
$Pt_{25}Ir_{20}Co_{55}/C$	Particles		600	1.27	[72]
13 nm Fe–Pt_3Ni	Octahedral	20.1	360	1.25	[73]
Pt@Nb–TiO_2	Membrane	14.74		0.16	[74]
Pt_2FeCo	Chemically	28	500		[75]
Pt_6FeCo	ordered	35	255		
Pd@PtNi core@shell octahedra	Core@shell octahedra	93	2500	2.7	[76]
Pt–Co–Sn	Particles		35.47	0.409	[77]
Octahedral Pd@$Pt_{1.8}Ni$ core–shell nanocrystals	Octahedral core–shell	178.01	790	0.45	[78]
Rh-doped Pt–Ni octahedral 0k	Octahedral		820	2.6	[79]
Rh-doped Pt–Ni octahedral 4k			1140	3.4	
Pt–Ni–Co octahedra	Octahedra		2800	6.1	[80]
Pt–Pd–Au 0k	Particles		1140	0.55	[81]
Pt–Pd–Au 30k			1471	0.9	
PtCu@PtCuNi dendrite@frame					
Dendrite@frame	33.8	2450	7.3	[82]	
Mo–PtNi/C	Octahedral				[83]
PtNi–NWs	Nanowires	80.7	4200	5.11	[84]
Chemically ordered Pt–Fe–Au	Chemically ordered	148.29	360	0.23	[85]

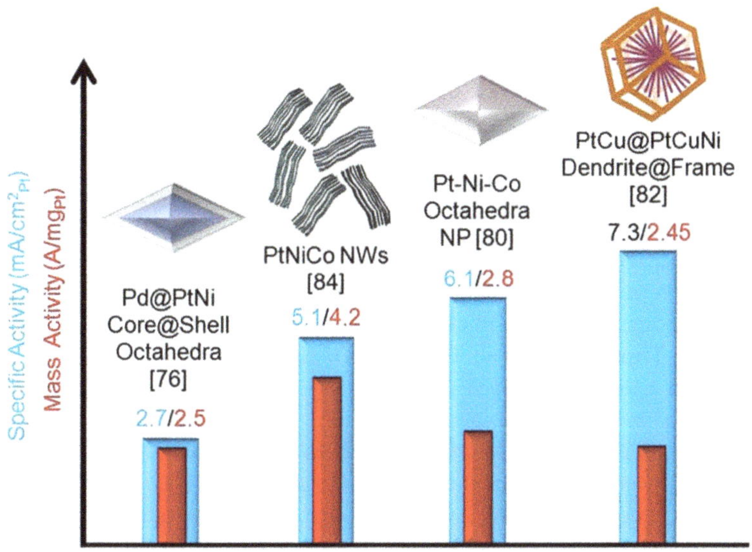

Figure 5.14. Comparative bar diagram of I_s and I_m of different geometries of Pt based ternary alloy electrocatalysts from previous reports.

5.5 Summary

This chapter provides a brief introduction to fuel cells and their classification, followed by a short introduction to and the working principle of polymer electrolyte membrane fuel cells (PEMFCs), and related technical challenges. Further, the mechanism of the oxygen reduction reaction (ORR) on platinum and platinum based alloy nanostructure surfaces and electrochemical measurements are shown explicitly. This chapter mainly investigates how the platinum and platinum based alloy nanostructures with different shapes and sizes influence the ORR in PEMFCs. Platinum based alloy nanostructures are receiving huge attention in research due to their higher sustainability in acid conditions in energy conversion devices. Recent progress made on platinum based alloy nanostructures (Pt based monometallic, binary, and ternary alloys) of materials with different geometries, with a special focus on the ORR activity in PEMFCs, including the composition, shape, and size of platinum based nanostructures and their electrochemical performance evaluation, are discussed briefly. Although recent improvements in platinum based alloy nanostructures for PEMFCs are exceptionally promising, there remain some challenges, such as the high cost, low availability, and rational design of low content Pt alloy nanostructures, and their long term durability, etc, which restrict their application in transferring on technology in the large scale. To overcome all the challenges and account of essentially sustainable electrocatalyst, an outside-the-box opinion is required.

Acknowledgements

The authors M L and R V M acknowledge FONDECYT 2021 Project No. 3200113 (FONDECYT-ANID) for financial support.

References

[1] Beena K B 2013 Carbon nanofiber based hybrid materials for PEMFC applications *PhD Thesis* University of Pune

[2] Grove W R 1839 XXIV. On voltaic series and the combination of gases by platinum *Philos. Mag. Ser.* **3** 127–30

[3] Vielstich W, Yokokawa H and Gasteiger H A 2009 *Handbook of Fuel Cells: Fundamentals Technology and Applications* (New York: Wiley)

[4] Steele B C and Heinzel A 2011 Materials for fuel-cell technologies *Materials for Sustainable Energy: A Collection of Peer-Reviewed Research and Review Articles from Nature Publishing Group* (Singapore: World Scientific) pp 224–31

[5] Strmcnik D, Tripkovic D, Van der Vliet D, Stamenkovic V and Marković N 2008 Adsorption of hydrogen on Pt (1 1 1) and Pt (1 0 0) surfaces and its role in the HOR *Electrochem. Commun.* **10** 1602–5

[6] Conway B and Tilak B 2002 Interfacial processes involving electrocatalytic evolution and oxidation of H_2, and the role of chemisorbed H *Electrochim. Acta* **47** 3571–94

[7] Innocente A and Angelo A 2006 Electrocatalysis of oxidation of hydrogen on platinum ordered intermetallic phases: kinetic and mechanistic studies *J. Power Sources* **162** 151–9

[8] Nørskov J K, Rossmeisl J, Logadottir A, Lindqvist L, Kitchin J R, Bligaard T and Jonsson H 2004 Origin of the overpotential for oxygen reduction at a fuel-cell cathode *J. Phys. Chem.* B **108** 17886–92

[9] Toda T, Igarashi H, Uchida H and Watanabe M 1999 Enhancement of the electroreduction of oxygen on Pt alloys with Fe, Ni, and Co *J. Electrochem. Soc.* **146** 3750–6

[10] Uribe F, Wilson M, Springer T, Gottesfeld S, Scherson D, Tryk D, Daroux M and Xing X 1992 Structural effects in electrocatalysis and oxygen electrochemistry *Workshop Electrochemical Society* p 11

[11] Mukerjee S, Srinivasan S, Soriaga M P and McBreen J 1995 Role of structural and electronic properties of Pt and Pt alloys on electrocatalysis of oxygen reduction an *in situ* XANES and EXAFS investigation *J. Electrochem. Soc.* **142** 1409–22

[12] Luo M, Sun Y, Zhang X, Qin Y, Li M, Li Y, Li C, Yang Y, Wang L and Gao P 2018 Stable high-index faceted Pt skin on zigzag-like PtFe nanowires enhances oxygen reduction catalysis *Adv. Mater.* **30** 1705515

[13] Garsany Y, Baturina O A, Swider-Lyons K E and Kocha S S 2010 Experimental methods for quantifying the activity of platinum electrocatalysts for the oxygen reduction reaction *Anal. Chem.* **82** 6321–8

[14] Patil I M, Lokanathan M, Ganesan B, Swami A and Kakade B 2017 Carbon nanotube/boron nitride nanocomposite as a significant bifunctional electrocatalyst for oxygen reduction and oxygen evolution reactions *Chem. Eur. J.* **23** 676

[15] Debe M K 2012 Electrocatalyst approaches and challenges for automotive fuel cells *Nature* **486** 43

[16] Gasteiger H A and Marković N M 2009 Just a dream—or future reality? *Science* **324** 48–9

[17] Gasteiger H A, Kocha S S, Sompalli B and Wagner F T 2005 Activity benchmarks and requirements for Pt, Pt-alloy, and non-Pt oxygen reduction catalysts for PEMFCs *Appl. Catal.* B **56** 9–35

[18] Higuchi E, Uchida H and Watanabe M 2005 Effect of loading level in platinum-dispersed carbon black electrocatalysts on oxygen reduction activity evaluated by rotating disk electrode *J. Electroanal. Chem.* **583** 69–76

[19] Kongkanand A, Kuwabata S, Girishkumar G and Kamat P 2006 Single-wall carbon nanotubes supported platinum nanoparticles with improved electrocatalytic activity for oxygen reduction reaction *Langmuir* **22** 2392–6

[20] Wang C, Daimon H, Onodera T, Koda T and Sun S 2008 A general approach to the size-and shape-controlled synthesis of platinum nanoparticles and their catalytic reduction of oxygen *Angew. Chem. Inter. Edn* **47** 3588–91

[21] Lim B, Lu X, Jiang M, Camargo P H, Cho E C, Lee E P and Xia Y 2008 Facile synthesis of highly faceted multioctahedral Pt nanocrystals through controlled overgrowth *Nano Lett.* **8** 4043–7

[22] Koenigsmann C, Zhou W, Adzic R R, Sutter E and Wong S S 2010 Size-dependent enhancement of electrocatalytic performance in relatively defect-free, processed ultrathin platinum nanowires *Nano Lett.* **10** 2806–11

[23] Lee S W, Chen S, Suntivich J, Sasaki K, Adzic R R and Shao-Horn Y 2010 Role of surface steps of Pt nanoparticles on the electrochemical activity for oxygen reduction *J. Phys. Chem. Lett.* **1** 1316–20

[24] Shao M, Peles A and Shoemaker K 2011 Electrocatalysis on platinum nanoparticles: particle size effect on oxygen reduction reaction activity *Nano Lett.* **11** 3714–9

[25] Saha M S, Zhang Y, Cai M and Sun X 2012 Carbon-coated tungsten oxide nanowires supported Pt nanoparticles for oxygen reduction *Int. J. Hydrogen Energy* **37** 4633–8

[26] Wang R, Higgins D C, Hoque M A, Lee D, Hassan F and Chen Z 2013 Controlled growth of platinum nanowire arrays on sulfur doped graphene as high performance electrocatalyst *Sci. Rep.* **3** 2431

[27] Wang C, Ma L, Liao L, Bai S, Long R, Zuo M and Xiong Y 2013 A unique platinum-graphene hybrid structure for high activity and durability in oxygen reduction reaction *Sci. Rep.* **3** 2580

[28] Ma L, Wang C, Xia B Y, Mao K, He J, Wu X, Xiong Y and Lou X W 2015 Platinum multicubes prepared by Ni^{2+}-mediated shape evolution exhibit high electrocatalytic activity for oxygen reduction *Angew. Chem. Inter. Ed.* **54** 5666–71

[29] Temmel S E, Fabbri E, Pergolesi D, Lippert T and Schmidt T J 2016 Investigating the role of strain toward the oxygen reduction activity on model thin film Pt catalysts *ACS Catal.* **6** 7566–76

[30] Stamenkovic V, Mun B S, Mayrhofer K J, Ross P N, Markovic N M, Rossmeisl J, Greeley J and Nørskov J K 2006 Changing the activity of electrocatalysts for oxygen reduction by tuning the surface electronic structure *Angew. Chem.* **118** 2963–7

[31] Stamenkovic V R, Fowler B, Mun B S, Wang G, Ross P N, Lucas C A and Marković N M 2007 Improved oxygen reduction activity on Pt_3Ni (111) via increased surface site availability *Science* **315** 493–7

[32] Stamenkovic V R, Mun B S, Arenz M, Mayrhofer K J, Lucas C A, Wang G, Ross P N and Markovic N M 2007 Trends in electrocatalysis on extended and nanoscale Pt-bimetallic alloy surfaces *Nat. Mater.* **6** 241

[33] Peng Z and Yang H 2009 Synthesis and oxygen reduction electrocatalytic property of Pt-on-Pd bimetallic heteronanostructures *J. Am. Chem. Soc.* **131** 7542–3

[34] Wang C, van der Vliet D, Chang K-C, You H, Strmcnik D, Schlueter J A, Markovic N M and Stamenkovic V R 2009 Monodisperse Pt_3Co nanoparticles as a catalyst for the oxygen reduction reaction: size-dependent activity *J. Phys. Chem.* C **113** 19365–8

[35] Sasaki K, Naohara H, Cai Y, Choi Y M, Liu P, Vukmirovic M B, Wang J X and Adzic R R 2010 Core-protected platinum monolayer shell high-stability electrocatalysts for fuel-cell cathodes *Angew. Chem. Inter. Edn* **49** 8602–7

[36] Kang Y and Murray C B 2010 Synthesis and electrocatalytic properties of cubic Mn–Pt nanocrystals (nanocubes) *J. Am. Chem. Soc.* **132** 7568–9

[37] Kim J, Lee Y and Sun S 2010 Structurally ordered FePt nanoparticles and their enhanced catalysis for oxygen reduction reaction *J. Am. Chem. Soc.* **132** 4996–7

[38] Wu J, Gross A and Yang H 2011 Shape and composition-controlled platinum alloy nanocrystals using carbon monoxide as reducing agent *Nano Lett.* **11** 798–802

[39] Zhang J, Yang H, Fang J and Zou S 2010 Synthesis and oxygen reduction activity of shape-controlled Pt_3Ni nanopolyhedra *Nano Lett.* **10** 638–44

[40] Wu J, Zhang J, Peng Z, Yang S, Wagner F T and Yang H 2010 Truncated octahedral Pt_3Ni oxygen reduction reaction electrocatalysts *J. Am. Chem. Soc.* **132** 4984–5

[41] Guo S and Sun S 2012 FePt nanoparticles assembled on graphene as enhanced catalyst for oxygen reduction reaction *J. Am. Chem. Soc.* **134** 2492–5

[42] Choi S I, Choi R, Han S W and Park J T 2011 Shape-controlled synthesis of Pt_3Co nanocrystals with high electrocatalytic activity toward oxygen reduction *Chem. Eur. J.* **17** 12280–4

[43] Cui C, Gan L, Li H-H, Yu S-H, Heggen M and Strasser P 2012 Octahedral PtNi nanoparticle catalysts: exceptional oxygen reduction activity by tuning the alloy particle surface composition *Nano Lett.* **12** 5885–9

[44] Wu J, Qi L, You H, Gross A, Li J and Yang H 2012 Icosahedral platinum alloy nanocrystals with enhanced electrocatalytic activities *J. Am. Chem. Soc.* **134** 11880–3

[45] Kakade B A, Wang H, Tamaki T, Ohashi H and Yamaguchi T 2013 Enhanced oxygen reduction reaction by bimetallic CoPt and PdPt nanocrystals *RSC Adv.* **3** 10487–96

[46] Choi S-I, Xie S, Shao M, Odell J H, Lu N, Peng H-C, Protsailo L, Guerrero S, Park J and Xia X 2013 Synthesis and characterization of 9 nm Pt–Ni octahedra with a record high activity of 3.3 A/mg_{Pt} for the oxygen reduction reaction *Nano Lett.* **13** 3420–5

[47] Cui C, Gan L, Heggen M, Rudi S and Strasser P 2013 Compositional segregation in shaped Pt alloy nanoparticles and their structural behaviour during electrocatalysis *Nat. Mater.* **12** 765

[48] Chen C, Kang Y, Huo Z, Zhu Z, Huang W, Xin H L, Snyder J D, Li D, Herron J A and Mavrikakis M 2014 Highly crystalline multimetallic nanoframes with three-dimensional electrocatalytic surfaces *Science* **343** 1339–43

[49] Shao M, He G, Peles A, Odell J H, Zeng J, Su D, Tao J, Yu T, Zhu Y and Xia Y 2013 Manipulating the oxygen reduction activity of platinum shells with shape-controlled palladium nanocrystal cores *Chem. Commun.* **49** 9030–2

[50] Wang D, Xin H L, Hovden R, Wang H, Yu Y, Muller D A, DiSalvo F J and Abruña H D 2013 Structurally ordered intermetallic platinum–cobalt core–shell nanoparticles with enhanced activity and stability as oxygen reduction electrocatalysts *Nat. Mater.* **12** 81

[51] Alia S M, Larsen B A, Pylypenko S, Cullen D A, Diercks D R, Neyerlin K C, Kocha S S and Pivovar B S 2014 Platinum-coated nickel nanowires as oxygen-reducing electrocatalysts *ACS Catal.* **4** 1114–9

[52] Zhang C, Hwang S Y, Trout A and Peng Z 2014 Solid-state chemistry-enabled scalable production of octahedral Pt–Ni alloy electrocatalyst for oxygen reduction reaction *J. Am. Chem. Soc.* **136** 7805–8

[53] Alia S M, Pylypenko S, Neyerlin K C, Cullen D A, Kocha S S and Pivovar B S 2014 Platinum-coated cobalt nanowires as oxygen reduction reaction electrocatalysts *ACS Catal.* **4** 2680–6

[54] Gan L, Heggen M, Cui C and Strasser P 2015 Thermal facet healing of concave octahedral Pt–Ni nanoparticles imaged *in situ* at the atomic scale: implications for the rational synthesis of durable high-performance ORR electrocatalysts *ACS Catal.* **6** 692–5

[55] Zhao Y, Liu J, Zhao Y, Wang F and Song Y 2015 Pt–Co secondary solid solution nanocrystals supported on carbon as next-generation catalysts for the oxygen reduction reaction *J. Mater. Chem.* A **3** 20086–91

[56] Brandiele R, Durante C, Grądzka E, Rizzi G, Zheng J, Badocco D, Centomo P, Pastore P, Granozzi G and Gennaro A 2016 One step forward to a scalable synthesis of platinum–yttrium alloy nanoparticles on mesoporous carbon for the oxygen reduction reaction *J. Mater. Chem.* A **4** 12232–40

[57] Escudero-Escribano M, Malacrida P, Hansen M H, Vej-Hansen U G, Velázquez-Palenzuela A, Tripkovic V, Schiøtz J, Rossmeisl J, Stephens I E and Chorkendorff I 2016 Tuning the activity of Pt alloy electrocatalysts by means of the lanthanide contraction *Science* **352** 73–6

[58] Huang H, Li K, Chen Z, Luo L, Gu Y, Zhang D, Ma C, Si R, Yang J and Peng Z 2017 Achieving remarkable activity and durability toward oxygen reduction reaction based on ultrathin Rh-doped Pt nanowires *J. Am. Chem. Soc.* **139** 8152–9

[59] Pizzutilo E, Knossalla J, Geiger S, Grote J P, Polymeros G, Baldizzone C, Mezzavilla, Ledendecker M, Mingers A and Cherevko S 2017 The space confinement approach using hollow graphitic spheres to unveil activity and stability of Pt–Co nanocatalysts for PEMFC *Adv. Energy Mater.* **7** 1700835

[60] Lu Y, Thia L, Fisher A, Jung C-Y, Yi S C and Wang X 2017 Octahedral PtNi nanoparticles with controlled surface structure and composition for oxygen reduction reaction *Sci. China Mater.* **60** 1109–20

[61] Luo M, Sun Y, Zhang X, Qin Y, Li M, Li Y, Li C, Yang Y, Wang L and Gao P 2018 Stable high-index faceted Pt skin on zigzag-like PtFe nanowires enhances oxygen reduction catalysis *Adv. Mater.* **30** 1705515

[62] Li J, Xi Z, Pan Y-T, Spendelow J S, Duchesne P N, Su D, Li Q, Yu C, Yin Z and Shen B 2018 Fe stabilization by intermetallic $L1_0$-FePt and Pt catalysis enhancement in $L1_0$-FePt/Pt nanoparticles for efficient oxygen reduction reaction in fuel cells *J. Am. Chem. Soc.* **140** 2926–32

[63] Wang X X, Hwang S, Pan Y-T, Chen K, He Y, Karakalos S G, Zhang H, Spendelow J S, Su D and Wu G 2018 Ordered Pt_3Co intermetallic nanoparticles derived from metal-organic frameworks for oxygen reduction *Nano Lett.* **18** 4163–71

[64] Gocyla M, Kuehl S, Shviro M, Heyen H, Selve S, Dunin-Borkowski R E, Heggen M and Strasser P 2018 Shape stability of octahedral PtNi nanocatalysts for electrochemical oxygen reduction reaction studied by *in situ* transmission electron microscopy *ACS Nano* **12** 5306–11

[65] Kwon H, Kabiraz M K, Park J, Oh A, Baik H, Choi S-I and Lee K 2018 Dendrite-embedded platinum–nickel multiframes as highly active and durable electrocatalyst toward the oxygen reduction reaction *Nano Lett.* **18** 2930–6

[66] Wang R, Wang L, Zhou W, Chen Y, Yan H, Ren Z, Tian C, Shi K and Fu H 2018 Ni_2P entwined by graphite layers as a low-Pt electrocatalyst in acidic media for oxygen reduction *ACS App. Mater. Interfaces.* **10** 9999–10010
[67] Qin Y, Luo M, Sun Y, Li C, Huang B, Yang Y, Li Y, Wang L and Guo S 2018 Intermetallic hcp-PtBi/fcc-Pt core/shell nanoplates enable efficient bifunctional oxygen reduction and methanol oxidation electrocatalysis *ACS Catal.* **8** 5581–90
[68] Wang C, Van Der Vliet D, More K L, Zaluzec N J, Peng S, Sun S, Daimon H, Wang G, Greeley J and Pearson J 2010 Multimetallic Au/$FePt_3$ nanoparticles as highly durable electrocatalyst *Nano Lett.* **11** 919–26
[69] Chou S-W, Shyue J-J, Chien C-H, Chen C-C, Chen Y-Y and Chou P-T 2012 Surfactant-directed synthesis of ternary nanostructures: nanocubes, polyhedrons, octahedrons, and nanowires of PtNiFe. their shape-dependent oxygen reduction activity *Chem. Mater.* **24** 2527–33
[70] Wang C, Li D, Chi M, Pearson J, Rankin R B, Greeley J, Duan Z and Wang G; der Vliet Van; More K L 2012 Rational development of ternary alloy electrocatalysts *J. Phys. Chem. Lett.* **3** 1668–73
[71] Sasaki K, Naohara H, Choi Y, Cai Y, Chen W-F, Liu P and Adzic R R 2012 Highly stable Pt monolayer on PdAu nanoparticle electrocatalysts for the oxygen reduction reaction *Nat. Commun.* **3** 1115
[72] Loukrakpam R, Shan S, Petkov V, Yang L, Luo J and Zhong C-J 2013 Atomic ordering enhanced electrocatalytic activity of nanoalloys for oxygen reduction reaction *J. Phys. Chem.* C **117** 20715–21
[73] Li Y, Quan F, Chen L, Zhang W, Yu H and Chen C 2014 Synthesis of Fe-doped octahedral Pt_3Ni nanocrystals with high electro-catalytic activity and stability towards oxygen reduction reaction *RSC Adv.* **4** 1895–9
[74] Du Q, Wu J and Yang H 2013 Pt@Nb-TiO_2 catalyst membranes fabricated by electro-spinning and atomic layer deposition *ACS Catal.* **4** 144–51
[75] Tamaki T, Atsuhiko A M, Arumugam B, Kakade B K and Yamaguch T 2014 Highly active and durable chemically ordered Pt–Fe–Co intermetallics as cathode catalysts of membrane–electrode assemblies in polymer electrolyte fuel cells *J. Power Sources* **271** 346–53
[76] Choi S-I, Shao M, Lu N, Ruditskiy A, Peng H-C, Park J, Guerrero S, Wang J, Kim M J and Xia Y 2014 Synthesis and characterization of Pd@ Pt–Ni core–shell octahedra with high activity toward oxygen reduction *ACS Nano* **8** 10363–71
[77] Knani S, Chirchi L, Napporn W, Baranton S, Léger J and Ghorbel A 2015 Promising ternary Pt–Co–Sn catalyst for the oxygen reduction reaction *J. Electroanal. Chem.* **738** 145–53
[78] Zhao X, Chen S, Fang Z, Ding J, Sang W, Wang Y, Zhao J, Peng Z and Zeng J 2015 Octahedral Pd@$Pt_{1.8}Ni$ core–shell nanocrystals with ultrathin PtNi alloy shells as active catalysts for oxygen reduction reaction *J. Am. Chem. Soc.* **137** 2804–7
[79] Beermann V, Gocyla M, Willinger E, Rudi S, Heggen M, Dunin-Borkowski R E, Willinger M-G and Strasser P 2016 Rh-doped Pt–Ni octahedral nanoparticles: understanding the correlation between elemental distribution, oxygen reduction reaction, and shape stability *Nano Lett.* **16** 1719–25
[80] Zhao Z, Feng M, Zhou J, Liu Z, Li M, Fan Z, Tsen O, Miao J, Duan X and Huang Y 2016 Composition tunable ternary Pt–Ni–Co octahedra for optimized oxygen reduction activity *Chem. Commun.* **52** 11215–8

[81] Li J, Yin H-M, Li X-B, Okunishi E, Shen Y-L, He J, Tang Z-K, Wang W-X, Yücelen E and Li C 2017 Surface evolution of a Pt–Pd–Au electrocatalyst for stable oxygen reduction *Nat. Energy* **2** 17111

[82] Park J, Kanti Kabiraz M, Kwon H, Park S, Baik H, Choi S-I and Lee K 2017 Radially phase segregated PtCu@PtCuNi Dendrite@Frame nanocatalyst for the oxygen reduction reaction *ACS Nano* **11** 10844–51

[83] Jia Q, Zhao Z, Cao L, Li J, Ghoshal S, Davies V, Stavitski E, Attenkofer K, Liu Z and Li M 2018 Roles of Mo surface dopants in enhancing the ORR performance of octahedral PtNi nanoparticles *Nano Lett.* **18** 798–804

[84] Jiang K, Zhao D, Guo S, Zhang X, Zhu X, Guo J, Lu G and Huang X 2017 Efficient oxygen reduction catalysis by subnanometer Pt alloy nanowires *Sci. Adv.* **3** e1601705

[85] Zhu H, Cai Y, Wang F, Gao P and Cao J 2018 Scalable preparation of the chemically ordered Pt–Fe–Au nano-catalysts with high catalytic reactivity and stability for oxygen reduction reactions *ACS Appl. Mater. Interfaces* **10** 22156–66

IOP Publishing

Nanostructured Materials for Sustainable Energy and Environmental Remediation

R V Mangalaraja, R Udayabhaskar, P Sathishkumar and Joydeep Dutta

Chapter 6

Ultrathin, flexible hybrid transition metal oxide nanostructures for renewable energy storage devices

B Gnana Sundara Raj, S Anandan and R V Mangalaraja

In our current world, the challenge of energy scarcity has been exacerbated significantly because of the problems of the demand for oil, the consumption of non-renewable resources, the expanding global economy, and the rapid growth of the population causing environmental pollution. There is an urgent need to create clean and sustainable energy sources. Electrochemical energy storage (EES) devices have become the most notable energy storage devices in current renewable and sustainable nanotechnology, and address the increasing demand for power and energy density. Traditional supercapacitors, as promising electrochemical devices, have motivated research. Novel functional nanostructured materials have been developed to accomplish the future requirements of supercapacitors. In this regard, suitable electrode materials for supercapacitors are transition metal oxides, due to their various oxidation states and ions. In transition metal oxides, the hybrid metal oxides have different oxidation states, which facilitate different redox reactions. They have achieved higher supercapacitive performance than single component metal oxides, and they are the most favorable and low cost group of materials for supercapacitors. Here we discuss the current situation of the rapidly growing field of nanostructured hybrid transition metal oxides as electrodes for supercapacitor applications.

6.1 Introduction

The depletion of fossil fuels and increasing environmental pollution are pushing scientists and engineers to develop sustainable, clean, and eco-friendly alternative devices in the modern electronics industry to supply and store energy [1–5]. Over the

doi:10.1088/978-0-7503-5143-0ch6

past decades, tremendous efforts have been made in the establishment of electrochemical energy storage/conversion systems (EES), such as lithium-ion batteries (LIBs), electrochemical capacitors (ECs), and fuel cells (FCs) for applications in portable devices, electric vehicles, and stationary energy storage systems [6–8]. Key parameters for evaluating the performance of EESs and their potential for practical applications include power density, energy density, specific capacity, cycle life, and calendar life, as well as environmental protection and cost [9, 10]. Figure 6.1 illustrates the specific energy and specific power output of representative current electrochemical EESs [11, 12]. This shows that no single energy device can achieve both energy and power density values at the same interval. A conventional capacitor has a very high specific power, but at the same times a low energy density. Batteries and supercapacitors bridge the gap between conventional capacitors and fuel cells, revealing their superior energy storage capabilities. The supercapacitor maintains a higher power density than an LIB [13] and a higher energy density than a conventional capacitor. Figure 6.1 shows that ECs generally possess more specific power than batteries, while retaining a more reasonable specific energy [9].

Among the various electrochemical energy storage devices, supercapacitors, also known as ECs, are remarkably attractive due to their fast recharge capability, high power density, and long cycling life. Typically, an electrochemical capacitor device contains two electrodes, an electrolyte, and a separator that electrically separates the two electrodes, as shown in figure 6.2 [14].

Generally, supercapacitors can be divided into two categories defined by the charge storage mechanism. The first is electrical double layer capacitors (EDLCs) (figure 6.3 (a)), in which the charges are accumulated on the electrode/electrolyte surface onto large surface area electrodes [15]. EDLCs have a higher specific

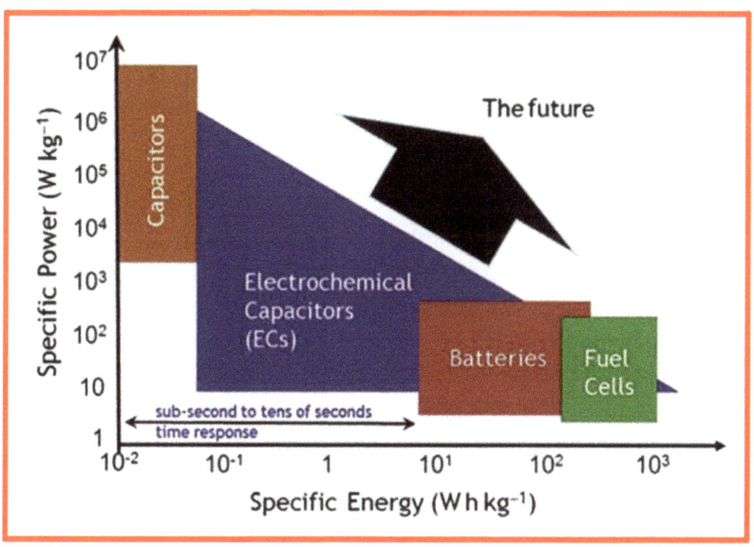

Figure 6.1. Ragone plot for electrochemical energy storage/conversion systems (EESs). (Reproduced with permission from [1]. Copyright 2011 Springer.)

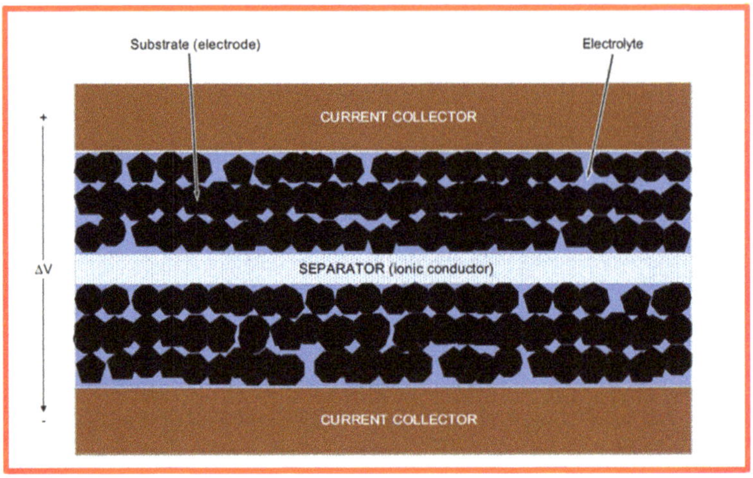

Figure 6.2. A schematic diagram of a symmetric supercapacitor. (Reproduced with permission from [2]. Copyright 2016 Springer.)

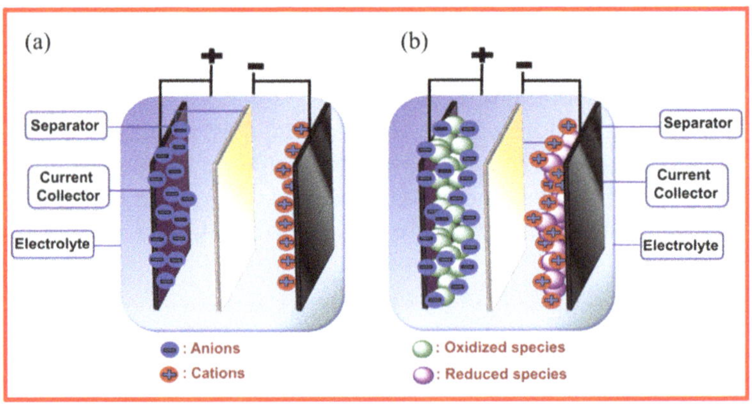

Figure 6.3. Schematic representation of (a) an EDLC and (b) a pseudocapacitive two-electrode cell. (Reproduced with permission from [3]. Copyright 2015 Springer.)

capacitance due to the more accessible surface and electronic conductivity of the electrode materials. From this point of view, the choice of electrode material for an EDLC must have a high specific surface area and good electrical conductivity.

For example, nanostructured carbon-based materials such as activated carbon, graphene, single-walled carbon nanotubes (SWCNTs), and reduced graphene oxide (RGO) are used as the active electrode materials. The double layer capacitance of carbon-based materials is controlled by their structure, in particular porosity and specific surface area [13, 16]. The second category of supercapacitors is pseudocapacitors or redox supercapacitors (figure 6.3(b)), in which the faster faradaic redox reaction takes up charge storage on the surface of the electroactive material. For pseudocapacitors, conducting polymers and transition metal oxides, for example, are used as the active electrode materials [17]. Supercapacitors can also be classified as symmetric and asymmetric supercapacitors based on the two electrodes of the device [9].

Although there have been excellent developments in the properties of carbon materials, their insufficient energy usually limits their practical application in EDLC-based supercapacitors. Therefore, numerous research attempts are focused on overcoming the low energy challenge of supercapacitors by designing other active electrode materials. Pseudocapacitive materials can be used in active materials with an alternative electrode for supercapacitors, which can achieve higher energy density and high specific capacitance than an EDLC [16]. The development of the capacitor can be modified by multiple oxidation states of the metal oxides, which allows a faster redox reaction at the electrode–electrolyte interface [18–21].

Recent developments have promoted hybrid transition metal oxides (HTMOs) as a group of reliable and low cost pseudocapacitive materials for supercapacitor applications. Due to the many redox reactions on HTMOs during electrochemical reactions they have a higher specific capacitance than single component metal oxides, and their synergistic effects enlarge the working potential window and increase the number of high electroactive sites during electrochemical reactions. It is worth noting that not only can the capacitance obtained by high faradaic redox reactions be improved, but the improvement of electronic conductivity and charge-transfer resistance can also be achieved by adding more metals to the metal oxides.

In this chapter, a detailed report is provided on the current developments in nanostructured HTMOs and graphene–HTMOs as electrodes for supercapacitors.

6.2 Fundamentals of nanostructured materials (NMs) for energy saving devices

Nanomaterials and nanostructures have played a key role in the recent development of many technologies. Nanomaterials differ from microscopic and aggregate materials in their specific dimensions not only in size, but they may also have new physical properties and offer new opportunities for various technological applications. Nanostructured materials are preferred as they provide a massive surface area for nanoscale ratios, different physical properties, reliable transport properties, and sealing effects created by the nanoscale dimensions. They have been studied consistently for energy related applications, such as lithium-ion batteries, supercapacitors, catalysts, hydrogen storage, solar cells, and thermoelectric systems. Nanostructured materials have several advantages in applications: (i) they provide a large area to enhance electrochemical reactions or molecular absorption, which takes place at the solid–liquid or solid–gas interface, (ii) they have optical effects that enhance optical absorption in solar cells, and (iii) the acceleration of electron or ion transport and electrolyte diffusion can be achieved using highly crystalline and/or porous structure nanomaterials [22].

Nanotechnology provides small sizes and dimensions, which offers a variety of benefits for the development of new technologies. For example, their small size makes nanoparticles potential carriers for delivering drugs to specific targets [23, 24]. As the size of electronic devices decreases, mobile phones promise to provide more functionality. As the number of defects decreases, the mechanical strength of nanomaterials is much higher than their bulk counterparts [25, 26]. However, the small size and large specific surface area of the nanomaterials can have detrimental effects in some applications. For example, nanowires or films with a thickness of several nanometers may have a lower conductivity than their total conductivity because the average free path of the most compressed electron often results from surface scattering [27]. The charge motion in a polycrystalline semiconductor is less than that of a crystal, and is much lower when the particle size decreases to nanometers [28].

Nanomaterials offer many benefits in energy storage and conversion applications. Their specific surface area, surface chemistry, and surface energy play a very important role in energy storage and transformation, including physical contact and/or chemical reactions at the surface or interface. Surface impacts are not limited to kinetics and rate, but to nuclear and consequent growth when involved in the thermodynamics and phase changes of heterogeneous reactions occurring at the interface. The small dimensions of nanomaterials can also provide very favorable mass, charge transfer, and heat, as well as accommodate dimensional changes associated with certain chemical reactions and phase changes. Nanotechnology presents new challenges in the use of energy storage and conversion. For example, a large specific surface area provides more sites for charge rectification at photovoltaics [29, 30] and small holes can control the diffusion of electrolyte ions in supercapacitors [31, 32].

The general advantages that nanostructured electrodes can offer are as follows [33]:

1. The electrode/electrolyte contact area per unit mass is significantly enlarged from the reduced dimensions of the electrode material, which provides higher ion absorption sites for double layer formation and charge-transfer reactions. In a few cases, the microporous (<2 nm) dimensions and their distribution can be modified to improve the areal or volumetric density of EDLCs and pseudocapacitance electrolyte–electrode interactions.
2. Micro-electrodes can indirectly reduce the ion and electronic diffusion distances, resulting in shorter diffusion times, higher energy density, and an improved high rate charge–discharge capacity.
3. Supercapacitors that control nanoscale material dimensions at the electrodes, and which show improved tolerance to stress and structural deformation, can then be cycled multiple times at high rates of stability.
4. Astonishing combinations of the properties of nano-sized electrode materials will allow supercapacitors to enter new markets. For example, supercapacitors that combine light weight, flexibility, transparency, and biodegradability would have applications in many fields.

Porous nanomaterials are used and commercialized in electric double layer capacitors (EDLCs) or supercapacitors. Capacitance is directly proportional to the total area, so nanopores are favorable for reaching a more specific area. On the other hand, small pores can penetrate and remove electrolyte ions, which can lead to lower energy densities. The impurities in microscopic materials may be unfavorable because they may react with the electrolyte to reduce cyclic stability. However, other impurities can increase the surface charge density, thus achieving greater efficiency. In general, nanostructures offer a larger surface area, making it easier for the electrolyte to access the active material, leading to shorter diffusion distances, and improved energy storage and performance in supercapacitors. The capacitance is enhanced by increasing the surface area by preventing accumulation in the case of activation processes or templating of graphene. However, since the pore structure must be carefully modified to allow for electrolyte ion diffusion, there is a limit to how much the surface can be magnified.

Various approaches have been proposed for energy storage and protection technologies through the design of anodes/cathodes/cells, the assembly of nanoparticles and nanocomposite materials, structural upgrades, and surface engineering/transformations. These approaches are used to improve the redox reaction rates of energy storage electrodes and to achieve faster kinetic mechanisms for higher power and faster charging [34–36]. Often, nano-sized particles/materials/composites and their configurations can control the rates of redox reactions that are directly present to enhance the electrode–electrolyte interface. In addition, nanomaterials can support reversible redox reactions, making them a promising device for clean energy storage and safety due to their high storage capabilities, safety, potential environmental friendliness, stability, and low cost [4, 35, 37, 38]. Nanoscale particles are used as a viable approach to achieve highly efficient energy storage and conservation technologies [36, 39–45].

Recently, different types of nanostructured materials (NMs) have been developed, including metal-based nanomaterials, carbon nanomaterials, nanostructured inorganic materials, and coordination polymers. They have key characteristics, such as high conductivity, excellent electro-functionality, and fast kinetics for environmentally friendly energy storage and conservation applications. Various NMs have been introduced by research communities to develop superior anode and cation products, such as nanomaterials, nanostructured organic–inorganic hybrids, carbon products, and polymers. In fact, such developments have led to significant growth in the field of energy storage and conservation research (figure 6.4) [8].

6.3 Materials with supercapacitor electrodes

The main component in pseudocapacitor devices is recognized as the electrode material, which affects the electrochemical performance of the supercapacitor. From this point of view, the selection and design of the electrode material for supercapacitor devices is essential. The electrode material must have good supercapacitive behavior according to the following requirements [46]:

i. High specific area, which enables specific capacitance [47–49]. Nanostructures offer a large area, which is important for fast ion/electron transfer and limited capacitance and rate capability. This ensures and facilitates adequate contact between the electrolyte and the active electrode materials, which leads to faster reaction kinetics, which affects the specific capacitance.

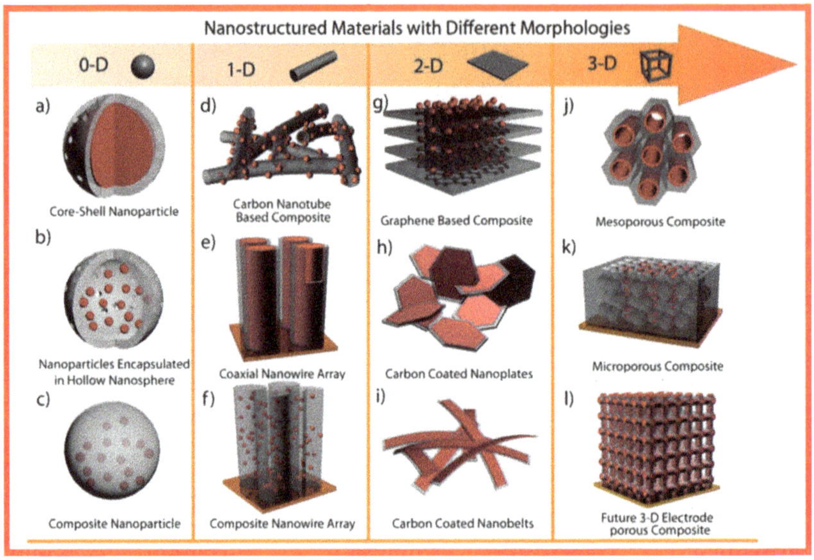

Figure 6.4. Different nanostructured materials based on their structural complexity. (Reproduced with permission from [4]. Copyright 2011 Springer.)

ii. High electronic conductivity, which is important for rate capability and energy density. Similarly high conductivity supercapacitive materials reduce the charge transfer resistance at the electrode–electrolyte interface [50, 51].
iii. Preferred electroactive sites that facilitate pseudocapacitance.
iv. High thermal stability and chemical stability [52].

These are some of the essential conditions that the electrode material must have. Transition metal oxide candidates have attracted a lot of attention considering their good capacitance values and cheap cost. Here we will consider only some of the commonly used hybrid transition metal oxide and carbon composite electrodes.

6.3.1 Supercapacitors based on hybrid transition metal oxides (HTMOs)

Recently, researchers have shown interest in HTMOs as electrode materials for supercapacitor applications due to their attractive electrochemical properties compared to single transition metal oxides. In the case of HTMOs, the synergistic effect between the multiple metal composition and the lower activation energy for electron movement leads to higher specific capacitance and electrical conductivity compared to a single TMO [53]. In this section, the composition and involvement of different HTMOs as supercapacitor electrodes are classified according to their main metal groups in the chemical composition.

Nickel and cobalt are preferred candidates for HTMOs because of their high redox activity [54–57]. Nickel cobalt HTMO has been widely reported to have better electrical performance than its single metal oxides. High faradaic redox reactions occur at the electrode–electrolyte interface in the previous case, based on the single metal oxides, nickel oxide, and cobalt oxide [58].

Nickel cobaltite ($NiCo_2O_4$), with the structural formula of [($Co^{2+}{}_{1-x}Co^{3+}{}_x$)($Co^{3+}Ni_x^{2+}Ni^{3+}{}_{1-x}$)$O_4$; where $0 < x < 1$] provides natural abundance, is eco-friendly, has extraordinary electrochemical functions, and is low cost. In particular, the tetrahedral lattice sites are found to be appropriate cations in the brackets ($Co^{2+}{}_{1-x}Co^{3+}{}_x$), whereas the octahedral lattice sites are found to be occupied by the remaining cations in the brackets ($Co^{3+}Ni^{2+}{}_xNi^{3+}{}_{1-x}$) [59, 60]. In summary, the lattice configuration of $NiCo_2O_4$ with the spinel generic formula of AB_2O_4 is illustrated in figure 6.5 [61], where O atoms are assigned in the cubic close-packed structure design, Co and Ni cations occupying the tetrahedral and octahedral lattice sites.

In general, the $NiCo_2O_4$ electrode operates below the potential range of 0–0.55 V on an alkaline electrolyte and its faradaic redox reactions are as follows [62]:

$$NiCo_2O_4 + OH^- + H_2O \leftrightarrow NiOOH + 2CoOOH + e^- \tag{6.1}$$

$$CoOOH + OH^- \leftrightarrow CoO_2 + H_2O + e^-. \tag{6.2}$$

Most researchers have found that $NiCo_2O_4$ has a variety of morphological and structural forms, such as nanoparticles, nanosheets, nanoplatelets, nanorods, microspheres, and more [63–71]. Each morphological system offers different supercapacitive performance. For example, BoopathiRaja *et al* [72] synthesized a hierarchical nanostructure $NiCo_2O_4$ sample with nanoflake (NCFs) and nanosheet (NCNs) like morphologies via a facile hydrothermal reaction, where nanoparticles served as the building blocks. A schematic representation of the synthesis of $NiCo_2O_4$ is shown in figure 6.6

Field emission scanning electron microscopy (FESEM) images of the NCN sample with different magnifications are shown in figure 6.7. The sample clearly shows that some irregular nanosheets are in the range of 2–3 μm. The FESEM images of the NCO sample clearly illustrate 3D hierarchical flower-like nanostructures, the thickness of

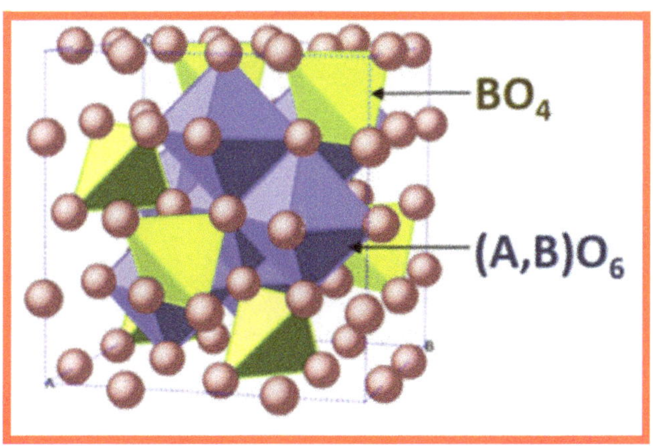

Figure 6.5. Spinel lattice structure of $NiCo_2O_4$.

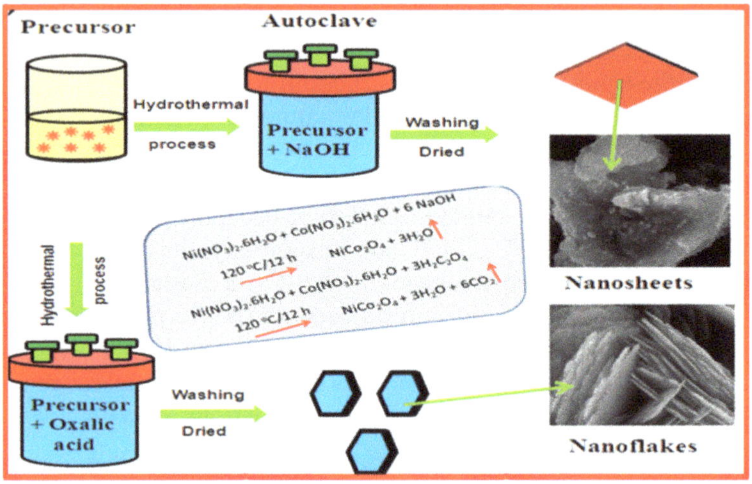

Figure 6.6. A schematic representation of the synthesis of $NiCo_2O_4$. (Reproduced with permission from [6]. Copyright 2020 Springer.)

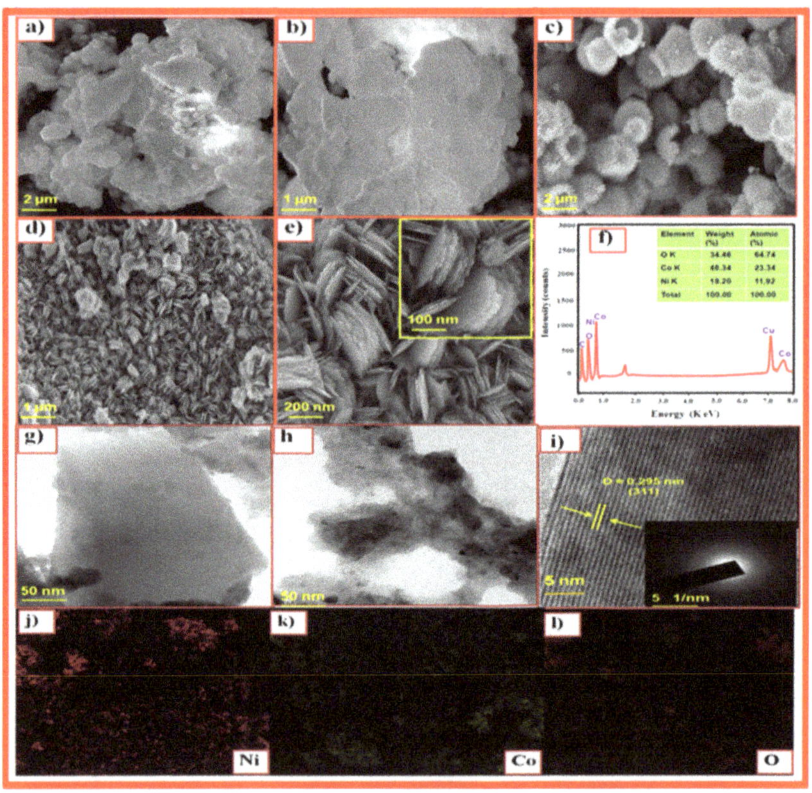

Figure 6.7. FESEM image of (a) and (b) NCN, and (c) NCO. (d) and (e) High resolution images of NCO with different magnifications. (f) EDS analysis of NCO. TEM images of (g) NCN and (h) NCO. (i) HRTEM image of NCO and its SAED pattern (inset). Elemental mapping of (j) Ni, (k) Co, and (l) O. (Reproduced with permission from [7]. Copyright 2020 Springer.)

the nanoflakes being approximately 15–20 nm. The energy dispersive x-ray spectroscopy (EDX) spectrum of the NCO sample basically shows the components of Ni, Co, and O from $NiCo_2O_4$. Furthermore, the morphologies of the sheets and scales were identified using TEM images. Clear, wrinkle type sheets and nanoflakes are identified in the corresponding images. The high resolution transmission electron microscopy (HRTEM) image of NCO clearly shows the latte edges with *d* values of 0.295 nm, which belong to the (311) plane of the cubic structure $NiCo_2O_4$. The corresponding selected area electron diffraction (SAED) pattern (inset) clearly shows a haloring pattern, indicating the polycrystalline nature. The uniform distribution of elements (Ni, Co, and O) in the samples is confirmed by element mapping.

The $NiCo_2O_4$ electrode samples express the specific capacitance of the NCO electrode, which is 1366 F g^{-1} at 2 A g^{-1}. On the positive side, at a high current density of 20 A g^{-1}, a specific capacitance of 566 F g^{-1} can be maintained. After 5000 cycles the specific capacitance of the NCN and NCO electrodes at 20 A g^{-1} was found to be 88.5% and 97.4%, respectively. This result suggests the presence of a unique nanoflakes structure that creates efficient access for the electrolyte to the surface of the material with the NCO electrode, which controls the ion diffusion path, resulting in better electrochemical performance of the electrode material.

Furthermore, a method to fabricate hollow bamboo-shaped $NiCo_2O_4$ (h-$NiCo_2O_4$) with ultrathin nanosheets is developed by Fang and co-authors [73] via a simple sacrificial template method. A detailed schematic procedure for fabricating the hollow $NiCo_2O_4$ is presented in figure 6.8. Two control experiments were conducted using only urea or NH_4F as alkaline precipitant, and the final obtained products were named as $NiCo_2O_4$-U and $NiCo_2O_4$-A. The $NiCo_2O_4$

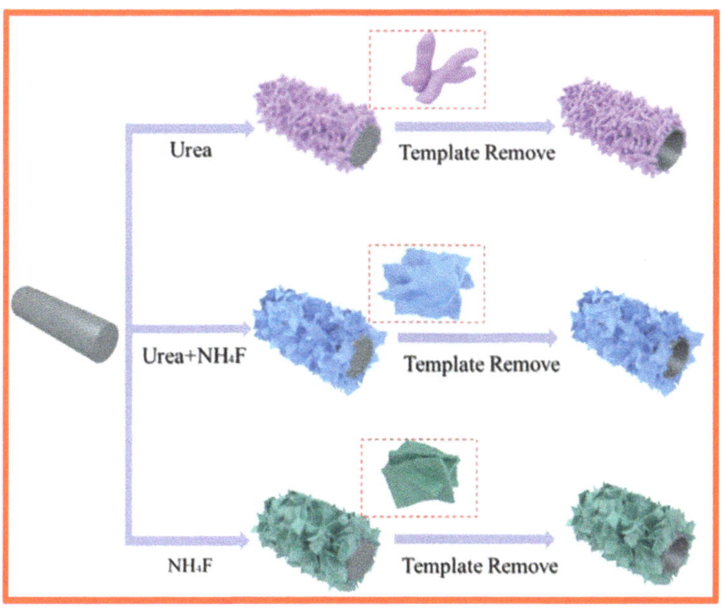

Figure 6.8. Schematic diagram of the preparation process for hollow bamboo-shaped $NiCo_2O_4$ with different alkaline precipitants. (Reproduced with permission from [8]. Copyright 2020 Springer.)

nanosheet is grown on the surface of template SBA-15 using a hydrothermal method with the mixed alkaline substances of urea and NH_4F, and then the template SBA-15 core is removed by NaOH etching. After the removing of the template, the bamboo shape is retained with thin $NiCo_2O_4$ nanosheets. In addition, the $NiCo_2O_4$ shell morphology can be controlled by adjusting the alkaline materials.

The SEM images in figure 6.9 show the SBA-15 template presenting a bamboo-shaped structure and the grown thin h-$NiCo_2O_4$ nanosheet. The obtained h-$NiCo_2O_4$ can retain the bamboo-shaped structure with a thin outer $NiCo_2O_4$ nanosheet. HRTEM was also used to analyse the hollow structure of the h-$NiCo_2O_4$. On this basis, the authors regulate the outer shell of $NiCo_2O_4$ by adjusting the component of the alkaline precipitant. Compared to the h-$NiCo_2O_4$, the width of the $NiCo_2O_4$-A nanosheet is much thicker and the surface is rougher. The thinner $NiCo_2O_4$ nanosheet of h-$NiCo_2O_4$ can effectively shorten the distance of the ion and charge transportation, which facilitates faster charge–discharge processes.

The h-$NiCo_2O_4$ has the highest specific capacity of 680.1 C g^{-1} at the current density of 1 A g^{-1}, followed by $NiCo_2O_4$-A (442 C g^{-1}), and $NiCo_2O_4$-U (436.7 C g^{-1}). Even if the current density is conducted at 10 A g^{-1}, the h-$NiCo_2O_4$ derived electrode can still hold at 618 C g^{-1}. This phenomenon is due to the outer shell structure of the fabricated $NiCo_2O_4$ [74]. The cycling test was carried out 5000 times for charging–discharging processes at a current density of 10 A g^{-1}, with h-$NiCo_2O_4$ still retaining 99.7% starting capacity. Moreover, it can also display an excellent coulombic efficiency of 99.72%. The above results reveal that the material possesses superior physical-chemical stability and redox reaction reversibility.

The results can be explained by the above excellent chemical properties of h-$NiCo_2O_4$ for the following reasons. (i) The thin outer shell $NiCo_2O_4$ nanosheet can provide more chemically active sites for the redox reaction to store charges. (ii) The thin $NiCo_2O_4$ nanosheet can effectively reduce the distance of charge transport and the hierarchical pore structure facilitates the permeation and reservoir of the electrolyte ions, which can ensure the fastest charging–discharging of the materials. (iii) The stable hollow structure can alleviate the volume changes of the

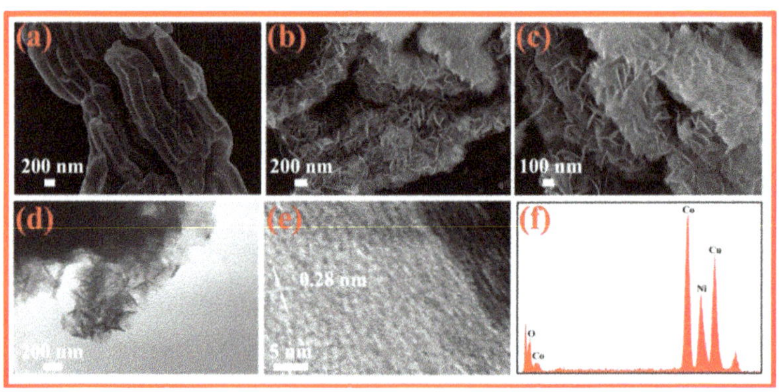

Figure 6.9. (a) SEM of the SBA-15 template. (b) and (c) Different SEM resolutions of h-$NiCo_2O_4$. (d) HRTEM of h-$NiCo_2O_4$. (e) Lattice fringe of h-$NiCo_2O_4$. (f) EDS spectrum of h-$NiCo_2O_4$. (Reproduced with permission from [9]. Copyright 2020 Springer.)

materials during the charging–discharging processes, and guarantee long-term cycling ability of the materials.

In addition to $NiCo_2O_4$, metal tungstate (MWO_4) [75, 76], metal molybdates ($MMoO_4$) [77–79], metal vanadate ($M_3V_2O_8$) [80, 81], metal ferrite (MFe_2O_4) [82], and metal manganites (MMn_2O_4) [83] have been identified as possible electrode materials for supercapacitor application. This is because of they are cost effective, naturally abundant, eco-friendly, highly stable, and have excellent electrochemical performances. The relevant electrochemical features of these materials are presented in table 6.1.

Although hybrid transition metal oxides have significant theoretical capacitance in combination with the redox reaction, they still suffer from low electrical conductivity, poor rate performance, low cycling stability, and inefficient use of the active substance during the electrochemical reaction. Therefore, the hybrid materials combine the morphological and electrochemical properties of metal oxides and carbonaceous materials into supercapacitor electrodes, combining the EDLC and faradaic properties of the active material for enhanced charge storage performance.

6.3.2 Supercapacitors based on graphene–hybrid transition metal oxides (G–HTMOs)

The combination of carbonaceous materials with various hybrid transition metal oxides is reported as a functional and effective approach to improve the

Table 6.1. A summary of the different hybrid transition metal oxides.

Morphology	Synthetic methods	Specific capacitance ($F\ g^{-1}$)	Current density	Cycling stability	Reference
$NiWO_4$ nanospheres	Chemical precipitation method	437.2	$0.1\ A\ g^{-1}$	44% (500 cycles)	[75]
$CoWO_4$ nanoparticles	Chemical precipitation reaction	378	$2\ mV\ s^{-1}$	95.5% (4000 cycles)	[76]
Honeycomb $NiMoO_4$	Electrodeposition method	1475	$1\ A\ g^{-1}$	87.9% (5000 cycles)	[77]
Nanoflake $NiMoO_4$	SILAR method	1853	$1\ A\ g^{-1}$	65% (2500 cycles)	[78]
$CoMoO_4$ microspheres	Hydrothermal method	384	$1\ A\ g^{-1}$	94.2 (1000 cycles)	[79]
$Ni_3V_2O_8$ nanoflakes	Chemical co-precipitation technique	1181	$0.625\ A\ g^{-1}$	73% (1000 cycles)	[80]
$NiCo_2V_2O_8$ nanoplates	Solvothermal method	1098.9	$4\ A\ g^{-1}$	68% (7000 cycles)	[81]
$NiFe_2O_4$ nanosheets	Hydrothermal method	240.9	$1\ A\ g^{-1}$	128% (2000 cycles)	[82]
Flower-like $CoMn_2O_4$ microspheres	Solvothermal method	188	$1\ A\ g^{-1}$	93% (1000 cycles)	[83]

Table 6.2. Summary for the development of graphene–HTMO nanocomposites [94–103] as working electrodes for supercapacitor systems.

Morphology	Synthetic methods	Specific capacitance	Current density	Cycling stability	Reference
$NiWO_4$–rGO	Solvothermal method	1031.3 F g^{-1}	0.5 A g^{-1}	100% (5000 cycles)	[94]
$CoWO_4$–N-rGO	Sonochemical method	450 F g^{-1}	2 A g^{-1}	97.1 (4000 cycles)	[95]
$NiMoO_4$ nanorods–graphene	Solvothermal method	670 F g^{-1}	0.3 A g^{-1}	88% (3000 cycles)	[96]
$CoMoO_4$ nanoplates–graphene	Hydrothermal method	1255.24 F g^{-1}	1 A g^{-1}	91.3% (3000 cycles)	[97]
$Ni_3V_2O_8$–graphene	Solvothermal method	748 F g^{-1}	0.5 A g^{-1}	71% (3000 cycles)	[98]
$Co_3V_2O_8$ micropencils–graphene	Solvothermal method	528.17 F g^{-1}	0.5 A g^{-1}	80% (5000 cycles)	[99]
$NiFe_2O_4$ nanocubes–rGO	Hydrothermal method	488 F g^{-1}	1 A g^{-1}	89.8 (10 000 cycles)	[100]
$CoFe_2O_4$–graphene	Hydrothermal method	166.5 C g^{-1}	0.5 A g^{-1}	79.3 (5000 cycles)	[101]
$NiMn_2O_4$ nanosheets–rGO	Co-precipitation	1243 F g^{-1}	3 A g^{-1}	80.8 (5000 cycles)	[102]
$CoMn_2O_4$–3D rGO	Hydrothermal method	1028 F g^{-1}	1 A g^{-1}	90.2 (3000 cycles)	[103]

electrochemical performance of supercapacitors. Recently, researchers have shown more interest in graphene, a sheet of sp^2 bonded carbon atoms in a honeycomb crystal structure, compared to other carbonaceous materials, such as activated carbon (AC) and carbon nanotubes (CNTs) [84]. This is mainly due to its attractive physical and chemical properties, such as a large theoretical surface, high electrical conductivity, excellent electrochemical stability, high flexibility, and excellent structural stability [85, 86].

The combination of a graphene sheet with nanostructured HTMO expands the available specific surface of the electrode active material. In addition the combined effects and well known interaction between the two phases contributes to the electron and ion diffusion process. Various synthesis approaches for 2D graphene sheets lead to the simple and controllable formation of a nanocomposite with unique nanostructures around a target application, such as supercapacitors. Freestanding 2D graphene nanosheets include 2D edge plane sites, which facilitate the ion absorption process, thus enhancing the specific power and specific energy of the G–HTMO nanocomposite [87, 88]. In this section, the synthesis and contribution of different G–HTMO nanocomposites as supercapacitor electrodes are discussed in detail and it is seen how they align well with the major metal groups in their chemical compositions.

Recently, graphene has been combined with metal cobaltite into nanocomposites, which has attracted a great deal of research interest due to their advanced physical and electrochemical properties compared to single metal cobaltite. For example, Jiu *et al* [89] grew $NiCo_2O_4$ (NCO) nanowires in a three-dimensional (3D) porous graphene aerogel (GA) using an easy two-step hydrothermal method.

SEM and TEM images were examined for the morphology and structure of the prepared samples. In the SEM images in figure 6.10 the interconnected 3D porous network structure of the graphene aerogel can be seen clearly; the walls of the holes are graphene sheets that are optionally cross-linked and interconnected. The morphology of the NCO–GA has NCO nanowires with diameters of 30–50 nm located densely on graphene aerogel sheets. TEM images of the NCO–GA show that the NCO nanowires are scattered on graphene sheets, which are compatible with the SEM images. A transparent lattice edge can be detected from high resolution TEM, and the latitude of 0.47 nm corresponds to the crystal plane of NCO (111). The well-defined diffraction rings in the SAED pattern can undoubtedly be indexed for NCO phase (220), (311), (400), (422), and (440) planes.

The electrochemical performance of NCO–GA (720 F g^{-1}) at a current density of 1 A g^{-1} shows higher capacitance than NCO (512 F g^{-1}). The increased capacitance

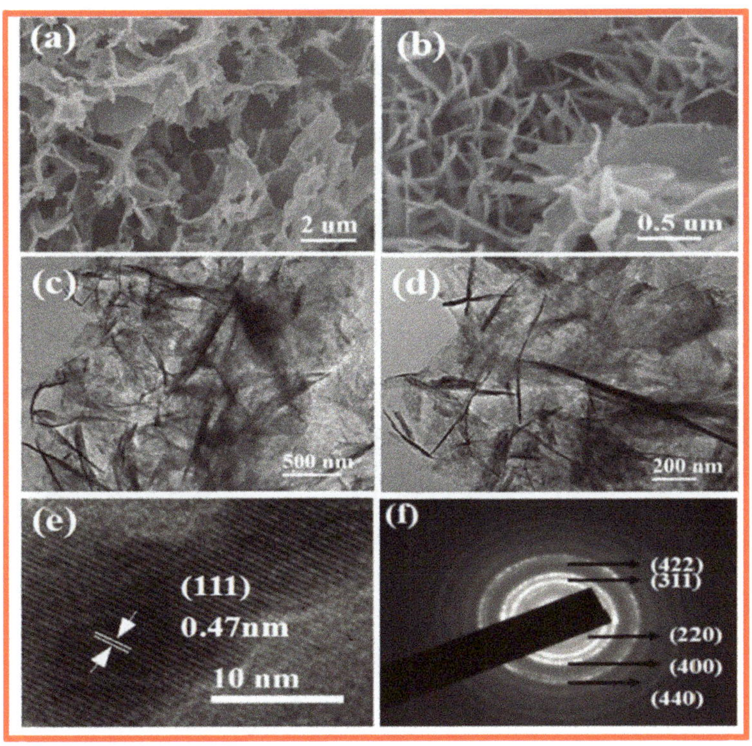

Figure 6.10. SEM images of (a) GA and (b) NCO–GA. (c) and (d) TEM images of NCO–GA at different magnifications. (e) HRTEM image and (f) SAED pattern of NCO–GA. (Reproduced with permission from [10]. Copyright 2019 Springer.)

of the NCO–GA electrode may have been contributed by the introduction of GA. The main reason can be generalized in several ways: (i) GA can be used as a deposition skeleton for the NCO, which inhibits the integration of the NCO and enhances the utilization of the electrode material; (ii) GA with its three-dimensional structure and better electrical properties can improve electron transfer and electrolyte transport, thereby leading to improved electrochemical properties; and (iii) the graphene structure of NCO–GA has good elasticity and thermal conductivity which can reduce the heat and pressure caused by electrochemical reactions during the charge–discharge process from the inside of the electrode [90–92]. For the cycling stability test of the NCO–GA electrode in 1 A g^{-1}, the specified capacitance is maintained at 84% after 1000 cycles.

Zhou and co-authors [93] created a 3D porous graphene (rGN) film based on a novel and green Cu_2O template-assisted 'integration engraving and precipitation' (figure 6.11).

The object's orientation processes are described in scheme 1. The morphology of the different specimens was examined using SEM and TEM images. Cu_2O forms uniform nanospheres with a diameter of about 700 nm, which is ideal for use as a template in designing the most successful nanogages (figure 6.12). When the Cu_2O was introduced to the graphene, the crushed graphene sheets coated with Cu_2O were glazed in several layers. Furthermore, some flocky lychee-like features appeared on the surface of the Cu_2O nanospheres due to the absorbed graphene. The 3D rGN film featured a porous structure with an aperture size of micrometers after removing the template. Such a pore configuration will have a large surface area and provide high quality use of the active material. The morphology of the 3D rGN–$NiCo_2O_4$ is such that $NiCo_2O_4$, with an average diameter of 830 nm, acquires the same spherical morphology as the Cu_2O nanospheres and is scatter throughout the rGN film, avoiding the accumulation of graphene sheets as much as possible.

The extensive features of $NiCo_2O_4$ show hollow hydrangea-like structures that contain numerous ultrasonic nanosheets perpendicular to the spherical surface. Apparently, these hollow $NiCo_2O_4$ hydrangeas form more pores in the film, which

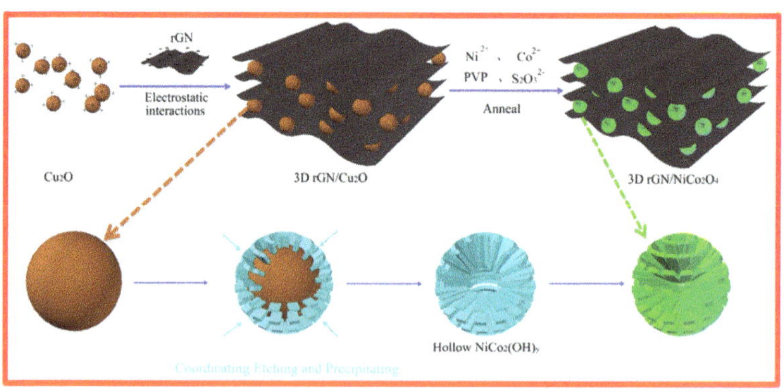

Figure 6.11. Schematic illustration of the formation of 3D rGN–$NiCo_2O_4$ film. (Reproduced with permission from [11]. Copyright 2020 Springer.)

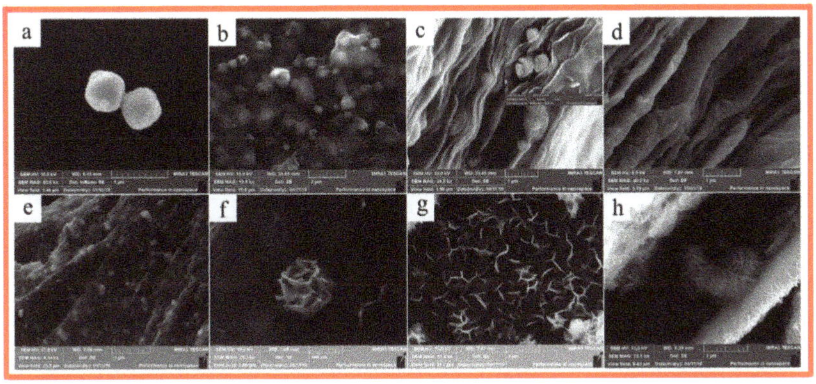

Figure 6.12. (a) SEM image of Cu_2O. (b) Surface and (c) cross-section SEM images of the rGN–Cu_2O film. (d) Cross-section SEM of the 3D rGN film. (e)–(h) Overview, surface, and cross-section SEM images of the 3D rGN–$NiCo_2O_4$ film. (Reproduced with permission from [12]. Copyright 2020 Springer.)

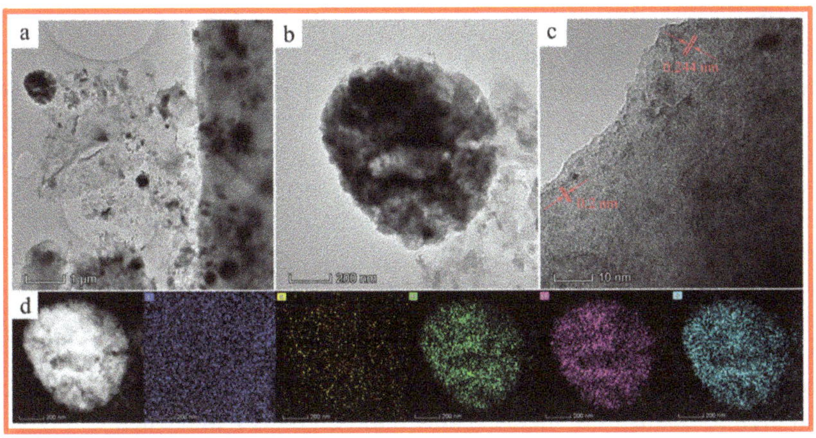

Figure 6.13. (a) and (b) TEM, (c) HRTEM, and (d) corresponding elemental mapping images of the 3D rGN–$NiCo_2O_4$ film. (Reproduced with permission from [13]. Copyright 2020 Springer.)

can store large amounts of electrolytes, increasing the reactive sites and accelerating the transport of ions, resulting in higher-rate discharge efficiency and increased energy storage performance.

TEM analysis was used to clarify the internal structure of the 3D rGN–$NiCo_2O_4$ film (figure 6.13). The apparent bright and dark variations fully stabilize the hollow structure of the $NiCo_2O_4$ hydrangeas. Closer observation again demonstrates that the nanosheets are the building blocks of the $NiCo_2O_4$ hydrangeas. The adjacent lattice edges of the $NiCo_2O_4$ are 0.244 nm and 0.20 nm, which fits well with the (400) and (311) planes in the HRTEM image. C, N, Ni, Co, and O elements distribution to ensure the fabrication of 3D rGN–$NiCo_2O_4$ film.

The specific capacitance value at the current density of 1 A g^{-1} for the 3D rGN–$NiCo_2O_4$ film is 708.36 F g^{-1}, respectively. The electrochemical stability of the 3D rGN–$NiCo_2O_4$ film is further characterized by the fact that 94.3% of the initial capacity is kept above 6000 charge–discharge cycles, i.e. the hybrid material has a good cycling life. The results demonstrate that graphene plays an important role in maintaining capacitive stability. On the one hand, strong Π–Π interactions between individual graphene sheets can produce integrity of the film for a greater number of cycles. Graphene, on the other hand, acts as a 'flexible confinement' to envelop hollow $NiCo_2O_4$ hydrangeas or Cu_2O nanospheres and prevent them from electrochemical dissolution. The 3D rGN–$NiCo_2O_4$ hybrid film has shown great potential for use as a flexible electrode for high performance energy storage devices.

In summary, these state-of-the-art examples clearly illustrate the synergistic effect between HTMO and graphene products. Therefore, combining graphene with HTMO is an attractive approach to improve capacitance and to exceed the limit of single HTMO (low electrical conductivity) for high performance supercapacitor applications.

6.4 Conclusions and future perspectives

In this chapter we conclude with the latest advances for supercapacitors, including their composition in nanostructured electrodes of hybrid transition metal oxides. Key advantages of hybrid transition metal oxides are their excellent electrical conductivity, stable structure, large electroactive centers, and environmentally friendly and harmless compatibility. In fact, nanostructures of various dimensions have unique features that give them the flexibility to retain the microscopic structure and properties of supercapacitor electrodes, while hybrid transition metal oxides provide a source of high specific capacitance and high energy density.

It is necessary to meet the requirements of practical applications to further improve the efficiency of specific capacitance, power density, energy density, rate efficiency, and long-term cycle stability. In recent years, despite the great development in hybrid transition metal oxide nanostructures and their supercapacitor applications, many important challenges still remain and they need to be explored.

First, its is necessary to control the morphology and dimensions of the hybrid transition metal oxide material because the shapes of the electrode materials have a significant impact on the speed of ion transfer and the capacity of the charge storage. This is a viable way to enhance electrochemically active sites, by designing advanced architectures with superficial and extraordinary structures such as 3D or hierarchical nanostructures, mesoporous, synergistic core–shell structures, and ordered nanostructured arrays. It is a feasible way to increase the electrochemically active sites for redox reactions or chemical absorption–desorption and to further increase the energy storage density.

Second, for future applications, a composition containing carbonaceous materials and other active nanomaterials to create the conductivity of the hybrid transition metal oxides will have a cohesive effect between the components by reducing the particle size, improving the specific surface area, promoting electron and proton transmission, stimulating porosity, blocking the particles from integration, protecting active materials from mechanical decay, improving cycling stability, and providing additional pseudocapacitance.

Third, one of the most effective ways to improve energy density is to create hybrid supercapacitors. An asymmetric hybrid supercapacitor can combine a pseudocapacitance or battery-type electrode with an EDLC-type electrode that can pass a relatively short static potential window and finite energy density of aqueous electrolytes and allow twice as much accessible operating voltage as EDLCs.

We look forward to promoting further improvements in this interesting area of research in order to make significant progress in supercapacitor applications in the future.

Acknowledgements

B G S R gratefully acknowledges the FONDECYT Postdoctoral Project No.: 3190090, Government of Chile, Santiago, for financial support through a grant.

References

[1] Yuan C, Wu H B, Xie Y and Lou X W 2014 Mixed transition-metal oxides: design, synthesis, and energy-related applications *Angew. Chem. Int. Ed.* **53** 1488–504

[2] Reddy A L M, Gowda S R, Shaijumon M M and Ajayan P M 2012 Hybrid nanostructures for energy storage applications *Adv. Mater.* **24** 5045–64

[3] Yang Z, Zhang J, Kintner-Meyer M C W, Lu X, Choi D, Lemmon J P and Liu J 2011 Electrochemical energy storage for green grid *Chem. Rev.* **111** 3577–613

[4] Arico A S, Bruce P, Scrosati B, Tarascon J M and vanSchalkwijk W 2005 Nanostructured materials for advanced energy conversion and storage devices *Nat. Mater.* **4** 366–77

[5] Simon P and Gogotsi Y 2008 Materials for electrochemical capacitors *Nat. Mater.* **7** 845–54

[6] Lin L, Ning H, Song S, Xu C and Hu N 2020 Flexible electrochemical energy storage: the role of composite materials *Compos. Sci. Technol.* **192** 108102

[7] Winter M, Barnett B and Xu K 2018 Before Li ion batteries *Chem. Rev.* **118** 11433–56

[8] Xu K 2014 Electrolytes and interphases in Li-ion batteries and beyond *Chem. Rev.* **114** 11503–618

[9] Chen D, Wang Q, Wang R and Shen G 2015 Ternary oxide nanostructured materials for supercapacitors: a review *J. Mater. Chem.* A **3** 10158–73

[10] Burke A 2000 Ultracapacitors: why, how, and where is the technology *J. Power Sources* **91** 37–50

[11] Rolison D R and Nazar L F 2011 Electrochemical energy storage to power the 21st century *MRS Bull.* **36** 486

[12] Darowicki K, Andrearczyk K, Slepski P, Sierczynska A, Lota G, Fic K and Lota K 2014 Determination of pseudocapacitance changes of nickel oxide NiO electrode with the use of dynamic electrochemical impedance spectroscopy *Int. J. Electrochem. Sci.* **9** 1702–14

[13] Low W H, Khiew P S, Lim S S, Siong C W and Ezeigwe E R 2019 Recent development of mixed transition metal oxide and graphene/mixed transition metal oxide based hybrid nanostructures for advanced supercapacitors *J Alloy Compd.* **775** 1324–56

[14] Gonzalez A, Goikolea E, Barrena J A and Mysyk R 2016 Review on supercapacitors: technologies and materials *Renew. Sustain. Energy Rev.* **58** 1189–206

[15] Achilleos D S and Hatton T A 2015 Surface design and engineering of hierarchical hybrid nanostructures for asymmetric supercapacitors with improved electrochemical performance *J. Colloid Interf. Sci.* **447** 282–301

[16] Noori A, El-Kady M F, Rahmanifar M S, Kaner R B and Mousavi M F 2019 Towards establishing standard performance metrics for batteries, supercapacitors and beyond *Chem. Soc. Rev.* **48** 1272–341

[17] Chen D, Wang Q, Wang R and Shen G 2015 Ternary oxide nanostructured materials for supercapacitors: a review *J. Mater. Chem.* A **3** 10158–73

[18] Lota G, Fic K and Frackowiak E 2011 Carbon nanotubes and their composites in electrochemical applications *Energy Environ. Sci.* **4** 1592–605

[19] Lee S W, Gallant B M, Lee Y, Yoshida N, Kim D Y, Yamada Y, Noda S, Yamada A and Shao-Horn Y 2012 Self-standing positive electrodes of oxidized few-walled carbon nanotubes for light-weight and high-power lithium batteries *Energy Environ. Sci.* **5** 5437–44

[20] Xu X, Shen J, Li N and Ye M 2014 Facile synthesis of reduced graphene oxide/$CoWO_4$ nanocomposites with enhanced electrochemical performances for supercapacitors *Electrochim. Acta* **150** 23–34

[21] Wu C, Cai J, Zhang Q, Zhou X, Zhu Y, Li L, Shen P and Zhang K 2015 Direct growth of urchin-like $ZnCo_2O_4$ microspheres assembled from nanowires on nickel foam as high-performance electrodes for supercapacitors *Electrochim. Acta* **169** 202–9

[22] Zhang Q, Uchaker E, Candelariaz S L and Cao G 2013 Nanomaterials for energy conversion and storage *Chem. Soc. Rev.* **42** 3127–71

[23] Singh R and Lillard J W Jr 2009 Nanoparticle-based targeted drug delivery *Exp. Mol. Pathol.* **86** 215–23

[24] Vasir J K, Reddy M K and Labhasetwar V D 2005 Nanosystems in drug targeting: opportunities and challenges *Curr.Nanosci.* **1** 47–64

[25] Terrones M 2003 Science and technology of the twenty-first century: synthesis, properties, and applications of carbon nanotubes *Annu. Rev. Mater. Res.* **33** 419–501

[26] Endo M, Strano M and Ajayan P 2008 Potential applications of carbon nanotubes *Top. Appl. Phys.* **111** 13–61

[27] Barati M and Sadeghi E 2001 Study of the ordinary size effect in the electrical conductivity of Bi nanowires *Nanotechnology* **12** 277

[28] Frank A J, Kopidakis N and Lagemaat J 2004 Electrons in nanostructured TiO_2 solar cells transport, recombination and photovoltaic properties *Coord. Chem. Rev.* **248** 1165–79

[29] Cahen D, Hodes G, Graetzel M, Guillemoles J F and Riess I 2000 Nature of photovoltaic action in dye-sensitized solar cells *J. Phys. Chem.* B **104** 2053–9

[30] Robertson N 2006 Optimizing dyes for dye-sensitized solar cells *Angew. Chem., Int. Ed.* **45** 2338–45

[31] Wang G, Zhang L and Zhang J 2012 A review of electrode materials for electrochemical supercapacitors *Chem. Soc. Rev.* **41** 797–828

[32] Fernandez J, Tennison S, Kozynchenko O, Rubiera F, Stoeckli F and Centeno T 2009 Effect of mesoporosity on specific capacitance of carbons *Carbon* **47** 1598–604

[33] Zhao X, Sanchez B M, Dobson P J and Grant P S 2011 The role of nanomaterials in redox-based supercapacitors for next generation energy storage devices *Nanoscale* **3** 839–55

[34] Kumar P, Kim K H, Bansal V and Kumar P 2017 Nanostructured materials: a progressive assessment and future direction for energy device applications *Coordin. Chem. Rev.* **353** 113–41

[35] Dai L, Chang W D, Baek B J and Lu W 2012 Carbon nanomaterials for advanced energy conversion and storage *Small* **8** 1130–66

[36] Dubal D, Ayyad O, Ruiz V and Gómez-Romero P 2015 Hybrid energy storage: the merging of battery and supercapacitor chemistries *Chem. Soc. Rev.* **44** 1777–90

[37] Winter M and Brodd J R 2004 What are batteries, fuel cells, and supercapacitors? *Chem. Rev.* **104** 4245–70
[38] Liu C, Li F, Ma P L and Cheng M J 2010 Advanced materials for energy storage *Adv. Mater.* **22** E28–62
[39] Chae H J, Ng C K and Chen Z G 2010 Nanostructured materials for the construction of asymmetrical supercapacitors *J. Power Energy* **224** 479–503
[40] Masquelier C and Croguennec L 2013 Polyanionic (phosphates, silicates, sulfates) frameworks as electrode materials for rechargeable Li (or Na) batteries *Chem. Rev.* **113** 6552–91
[41] Reddy M, Subba Rao G and Chowdari B 2013 Metal oxides and oxysalts as anode materials for Li ion batteries *Chem. Rev.* **113** 5364–457
[42] Liu C, Yu Z, Neff D, Zhamu A and Jang Z B 2010 Graphene-based supercapacitor with an ultrahigh energy density *Nano Lett.* **10** 4863–8
[43] Xie K and Wei B 2014 Materials and structures for stretchable energy storage and conversion devices *Adv. Mater.* **26** 3592–617
[44] Bruce G P, Scrosati B and Tarascon M J 2008 Nanomaterials for rechargeable lithium batteries *Angew. Chem. Int. Ed.* **47** 2930–46
[45] Liu R, Duay J and Lee B S 2011 Heterogeneous nanostructured electrode materials for electrochemical energy storage *Chem. Commun.* **47** 1384–404
[46] Kate R S, Khalate S A and Deokate R J 2018 Overview of nanostructured metal oxides and pure nickel oxide (NiO) electrodes for supercapacitors: a review *J Alloy Compd.* **734** 89–111
[47] Chang J, Park M, Ham D, Ogale S B, Mane R S and Han S 2008 Liquid-phase synthesized mesoporous electrochemical supercapacitors of nickel hydroxide *Electrochim. Acta* **53** 5016–21
[48] Meher S K, Justin P and Rao G R 2011 Microwave-mediated synthesis for improved morphology and pseudocapacitance performance of nickel oxide *Interfaces* **3** 2063–73
[49] Wang Y and Xia Y 2006 Electrochemical capacitance characterization of NiO with ordered mesoporous structure synthesized by template SBA-15 *Electrochim. Acta* **51** 3223–7
[50] Zhi M, Xiang C, Li J, Lia M and Wu N 2013 Nanostructured carbon–metal oxide composite electrodes for supercapacitors: a review *Nanoscale* **5** 72–8
[51] Justin P, Meher S K and Ranga Rao G 2010 Tuning of capacitance behavior of NiO using anionic, cationic, and nonionic surfactants by hydrothermal synthesis *J. Phys. Chem.* **114** 5203–10
[52] Zhi M, Xiang C, Li J, Lia M and Wu N 2013 Nanostructured carbonemetal oxide composite electrodes for supercapacitors: a review *Nanoscale* **5** 72–8
[53] Reddy M V, Subba Rao G V and Chowdari B V R 2013 Metal oxides and oxysalts as anode materials for Li ion batteries *Chem. Rev.* **113** 5364–457
[54] Huang Y Y and Lin L Y 2018 Synthesis of ternary metal oxides for battery-supercapacitor hybrid devices: influences of metal species on redox reaction and electrical conductivity *ACS Appl. Energy Mater.* **1** 2979–90
[55] Zhang J, Liu F, Cheng J P and Zhang X B 2015 Binary nickel-cobalt oxides electrode materials for high-performance supercapacitors: influence of its composition and porous nature *ACS Appl. Mater. Interfaces* **7** 17630–40
[56] Lin L Y and Lin L Y 2017 Material effects on the electrocapacitive performance for the energy-storage electrode with nickel cobalt oxide core/shell nanostructures *Electrochim. Acta* **250** 335–47
[57] Hong W L, Lin L Y and Lin L Y 2017 Growing sequence effects of core–shell nanostructure on morphology and electrocapacitive ability for energy-storage electrodes *Electrochim. Acta* **255** 309–22

[58] Gonzalez A, Goikolea E, Barrena J A and Mysyk R 2016 Review on supercapacitors: technologies and materials *Renew. Sust. Energ. Rev.* **58** 1189–206

[59] Changzhou Y, Jiaoyang L, Linrui H, Xiaogang Z, Laifa S and Wen L X 2012 Ultrathin mesoporous $NiCo_2O_4$ nanosheets supported on Ni foam as advanced electrodes for supercapacitors *Adv. Funct. Mater.* **22** 4592–7

[60] Kim J G, Pugmire D L, Battaglia D and Langell M A 2000 Analysis of the $NiCo_2O_4$ spinel surface with auger and x-ray photoelectron spectroscopy *Appl. Surf. Sci.* **165** 70–84

[61] Bitla Y *et al* 2015 Origin of metallic behavior in $NiCo_2O_4$ ferrimagnet *Sci. Rep.* **5** 15201

[62] Wu Z, Zhu Y and Ji X 2014 $NiCo_2O_4$-based materials for electrochemical supercapacitors *J. Mater. Chem.* A **2** 14759e–72

[63] Yan D, Wang W, Luo X, Chen C, Zeng Y and Zhu Z 2018 $NiCo_2O_4$ with oxygen vacancies as better performance electrode material for supercapacitor *Chem. Eng. J.* **334** 864–72

[64] Li Y, Hou X, Zhang Z, Hai Z, Xu H, Cui D, Zhuiykov S and Xue C 2018 $NiCo_2O_4$ particles with diamond-shaped hexahedron structure for high-performance supercapacitors *Appl. Surf. Sci.* **436** 242–51

[65] Saraf M, Natarajan K and Mobin S M 2017 Multifunctional porous $NiCo_2O_4$ nanorods: sensitive enzymeless glucose detection and supercapacitor properties with impedance spectroscopic investigations *New J. Chem.* **41** 9299–313

[66] Huang T, Liu B, Yang P, Qiu Z and Hu Z 2018 Facilely synthesized $NiCo_2O_4$ nanoparticles as electrode material for supercapacitors *Int. J. Electrochem. Sci.* **13** 6144–54

[67] Zhang G and Lou X W 2013 Controlled growth of $NiCo_2O_4$ nanorods and ultrathin nanosheets on carbon nanofibers for high-performance supercapacitors *Sci. Rep.* **3** 1470

[68] Chen H, Jiang G, Yu W, Liu D, Liu Y, Li L and Huang Q 2017 Electrospun carbon nanofiberic coated with ambutan-like $NiCo_2O_4$ microspheres as electrode materials *MRS Commun.* **7** 90–6

[69] Zhicheng J, Guangyao M, Yulong Z, Zheng X, Yinghuai Q and Yitai Q 2015 A facile method for synthesis of porous $NiCo_2O_4$ nanorods as a high-performance anode material for Li-ion batteries *Part. Part. Syst. Char.* **32** 1012–9

[70] Sahoo S, Ratha S and Rout C 2015 Spinel $NiCo_2O_4$ nanorods for supercapacitor applications *Am. J. Eng. Appl. Sci.* **8** 371–9

[71] Jiang S, Sun Y, Dai H, Ni P, Lu W, Wang Y, Li Z and Li Z 2016 A facile enhancement in battery-type of capacitive performance of spinel $NiCo_2O_4$ nanostructure via directly tuning thermal decomposition temperature *Electrochim. Acta* **191** 364–74

[72] BoopathiRaja R and Parthibavarman M 2020 Reagent induced formation of $NiCo_2O_4$ with different morphologies with large surface area for high performance asymmetric supercapacitors *Chem. Phys. Lett.* **755** 137809

[73] Fang J, Kang C, Fu L, Li S and Liu Q 2020 Fabrication of hollow bamboo-shaped $NiCo_2O_4$ with controllable shell morphologies for high performance hybrid supercapacitors *J. Alloy Compd.* **849** 156317

[74] Xiang C, Wang Q, Zou Y, Huang P, Chu H, Qiu S, Xu F and Sun L 2017 Simple synthesis of graphene-doped flower-like cobalt–nickel–tungsten–boron oxides with self-oxidation for high-performance supercapacitors *J. Mater. Chem.* **5** 9907–16

[75] Huang Y *et al* 2019 Amorphous $NiWO_4$ nanospheres with high-conductivity and capacitive performance for supercapacitors *J. Phys. Chem.* C **123** 30067–76

[76] Adib K, Nasrabadi M R, Rezvani Z, Pourmortazavi S M, Ahmadi F, Naderi H R and Ganjali M R 2016 Facile chemical synthesis of cobalt tungstates nanoparticles as high performance supercapacitor *J. Mater. Sci.: Mater. Electron.* **27** 4541–50

[77] Kumbhar V S, Nguyen V Q, Lee Y R, Lokhande C D, Kim D H and Shim J J 2018 Electrochemically growth-controlled honeycomb-like $NiMoO_4$ nanoporous network on nickel foam and its applications in all-solid-state asymmetric supercapacitors *New J. Chem.* **42** 14805–16

[78] Chavan H S *et al* 2018 Nanoflake $NiMoO_4$ based smart supercapacitor for intelligent power balance monitoring *Sol. Energ. Mat. Sol.* C **185** 166–73

[79] Li W, Wang X, Hu Y, Sun L, Gao C, Zhang C, Liu H and Duan M 2018 Hydrothermal synthesized of $CoMoO_4$ microspheres as excellent electrode material for supercapacitor *Nanoscale Res. Lett.* **13** 120

[80] Liu M C *et al* 2014 Synthesis and characterization of $M_3V_2O_8$ (M = Ni or Co) based nanostructures: a new family of high performance pseudocapacitive materials *J. Mater. Chem.* A **2** 4919–26

[81] Zhang J *et al* 2017 Facile synthesis of 3D porous $Co_3V_2O_8$ nanoroses and 2D $NiCo_2V_2O_8$ nanoplates for high performance supercapacitors and their electrocatalytic oxygen evolution reaction properties *Dalton Trans.* **46** 3295–302

[82] Gao X, Wang W, Bi J, Chen Y, Hao X, Sun X and Zhang J 2019 Morphology-controllable preparation of $NiFe_2O_4$ as high performance electrode material for supercapacitor *Electrochim. Acta* **296** 181–9

[83] Ren L, Chen J, Wang X, Zhi M, Wu J and Zhang X 2015 Facile synthesis of flower-like $CoMn_2O_4$ microspheres for electrochemical supercapacitors *RSC Adv.* **5** 30963–9

[84] Ma L, Shen X, Zhou H, Ji Z, Chen K and Zhu G 2015 High performance supercapacitor electrode materials based on porous $NiCo_2O_4$ hexagonal nanoplates/reduced graphene oxide composites *Chem. Eng. J.* **262** 980–8

[85] Xu Y, Wei J, Tan L, Yu J and Chen Y 2015 A facile approach to $NiCoO_2$ intimately standing on nitrogen doped graphene sheets by one-step hydrothermal synthesis for supercapacitors *J. Mater. Chem.* A **3** 7121–31

[86] Al-Rubaye S, Rajagopalan R, Dou S X and Cheng Z X 2017 Facile synthesis of reduced graphene oxide wrapped porous $NiCo_2O_4$ composite with superior performance as an electrode material for supercapacitors *J. Mater. Chem.* A **5** 18989–97

[87] Li X and Zhi L 2018 Graphene hybridization for energy storage applications *Chem. Soc. Rev.* **47** 42–6

[88] Bulbula S T, Lu Y, Dong Y and Yang X Y 2018 Hierarchically porous graphene for batteries and supercapacitors *New J. Chem.* **42** 5634–55

[89] Jiu H, Jiang L, Gao Y, Zhang Q and Zhang L 2019 Synthesis of three-dimensional graphene aerogel-supported $NiCo_2O_4$ nanowires for supercapacitor application *Ionics* **25** 4325–31

[90] Lv H P, YuanY X Q J, Liu H M, Wang Y G and Xia Y Y 2018 Carbon quantum dots anchoring MnO_2/graphene aerogel exhibits excellent performance as electrode materials for supercapacitor *J. Power Sources* **398** 167–74

[91] Tian Z, Wang X M, Li B X, Li H J and Wu Y C 2019 High rate capability electrode constructed by anchoring $CuCo_2S_4$ on graphene aerogel skeleton toward quasi-solid-state supercapacitor *Electrochim. Acta* **298** 321–9

[92] Wang H Y, Li R Y, Li M and Li Z J 2018 Flower-like Fe_2O_3@multiple graphene aerogel for high-performance supercapacitors *J. Alloys Compd.* **742** 759–68

[93] Zhou Y, Huang Z, Liao H, Li J, Wang H and Wang Y 2020 3D porous graphene/$NiCo_2O_4$ hybrid film as an advanced electrode for supercapacitors *Appl. Surf. Sci.* **534** 147598

[94] Xu X, Pei L, Yang Y, Shen J and Ye M 2016 Facile synthesis of $NiWO_4$/reduced graphene oxide nanocomposite with excellent capacitive performance for supercapacitors *J. Alloys Compd.* **654** 23–31

[95] Naderi H R, Sobhani-Nasab A, Rahimi-Nasrabadi M and Ganjali M R 2017 Decoration of nitrogen-doped reduced graphene oxide with cobalt tungstate nanoparticles for use in high-performance supercapacitors *Appl. Surf. Sci.* **423** 1025–34

[96] Ezeigwe E R, Khiew P S, Siong C W, Kong I and Tan M T T 2017 Synthesis of $NiMoO_4$ nanorods on graphene and superior electrochemical performance of the resulting ternary based composites *Ceram. Int.* **43** 13772–80

[97] Jiang Y, Zheng X, Yan X, Li Y, Zhao X and Zhang Y 2017 3D architecture of a graphene/ $CoMoO_4$ composite for asymmetric supercapacitors usable at various temperatures *J. Colloid Interf. Sci.* **493** 42–50

[98] Low W H, Khiew P S, Lim S S, Siong C W and Ezeigwe E R 2018 Facile solvothermal designing of graphene/$Ni_3V_2O_8$ nanocomposite as electrode for high performance symmetric supercapacitor *J. Alloys Compd.* **768** 995–1005

[99] Low W H, Siong C W, Chia C H, Lim S S and Khiew P S 2019 A facile synthesis of graphene/$Co_3V_2O_8$ nanocomposites and their enhanced charge storage performance in electrochemical capacitors *J. Sci.: Adv. Mater. Dev.* **4** 515–23

[100] Zhang X *et al* 2019 $NiFe_2O_4$ nanocubes anchored on reduced graphene oxide cryogel to achieve a 1.8 V flexible solid-state symmetric supercapacitor *Chem. Eng. J.* **360** 171–9

[101] Liu P, Chen H, Chang X, Xue Y, Zhou J, Zhao Z, Lin H and Han S 2017 Novel method of preparing $CoFe_2O_4$/graphene by using steel rolling sludge for supercapacitor *Electrochim. Acta* **231** 565–74

[102] Li L, Hu H and Ding S 2018 Facile synthesis of ultrathin and perpendicular $NiMn_2O_4$ nanosheets on reduced graphene oxide as advanced electrodes for supercapacitors *Inorg. Chem. Front.* **5** 1714–20

[103] Zhang C *et al* 2019 Heterostructural three-dimensional reduced graphene oxide/$CoMn_2O_4$ nanosheets toward a wide-potential window for high-performance supercapacitors *ACS Appl. Energy Mater.* **2** 5219–30

Chapter 7

Graphene based nanocomposites for energy conversion: the oxygen reduction, oxygen evolution, and hydrogen evolution reactions

Periyasamy Gnanaprakasam, Periyasamy Sundaresan, Muthukumar Abinaya, R V Mangalaraja, Ruey Shin Juang and Baskaralingam Vaseeharan

Graphene is an allotrope of carbon with a two-dimensional structure that has drawn significant attention worldwide due to its unique physicochemical and structural properties. Graphene based research has been done in almost all fields of basic science and engineering. Heteroatom doping on graphene creates a bandgap which is very useful for different applications. In addition, the incorporation of heteroatoms such as nitrogen, boron, fluorine, and phosphorous provides valuable properties in graphene nanosheets. Since graphene and doped graphene have a variety of important scientific applications, it is necessary to examine their features through the recent literature. This chapter briefly discusses the synthesis methods of graphene and heteroatom doped graphene composites, their electrocatalytic behaviors in various energy conversion systems, and future perspectives. In the application section, we discuss the recent reports of graphene based composites as efficient electrocatalysts for oxygen reduction reactions (ORRs), oxygen evolution reactions (OERs), and hydrogen evolution reactions (HERs). We hope this chapter will be helpful for those who are researching the synthesis methods of graphene and its heteroatom doped composites and their various energy conversion applications.

7.1 Introduction

Graphene, a two-dimensional sp^2-bonded carbon sheet packed in a hexagonal lattice one atom thick, is one of the thinnest materials in the world. Its extraordinary electronic, physical, electrochemical, and mechanical properties have inspired

doi:10.1088/978-0-7503-5143-0ch7

researchers in the fields of chemistry, physics, materials science, and biology, etc [1–3]. Graphene has excellent electron mobility at room temperature (250 000 cm^2 V^{-1} s^{-1}) [4] and the highest electrical conductivity (10^6 S cm^{-1}) [5]. It has exceptional mechanical properties with a Young's modulus of 1 TPa [6] and superior thermal conductivity (5000 W m^{-1} K^{-1}) [7], and has the highest theoretical surface area of 2630 m^2 g [8]. Generally, graphene sheets absorb only 2.3% of light in broad wavelengths, so the single layer of graphene is translucent and suitable for optoelectronic applications [9]. Due to the aforementioned unique properties, graphene has been used as a potential material in various fields such as optoelectronics and electronics, and in biomedical and electrochemical applications [10, 11].

Despite its unique physicochemical properties, pristine graphene sheets need to be modified for desired applications [12]. For example, graphene nanosheets absorb very little visible light and do not have a bandgap, which makes pristine graphene an unsuitable material for harvesting solar energy [13]. The capacitance of graphene is very low, and is restrained by the electronic double-layer [14]. Moreover, the aggregation of graphene layers leads to a reduction of the surface area and vanishing of the special properties of graphene. Consequently, the modification of pristine graphene with other atoms, functional groups, or metal nanoparticles will decrease the aggregation and retain its unique properties [15]. In this context, numerous graphene based composites have been prepared and show improved properties compared to pure graphene. Pristine graphene and its composites can be synthesized using various methods, such as exfoliation of graphite using ultrasonication [16], chemical vapor deposition (CVD) [17], exfoliation and reduction from graphite oxide [18], micromechanical exfoliation of graphite with sticky tape [19], epitaxial growth on single-crystal SiC [20], organic coupling reactions, and unzipping of carbon nanotubes [21]. Among these methods, oxidation exfoliation followed by solution reduction of graphite is known as the most suitable approach for the scalable production of graphene (reduced graphene oxide) [18]. Graphene nanosheets can be modified easily with functional groups including organic molecules, polymers, heteroatoms (e.g. nitrogen, fluorine, and boron) or inorganic molecules [22, 23]. Most recent works have focused on graphene and graphene based composites engineered for energy conversion and storage, primarily involving the oxygen reduction reactions (ORR) of fuel cells, oxygen evolution reactions (OERs), hydrogen evolution reactions (HERs), field-effect transistors (FETs), lithium ion batteries (LIBs), and supercapacitors. The early results reveal the enormous potential of graphene based composites in the field of renewable energy [24–27].

7.2 Synthesis of graphene

7.2.1 Top-down approaches

7.2.1.1 Exfoliation methods

Figure 7.1 shows a flow chart of various synthesis methods for graphene nanosheets. The exfoliation synthesis approach is an easy way to produce high quality graphene nanosheets with various solvents. This method avoids extreme reaction conditions (e.g. toxic chemicals, high temperature) and is also very easy and suitable for

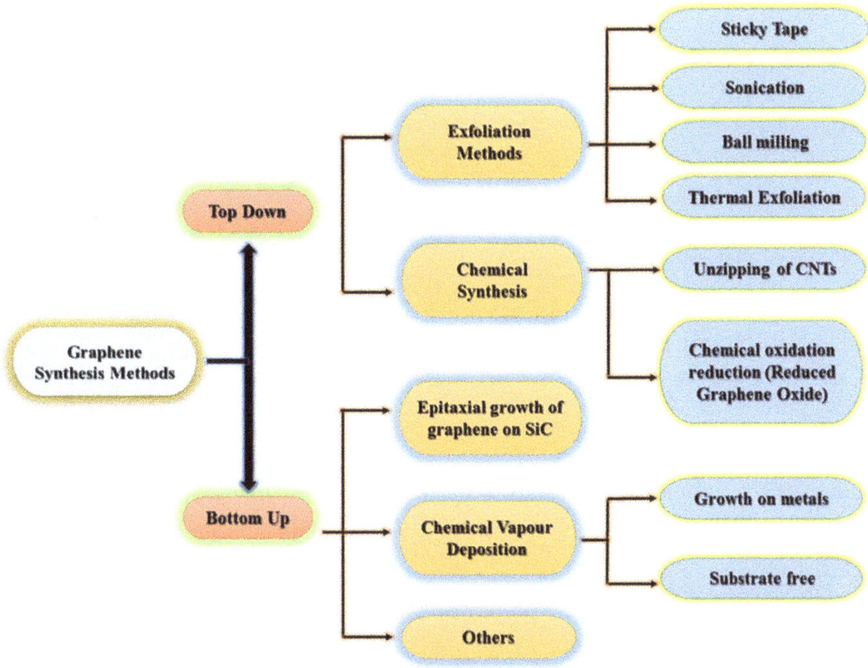

Figure 7.1. Flow chart of various synthesis methods of graphene.

large-scale synthesis of pure graphene. Generally, the presence of lattice defects and the disturbance of sp^2 hybridization of the graphene nanosheets will greatly influence the physicochemical properties of the graphene. Hence, the process of exfoliating graphite into graphene in various solvents could effectively minimize lattice defects and results in graphene of high quality [28, 29].

The micromechanical exfoliation or 'Scotch tape' method involves graphite exfoliation using sticky tape to peel apart the layers, and it was the first experimental method that was used to isolate graphene. Replicated cleavages lead to the formation of few-, bi-, and mono-layer graphene. Because of the very limited graphite processing, the as-prepared sheets are of high quality, but the technique is slow, labor intensive, and non-scalable [19]. Highly oriented graphene could be prepared by solution phase exfoliation of expanded graphite (EG) in N-methyl-2-pyrrolidone (NMP) using ultrasonication [30]. Graphene nanosheets in water can be prepared using 7,7,8,8-tetracyanoquinodimethane (TCNQ) as a stabilizing agent and EG as a starting material (figure 7.2(A) and (B)). The dissolved TCNQ is diffused effectively into the interlayers of EG which reduces the π–π interaction and increases the interlayer distance [31].

The functionalized graphene can also be synthesized in polar solvents, such as DMSO and N,N-dimethylformamide (DMF). For example, a study reported that 4-bromophenyl-modified EG was synthesized from the precursors of 4-bromoaniline and EG in the presence of DMF using ultrasonication, resulting in the successful exfoliation of functionalized EG and producing a thin layer of functionalized

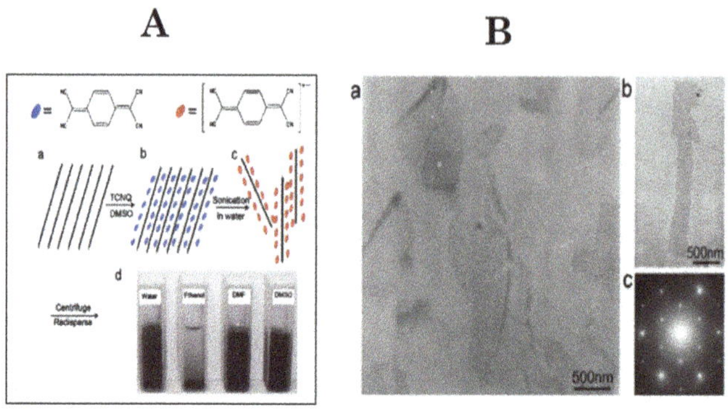

Figure 7.2. (A) Schematic illustration of aqueous graphene dispersions stabilized with TCNQ anions. (a) Pristine expanded graphite by heating expandable graphite at 1000 C. (b) TCNQ insertion into the intercalated graphite with the aid of DMSO. (c) After durative sonication, TCNQ anion stabilized graphene in water. (d) A photograph of TCNQ anion adsorbed graphene dispersed in different solvents: water, ethanol, DMF, and DMSO. (B) TEM and ED images of graphene sheets: (a) several isolated graphene sheets; (b) a typical graphene ribbon; and (c) the corresponding SAED. (Reproduced with permission from [31]. Copyright 1996 Royal Society of Chemistry)

graphene with greater dispersibility in DMF than the unmodified graphene [32]. In another study, readily soluble EG was prepared in a one-step exfoliation process using fluorinated graphite intercalation compound using the single insertion and thermal expansion method, and it was noted that the interlayer distance of the functionalized EG was 17% higher than that of the EG produced using standard methods [33].

In addition to the ultrasonication based methods, the solvothermal approach is also known as a familiar liquid-phase exfoliation method to prepare quality graphene from graphite. A report suggested that oleyl amine was used as the solvent to exfoliate single-layer graphene membranes from the starting material of graphite, the resultant graphene layers were identified as more stable from re-aggregation because of the stabilizing effect of oleyl amine [34]. In another study, acetonitrile was used as the solvent to produce mono-layer and bilayer graphene from EG using a solvothermal exfoliation process. The solvothermal method helped to inject acetonitrile molecules into the graphite interlayer, thereby after sonication mono- and bilayer graphene was cleaved from the EG and firmly dispersed in acetonitrile solvent [35].

7.2.1.2 Synthesis of graphene oxide and its reduction

The synthesis of graphene (reduced graphene oxide) via exfoliation and reduction of graphite oxide (GO) is identified as a suitable method for large-scale synthesis. Reduced graphene oxide could be synthesized easily via the oxidation of graphite using strong oxidants and concentrated acids, in the Brodie [36], Straudenmaier [37], or Hummers methods [38]. The Hummers method is a widely used approach. After the Hummers method, the resultant GO will have more oxygen functional groups, such as hydroxyl, epoxy on basal plane and/or carbonyl and carboxyl groups on the edges of the graphene sheets, making the GO more hydrophilic, and they offer major benefits not only for better dispersion in certain solvents but also for more functionalization, which will make GO much easy to use in many applications. In addition, the existence of oxygen containing groups between layers also results in graphite oxide with higher interlayer spacing (6–12 Å) than graphite (3.4 Å) [39].

Further, the conversion of graphite oxide to graphene could be achieved with the help of reducing agents, microwave irradiation, or ultrasonication. Even after the reduction process, a small amount of oxygen containing groups will remain in the nanosheets. The resulting material is commonly called reduced graphene oxide (rGO) rather than graphene, because there has not been complete reduction [40]. Graphite oxide can be exfoliated more easily than graphite using ultrasonication in water and polar solvents. The exfoliated graphite oxide could be reduced easily into reduced graphene oxide using chemical, thermal, or ultrasonication treatments. Although graphene oxide is electrically insulating because of its functional groups, the efficient reduction of oxygen functional groups using effective reducing agents will make graphene nanosheets electrical conductors. However, the reduction of oxygen functional groups leads to the formation of a graphitic lattice, and because of the harsh oxidation process during the synthesis of graphite oxide may result in more defects on the graphene nanosheets and their properties will not match those of graphene [41].

7.2.1.3 Unzipping of carbon nanotubes

The synthesis of graphene and few-layer graphene nanoribbons (GNRs) by unzipping of single carbon nanotubes (CNTs) or multi-walled carbon nanotubes (MWCNTs) using suitable chemicals (e.g. oxidizing agents) [42] or plasma etching [43] and laser ablation processes [44] has received interest in recent years. The high-resolution transmission electron microscopy (HRTEM) and atomic force microscopy (AFM) images in figure 7.3 clearly show the conversion of MWCNTs synthesized by various methods such as CVD and arc discharge, and the nitrogen and boron functionalized MWCNTs into the respective graphene nanoribbons using a laser ablation method. The diameter of the unzipped GNRs depends on the diameter of the tubes. The unzipped GNRs are one-dimensional materials with properties that depend on their dimensions and the configuration of carbon atoms (zigzag/armchair) on the edges of the nanoribbons [45]. The unzipping of CNTs occurs through the breaking of C–C bonds, which usually starts from a defect site and leads to irregular unzipping [46]. Recently the synthesis of regularly oriented GNRs was performed by unzipping of flattened nanotubes, where attacking takes place preferentially along bent edges. GNR production from CNTs is attracting researchers in the field of energy conversion because graphene is often known as the better alternative to CNTs [47]. Interestingly, two recent independent researchers demonstrated the unzipping of MWCNTs to GNRs via a very elegant approach, which is focused on the longitudinal unzipping of MWCNTs. These procedures are very attractive, easy, and cheap, and lead to defined shapes of GNRs [42, 48].

7.2.2 Bottom-up approaches

7.2.2.1 Chemical vapor deposition (thermal CVD, plasma enhanced CVD)

Chemical vapor deposition is one of the appealing methods for preparing high quality graphene films on the surface of metal substrates by thermal decomposition

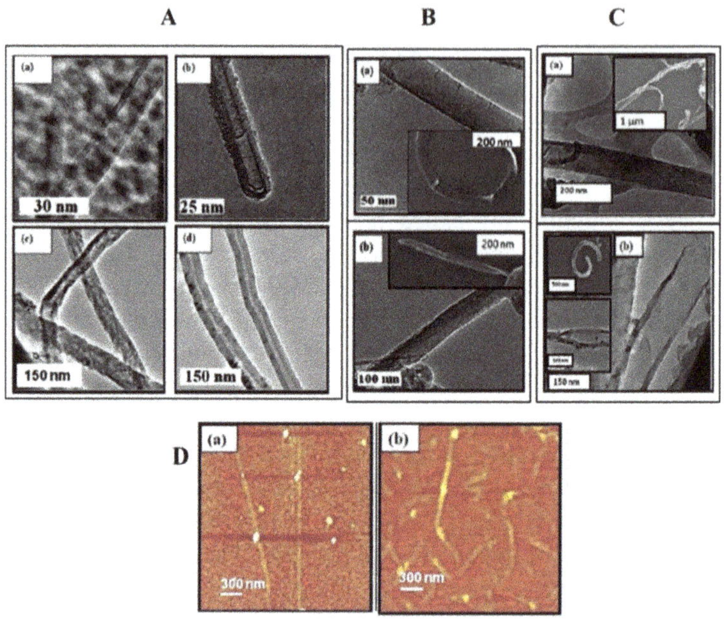

Figure 7.3. (A) TEM images of MWNCTs: (a) CVD-MWCNTs; (b) arc-MWCNTs; (c) N-MWCNTs; and (d) B-MWCNTs. (B) TEM and FESEM images of GNRs obtained by laser irradiation of MWCNTs: (a) CVD-MWCNTs irradiated at 250 mJ and (b) arc-MWCNTs irradiated at 350 mJ; corresponding FESEM images are shown as insets. (C) TEM and FESEM images of N- and B-doped GNRs obtained by laser irradiation of doped MWCNTs: (a) N-MWCNTs irradiated at 250 mJ, the inset shows the corresponding FESEM image; (b) B-MWCNT sample irradiated at 250 mJ, the top inset shows the corresponding FESEM image and the bottom inset shows a TEM image of a partially opened B-MWCNT irradiated at 200 mJ. (D) AFM images of GNRs obtained by laser irradiation of (a) CVD-MWCNTs irradiated at 250 mJ and (b) N-MWCNTs irradiated at 250 mJ laser energy. (Reproduced with permission from [44]. Copyright 2009 Royal Society of Chemistry.)

of carbon containing gases at higher temperatures. The growth of CVD graphene can be classified into either segregation or surface catalyzed methods, which depend on the metal substrate. In the segregation process, the carbon is dissolved from the bulk metal to the surface of a metal and forms graphene composites, which typically happens by cooling, because of the reduced carbon solubility in metals at lower temperatures. The rate of cooling and the amount of dissolved carbon determines the number of graphene layers produced by the segregation process. In the surface catalyzed method, the metal surface is the site of the reaction where both carbon species decomposition and graphene formation occur at the same place, which leads to the formation of a graphene mono-layer. The reaction will be completed once the surface is completely covered [49]. Graphene has been grown on a wide range of metals such as Cu [50], Co [51], Fe [52], Rh [53], Ru [54], Ni [17], Pt [55], Pd [56], Au [57], Ir [58], and many alloys (Ni–Mo [59], Au–Ni [60], Co–Ni [61]). Hitherto, most attention has been paid to copper and nickel, in particular the polycrystalline copper foils from which the graphene films of diagonal lengths of 30 inches have being grown and transferred. It is thought that graphene growth on copper occurs via the surface catalyzed mechanism, resulting in the formation of mono-layer graphene for various reaction conditions, whereas the graphene growth on nickel occurs through the segregation process, that leads to the formation of multi-layer graphene and is more difficult to control. The CVD growth of graphene on metals has some disadvantages, i.e. extreme growth conditions, particularly the requirement for ultra-high vacuum (UHV) conditions for some metals (Cu) [62]. Maintaining these conditions is quite difficult and patches of multi-layer graphene films have been obtained in the case of failure of maintaining UHV conditions. There have been also some attempts to reduce the temperature needed for the reaction, in particular such research is being carried out using carbon sources with lower temperatures of decomposition [63].

7.2.2.2 Epitaxial growth on silicon carbide

The synthesis of graphene from silicon carbide (SiC) is carried out through the selective sublimation of silicon from the surface of SiC followed by the simultaneous graphitization of the remaining carbon atoms. This reaction happens at temperatures above 1000 °C and is usually carried out under UHV conditions [20]. The presence of a meager amount of disilane or argon atmospheres reduces the rate of silicon sublimation and allows the reaction to occur at higher temperatures, resulting in the formation of high quality graphene [64]. In addition to thermal treatment, pulsed electron irradiation is also used to induce preferential Si sublimation. Usually, the hexagonal phase of SiC is used for the synthesis of graphene, but recently the cubic phase of SiC has also been used as the starting material. The demand for high temperature is one of the disadvantages of graphene growth on SiC, which could be solved with the help of a Ni catalyst. The usage of the nickel catalyst allows the growth reaction to occur at a lower temperature (700 °C–800 °C), where the surface of the SiC substrate is covered with nickel before the annealing process, then the graphene films will form over the nickel layer (figure 7.4(A) and (B)) [65]. Epitaxial growth of graphene on SiC is usually chosen for fabricating electronic devices or components. In that case it is not necessary to remove the graphene layer from the substrate. Although the interaction between the grown graphene and SiC surface is strong, the transfer of graphene from SiC to another surface has been demonstrated. SiC is an easily available material but because of its high cost, the transfer of graphene from SiC will reduce the cost of production and will be helpful for the use of graphene in multiple commercial applications [66].

7.3 Heteroatom doped graphene

Heteroatom doping on graphene with nitrogen, fluorine, or boron could change the electronic structure and create a bandgap which would make it a p-type or n-type material. Such doped graphene materials have unique properties and have attracted much attention, in particular in the fields of fuel cells, water-splitting, supercapacitors, and batteries [22]. For example, N-doped graphene is a better electron donor than pristine graphene, which can be used as a electrocatalyst for ORR, OER, and so on. The B-doped graphene could be produced through a chemical approach or CVD

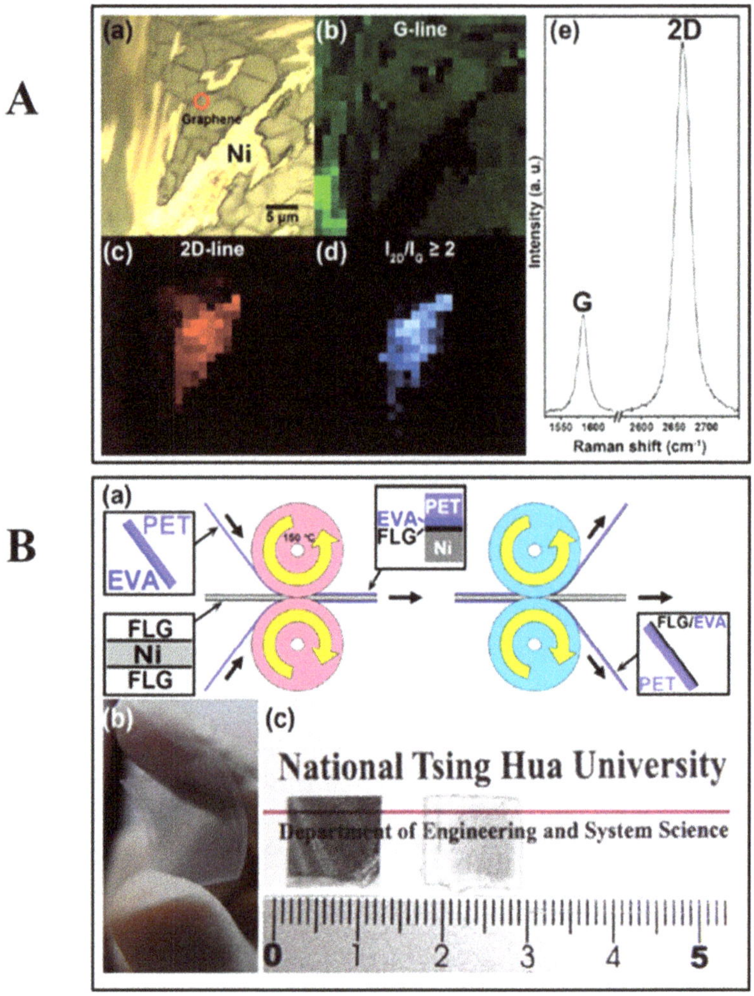

Figure 7.4. (A) (a) to (d) Optical microscope and Raman images of as-grown graphene on Ni foil processed with a slow cooling rate of 0.3 °C s^{-1}. (e) Raman spectra taken at the position marked with a circle, indicating the presence of single-layer graphene to bilayer graphene. (B) A diagram to show the roll-to-roll process for the transfer of few-layer graphene (FLG) from Ni foil to ethylenevinyl acetate copolymer –polyethylene terephthalate (EVA–PET) substrates. (b) A 2 × 3 cm^2 transparent flexible FLG–eEVA–PET sample. (c) Two 1 × 1 cm^2 transferred samples with different transmittances. (Reproduced with permission from [65]. Copyright 2009 Elsevier.)

method that will have high conductivity, p-type characteristics, and transparency, that leads to the use of the resultant material as an electrode in photovoltaic devices [67]. The synthesis, properties, and catalytic activities of nitrogen, boron, and fluorine-doped graphene will be discussed in the following sections.

7.3.1 Synthesis of N-doped graphene

Numerous methods have been used to synthesize nitrogen-doped (N-doped) graphene, which include thermal annealing [68], plasma treatment [69], CVD [70], arc discharge [71], and liquid-phase reaction [72]. Rao *et al* prepared N-doped graphene by arc discharge between graphite electrodes in the presence of ammonia, helium, and hydrogen or pyridine vapors. The obtained graphene contains few layers with a nitrogen content of ~1% as calculated using electron-energy loss spectroscopy (EELS). Wei *et al* synthesized N-doped graphene with ~9% of nitrogen content via a CVD process on a Si substrate (at 800 °C) using Cu metal catalyst in the presence of methane, argon, ammonia, and hydrogen. The N-doped graphene that was prepared under nitrogen plasma conditions had 2.4% of nitrogen content [73]. Dai *et al* prepared N-doped graphene with 5% nitrogen content by thermal annealing of graphene oxide (GO) with ammonia, in the temperature range of 300 °C to 1100 °C. The degree of doping at the edges of the graphene depends on the temperature and oxygen functionality. N-doped graphene was also prepared at different temperatures by heating GO with melamine or urea as the source of nitrogen. In this process, the GO was ground finely at various mass proportions with melamine or urea, and heated at high temperatures. The reaction temperature and the concentration of nitrogen source determined the extent of doping. GO contains oxygen functionalities (epoxide, hydroxyl, carbonyl, carboxyl, and phenol) which were amidated by the ammonia that formed through the decomposition of melamine or urea at 300 °C and then was converted into pyridinic and pyrrolic nitrogen at the edges of the graphene. At high temperatures nitrogen will be incorporated into the graphene lattice and become graphic nitrogen [74]. N-doped graphene can also be synthesized in lower temperature conditions using solvothermal and hydrothermal methods. By reacting tetrachloromethane with lithium nitride, Bao *et al* synthesized N-doped graphene with 13% nitrogen content (figure 7.5(A) and (B)). Rao *et al* recently achieved the highest level of 15.6% of nitrogen doping by treating GO with urea for a few seconds under microwave conditions (900 W). Here, GO was reduced

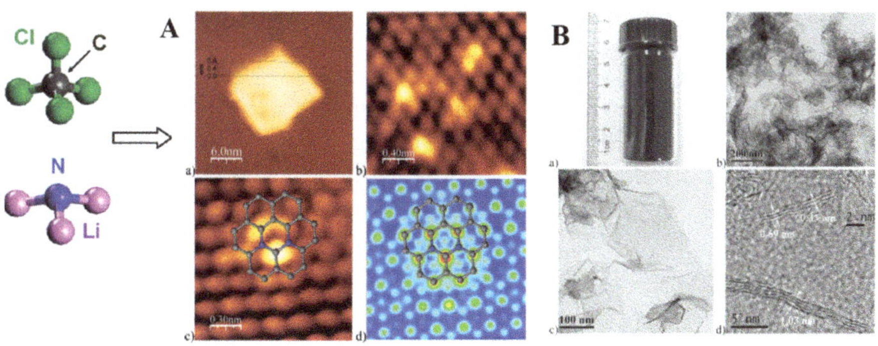

Figure 7.5. (A) STM images of NG-2. (a) Isolated bilayer N-doped graphene, the black curve on top showing the height measurement across this bilayer. (b) and (c) High resolution images with defects arranged in different configurations, measured at a bias = 0.5 V, current I = 53.4 pA, and bias = 0.9 V, I = 104 pA, respectively. (d) Simulated STM image for (c). The inserted schematic structures represent N-doping graphene, with the hexagons in gray balls highlighting the atomic network of graphene and deep blue balls labeling N atoms. (B) Vial containing NG-1 (1.2 g) obtained with a 40 ml autoclave per batch in the laboratory. (b) and (c) TEM images of NG-2 at different magnifications. (d) HRTEM images of NG-1 showing the cross sections of single (inserted image) to four layer graphene. (Reproduced with permission from [72]. Copyright 2013 American Chemical Society.)

simultaneously and also the nitrogen was incorporated into the lattice of the graphene [72]. Long *et al* demonstrated N-doped graphene with 5% nitrogen by reducing GO with hydrazine hydrate and ammonia using the wet chemical reaction between dicyandiamide and GO at a heating of 180 °C for 12 h. The milling of pure graphite in the presence of ammonia (4.49%) or nitrogen (14.84%) was reported. Ball-milling can also be used effectively to manufacture N-doped graphene in large quantities [75].

7.3.2 Synthesis of B-doped graphene

Han *et al* developed a solvothermal process for the synthesis of boron-doped (B-doped) graphene platelets using the precursors GO and borane–tetrahydrofuran complex [76]. The synthesis of B-doped graphene oxide from a suspension of GO and H_3BO_3 in DMF by annealing was demonstrated by Khai *et al* [77]. Sheng *et al* prepared B-doped graphene via a catalyst free thermal annealing approach in the presence of boron oxide. The successful doping of boron atoms on the graphene surface was confirmed by x-ray photoelectron spectroscopy and the atomic percentage was estimated as 3.2% [78]. Kim *et al* reported a B-doped single-layer graphene prepared by mechanical exfoliation using B-doped graphite as a precursor, which was synthesized by thermally treating graphite with a boron compound at 2450 °C. The amount of boron was measured as 0.22% on the graphene surface [79]. An arc discharge method between the boron-filled graphite electrodes or graphite electrodes in the presence of $H_2 + B_2H_6$ produced two to four layers of B-doped graphene with a boron content of 1%–3% [80]. B-doped graphene can be synthesized directly on Cu substrates using ethanol and boron powder as precursors, at 950 °C under the CVD method. The thus synthesized sample was mainly a few layers (~3–5 layers) thick, containing a boron content of 0.5%. A CVD experiment was performed for the synthesis of B-doped single-layer graphene on polycrystalline copper foils using diborane and methane, as the precursors for boron and carbon, respectively. The content of boron was 1.5% [81].

Phenylboronic acid [82] and triethylboron [83] might also be used as a source of boron. Using a microwave plasma, the decomposition of trimethylboron resulted in the formation of B-doped graphene with tunable bandgap and transport properties through controllable doping. By regulating the ion reaction time, the boron content was changed to between 0% and 13% [84]. B-doped graphene can also be prepared by a reductive coupling reaction of rapid Wurtz-type at lower temperatures without using transition metal catalysts. This process involves the reaction between CCl_4 and potassium in the presence of a sufficient amount of borontribromide at 150 °C–210 °C for 10 min, that had considerable advantages compared to the solvothermic method [85].

7.3.3 Synthesis of F-doped graphene

The synthesis methods for fluorinated graphene or F-doped graphene are generally classified into two categories: exfoliation and fluorination methods. Sonochemical exfoliation, Hummer's exfoliation, and thermal exfoliation are examples of exfoliation methods. Fluorination methods include hydrothermal fluorination, plasma fluorination, photochemical/electrochemical synthesis, and direct gas fluorination [86, 87].

Nair *et al* prepared fluorographene using graphene film and XeF_2 gas at 70 °C by direct gas fluorination. The mild and controllable fluorination using XeF_2 gas is one of the commonly used techniques for preparing fluorographene with various fluorinated structures [88]. Baraket *et al* used an electron beam generated plasma for synthesizing fluorinated graphene with Ar/SF_6 and noticed that the C–F bonds in the fluorinated graphene could be reduced to original C–C bonds by removing F atoms by annealing (500 °C) [89]. Wang *et al* used a hydrothermal process as a practical method for fluorinating the distributed GO using hydrofluoric acid (HF). Note that during the hydrothermal reaction, some oxygen containing groups were replaced by F atoms. Furthermore, the ratio of F/C can be altered by varying the temperature, time, and concentration of HF (figure 7.6(A) and (B)) [90]. Using UV irradiation on GO dispersion in HF at room temperature, Gong *et al* prepared fluorinated rGO [91]. Bruna *et al* reported the synthesis of oxy-fluorinated graphene via an electrochemical process [92]. Zborˇil *et al* demonstrated fluorinated graphene with a F/C ratio of 1.00 using a single-step liquid-phase exfoliation. In this process, commercial fluorinated graphite dispersed in sulfolane was exfoliated using an ultrasonic bath of 135 W at 50 ° C for 1 h to obtain fluorinated graphene [93]. Fluorinated graphene can also be prepared by a thermal exfoliation method from fluorinated graphite. Dubois *et al* synthesized fluorographene by means of thermal exfoliation of fluorinated HOPG in

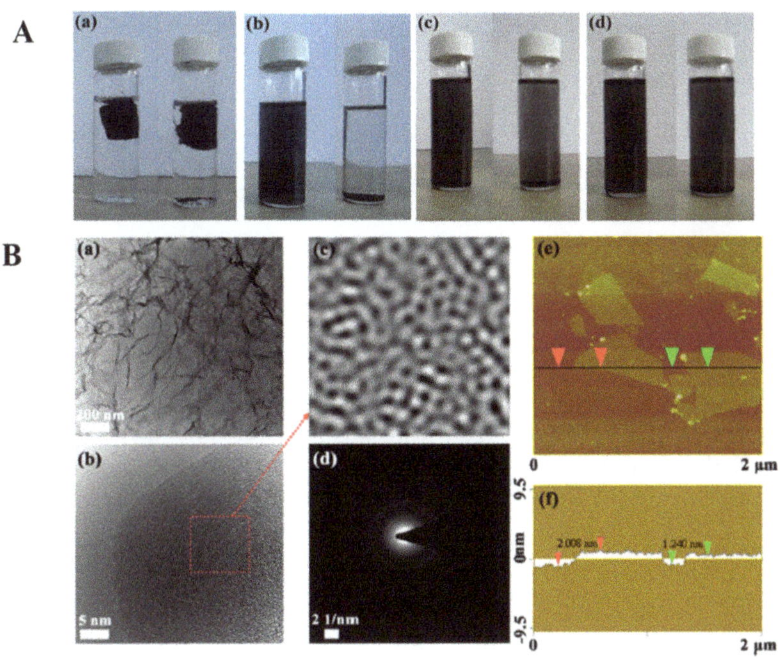

Figure 7.6. (A) Optical photographs of fluorinated graphene sheets (FGS) dispersed in (a) water, (b) ethanol, (c) DMF, and (d) NMP. (a) and (b) are FGS just dispersed by ultrasonics, and (b)–(d) are the FGS dispersions after 24 h. (B) (a) TEM and (b) HRTEM micrographs of the synthesized FGS. (c) Magnified image of the red section in panel (b). (d) Selected area electron diffraction pattern of the FGS sample. (e)–(f) AFM image and thickness of the FGS. (Reproduced with permission from [90]. Copyright 2012 Elsevier.)

the presence of F_2. Fluorographene was acquired by rapid removal of inter-laminar fluorinated HOPG species with an acute increase in temperature followed by color changes from greyish to black [94]. Using a modified Hummer process, Ajayan *et al* demonstrated a method for synthesizing fluorinated GO. The results of magic-angle spinning (MAS) 13C NMR showed two forms of fluorinated GO: partially fluorinated GO (FGO) and strongly fluorinated GO (HFGO), respectively. The hydrophilicity of FGO was identical to GO, whereas HFGO was fairly hydrophobic [95].

7.4 Energy conversion applications of graphene composites

7.4.1 Oxygen reduction reaction

A fuel cell is a device that directly transforms chemical energy into electrical energy, with high performance and low emissions. The ORR proceeds under both acidic and alkaline conditions, via either a direct four-electron ($4e^-$) pathway or a two-step, two-electron ($2e^-$) pathway [96].

Direct $4e^-$ pathway:

$$O_2 + 2H_2O + 4e^- \rightarrow 4OH^- \quad \text{(alkaline)} \tag{7.1}$$

or

$$O_2 + 4H^+ + 4e^- \rightarrow 2H_2O \quad \text{(acidic)} \tag{7.2}$$

Two-step $2e^-$ pathway:

$$O_2 + H_2O + 2e^- \rightarrow HO_2^- + OH^- \quad \text{(alkaline)} \tag{7.3}$$

$$HO_2 + H_2O + 2e^- \rightarrow 3OH^- \quad \text{(alkaline)} \tag{7.4}$$

or

$$O_2 + 2H^+ + 2e^- \rightarrow H_2O_2 \quad \text{(acidic)} \tag{7.5}$$

$$H_2O_2 + 2H^+ + 2e^- \rightarrow H_2O \quad \text{(acidic).} \tag{7.6}$$

The ORR activity can be calculated by the onset potential, the kinetic limiting current density (j_{kin}), the electron transfer number (n), the Tafel slope, etc. Specifically, the n values of O_2 molecule that are involved in the ORR cycle can be determined using the following Koutecky–Levich (K–L) plots equations:

$$\frac{1}{j} = \frac{1}{j_{kin}} + \frac{1}{j_{lim}} = \frac{1}{j_{kin}} + \frac{1}{B.\ \omega^{1/2}}, \tag{7.7}$$

$$B = 0.\ 62nFD_O^{2/3}\upsilon^{-1/6}C_O, \tag{7.8}$$

$$j_{kin} = nFkC_O, \tag{7.9}$$

where j is the measured current density, j_{lim} is the diffusion limiting current density, and F is the Faraday constant ($F = 96\,485$ C mol^{-1}), B is the reciprocal of the slope,

D_O is the diffusion coefficient of O_2, C_O is the concentration of O_2, ν is the kinematic viscosity of the electrolyte, ω is the electrode rotating speed in rpm, and k is the electron transfer rate constant [96].

Platinum (Pt) has been considered for a long time as the perfect catalyst for both anodic oxidation reactions and cathode reduction reactions. However, for the cathodic ORR, the use of Pt catalyst is impeded by its disadvantages, such as low abundance, high cost, CO poisoning effect, and susceptibility to crossover impacts. Non-precious metals (e.g. metal chalcogenides) and metal-free electrocatalysts for ORR were investigated largely to minimize the use of Pt and the cost of a fuel cell [97]. Graphene and related materials have recently gained significant attention as metal-free catalysts for fuel cells, in addition to their use as catalyst supports [12, 97]. Heteroatom, such as N, B, F, S, P, etc, doped graphene have been considered to be more suitable materials for the ORR. Previous reports reveal that when some of the carbon atoms in the graphene network are replaced by heteroatoms, such as boron or nitrogen, the electrochemical properties of the doped graphene are considerably enhanced [12].

Gong *et al* demonstrated vertically aligned nitrogen-doped carbon nanotubes (VA-NCNTs) that could catalyze the four-electron ORR process effectively with much greater catalytic activity, smaller overpotential, lower crossover sensitivity, and greater long-term operating stability than the commercial catalyst platinum/carbon black. The performance of VA-NCNTs as catalysts for the ORR has inspired N-doped graphene research for the same application. Indeed, nitrogen-doped graphene films produced by CVD in the presence of ammonia provided excellent performance for ORR, similar to the VA-NCNTs in an alkaline medium with the same nitrogen content [98].

N-doped graphene synthesized by other methods, such as nitrogen plasma treatment, edge functionalization of graphene, thermal treatment of graphene with ammonia, and solvothermal treatment between graphene with tetrachloromethane and lithium nitride, also show better electrocatalytic activities for the ORR [72]. Recently, by annealing GO in the presence of nitrogen, boric acid, and ammonia, nitrogen and boron co-doped graphene (BCN-graphene) was successfully synthesized by Wang *et al.* The resultant material has significantly improved ORR electrocatalytic efficiency, much better than that of a commercial platinum/carbon catalyst, because of a poorly understood synergistic effect associated with nitrogen and boron codoping of graphene (figure 7.7) [99].

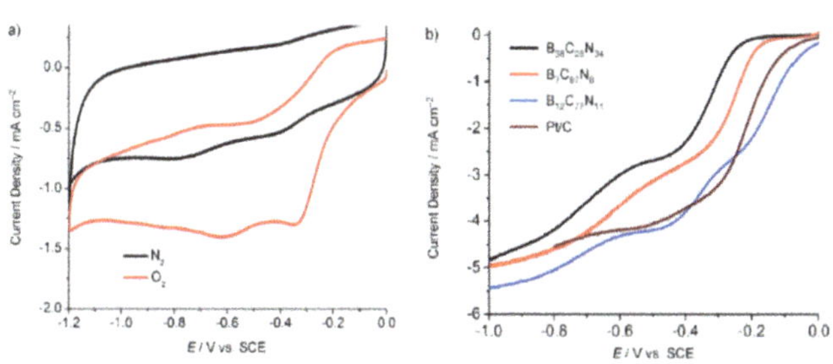

Figure 7.7. (a) Cyclic voltammetry (CV) curves of ORR on BCN ($B_{12}C_{77}N_{11}$) graphene in nitrogen- and oxygen-saturated 0.1 M KOH solutions at a scan rate of 50 mV s^{-1}. (b) Linear sweep voltammetry (LSV) curves of ORR on BCN-graphene with different compositions in oxygen-saturated 0.1 M KOH solution at 10 mV s^{-1} and compared with the commercial Pt/C electrocatalyst. (Reproduced with permission from [99]. Copyright 2012 Wiley.)

Interestingly, Kakaei *et al* synthesized porous fluorine-doped graphene nanosheets via thermal treatment of graphene oxide and NaF (sodium fluoride) in H_2SO_4 solution phase for used as efficient metal-free electrocatalysts for oxygen reduction in a gas diffusion electrode. The synthesized materials showed highest electrocatalytic activity towards the ORR along with long-term stability, that can be attributed to the high surface area and the presence of F on the surface of the rGO [100].

7.4.2 Oxygen evolution reaction

Among the electrochemical energy conversion devices such as fuel cells, metal–air batteries, and water-splitting, the latter is an energy conversion system that has the potential to produce energy in a completely green way. The oxygen evolution reaction (OER) is the core reaction for water-splitting [101]. The OER can be demonstrated under basic or acidic and neutral solutions as follows:

in alkaline solutions:

$$4OH^- \rightarrow 2H_2O + O_2 + 4e^-,$$

in acidic or neutral solutions:

$$2H_2O \rightarrow 4H^+ + O_2 + 4e^-.$$

The effectiveness of electrocatalysts is calculated by the strength of the binding of the intermediates (HO*, O*, and HOO*) on the surface of the electrode. Electrocatalytic efficiency can be assessed by comparing their catalytic activity, faradaic efficiencies, and stability. In particular, the activity is determined by the overpotential required to attain the current density of 10 mA cm^{-2} per geometrical area. Generally, the OER has very sluggish reaction kinetics because of the multiple coupling reactions between the protons and electrons to produce molecular oxygen, which requires the use of an efficient electrocatalyst to improve the rate of reaction. As of now, Ir and Ru based materials are known as efficient catalysts for the OER [101]. Numerous attempts have been made to replace Ir and Ru with low-cost transition metal composites (Fe, Co, Ni, etc) [102]. In addition, heteroatom-doped graphene has been identified as a better alternative to Ir and Ru for the OER, because of their lower production cost and the presence of highly active centers. Apart from these, heteroatom doped graphene and its composites with non-metal species such as N-doped CNTs and graphitic carbons are also identified as better catalysts for the OER [103]. For example, N-doped graphene prepared from a polymer rich in nitrogen displayed excellent OER performance in alkaline solution. The catalyst has an overpotential of 380 mV for obtaining a current density of 10 mA cm^{-2}. The enhanced OER activity is because of the presence of pyridinic and pyrrolic nitrogen present in the N-doped graphene [104]. Qiao *et al* reported a three-dimensional hybrid paper of N-doped graphene–$NiCo_2O_4$ (PNG–NiCo) as a catalyst for the OER with uniform pores. The 3D hybrid paper demonstrated impressive catalytic activity for the OER. The PNG–NiCo has an overpotential of 373 mV to attain a current density of 5 mA cm^{-2} (figure 7.8). The strong synergy

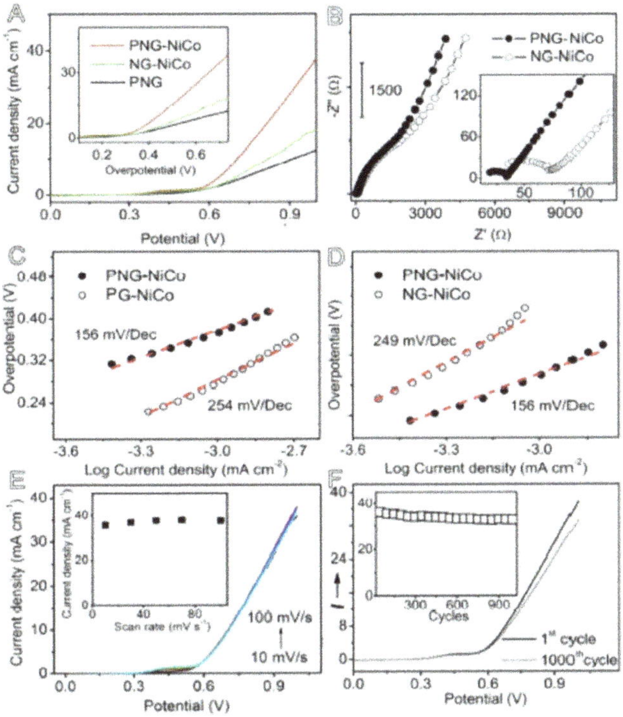

Figure 7.8. (a) LSV plots of PNG–NiCo, NG–NiCo, and PNG at 50 mV s^{-1} in 0.1 M KOH; the inset shows the corresponding data re-plotted as the current density versus overpotential. (b) EIS spectra of PNG–NiCo and NG–NiCo. Tafel plots for PNG–NiCo in comparison to those for (c) PG–NiCo and (d) GNiCo. (e) LSV plots of PNG–NiCo at different scan rates; the inset shows the corresponding data re-plotted as the current density (at 1.0 V versus Ag/AgCl) versus scan rates. (f) LSV plots for PNG–NiCo before and after CV testing of 1000 cycles; the inset shows the relationship between current density (express as mA cm^{-2}, at 1.0 V versus Ag/AgCl) and cycle numbers. (Reproduced with permission from [105]. Copyright 2013 American Chemical Society.)

between the $NiCo_2O_4$ and N-doped graphene resulted in the enhancement of catalytic activity for the OER [105]. N-doped CNTs have received exceptional interest due to their environmental friendliness, high durability, low-cost, and abundant catalytic sites. The superior coupling effects between graphene and the CNTs will avoid the leaching out of catalyst from the electrode surface and results in a higher catalytic durability for the OER. Further, 3D N-doped hybrid carbon film made by incorporating N-doped graphene and g-C_3N_4 ultrathin nanosheets on cellulose-based fiber papers was also explored. The N-doped graphene demonstrates high electrical conductivity and fast charge transport, and the graphitic carbon nanosheets possesses higher active sites for electrocatalysis. Due to their strong synergistic effects between cellulose fibers and the synthesized carbon material, the resultant framework shows an excellent catalytic activity, in particular the material shows competitive catalytic activity and excellent durability compared to the commercial catalyst for the OER (IrO_2) [106]. Hashimoto *et al* developed an

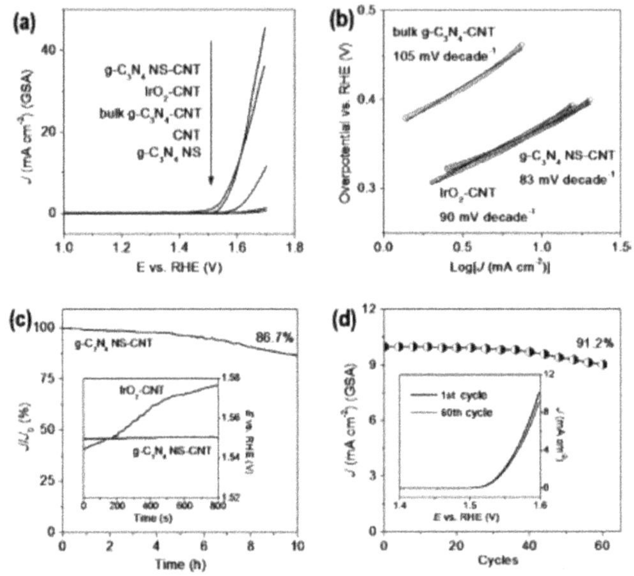

Figure 7.9. (a) LSV and (b) Tafel plots for g-C_3N_4 NS–CNT, IrO_2–CNT, bulk g-C_3N_4–CNT, purified oxidized CNTs, and g-C_3N_4 NSs on a rotating disc electrode (RDE) (1500 rpm) in an O_2-saturated 0.1 M KOH solution (scan rate: 5 mV s^{-1}). The calculation of the OER current density was based on the geometric surface area (GSA). (c) Chronoamperometric response at a constant potential of 1.54 V; the inset shows the chronopotentiometric response of g-C_3N_4 NS–CNTs compared to that of IrO_2–CNT at a constant current density of 3.0 mA cm^{-2}. (d) A curve of the current density of g-C_3N_4 NS–CNT at 1.60 V versus scan cycles; the inset shows the LSVs of g-C_3N_4 NS–CNTs before and after 60 scan cycles (scan rate: 5 mV s^{-1}). (Reproduced with permission from [107]. Copyright 2014 Wiley.)

N-doped graphite electrocatalyst which had an overpotential of 0.38 V to obtain the current density of 10 mA cm^{-2}. The pyridinic N and graphitic N atoms of the catalyst act as the active centers for OER and the same was confirmed by electrochemical and physical characterization [104].

Recently, Quio *et al* and others demonstrated that the composites of graphene and carbon are ideal candidates for non-metal electrocatalysts for the OER, such as g-C_3N_4, surface-oxidized carbon nanotubes, 3D N- and P-doped porous carbons, and heteroatom-doped graphene. These carbon composite materials can be acquired with the larger active centers desirable for OER kinetics, the high porosity that is required for the easy transport of reactants or products, and a high surface area ensuring the overpotential of 0.3 V for obtaining a current density of 10 mA cm^{-2} (figure 7.9) [107].

7.4.3 Hydrogen evolution reaction

The complete HER's reaction pathways and the equations used to calculate the reaction kinetics are given briefly as follows: in acidic media, the first step of the HER is the adsorption of hydrogen atoms on the surface of the electrode by moving a proton from the electrolyte and coupled with one electron from the electrode surface (the Volmer reaction). There are two possibilities to carry out the second step of the reaction: the first one is the Tafel reaction pathway where two adjacent H* will interact that will form a single H_2 molecule and the second one is the Heyrovsky pathway, where the adsorbed hydrogen (H*) is combined with one proton and electron to form a H_2 molecule. In an alkaline medium, the water dissociation step is an additional step, which will make an extra energy barrier and affects the entire reaction mechanism and rate of the reaction [103].

In acidic media:
Step 1:

$$H^+ + e^- + * \rightarrow H^* \quad \text{(Volmer)}.$$

Step 2:

$$H^+ + e^- + H^* \rightarrow H_2 \quad \text{(Heyrovsky)}$$

or

$$2H^* \rightarrow H_2 \quad \text{(Tafel)}.$$

In alkaline media:
Step 1:

$$H_2O + e^- \rightarrow H^* + OH^- \quad \text{(Volmer)}.$$

Step 2:

$$H_2O + e^- + H^* \rightarrow H_2 + OH^- \quad \text{(Heyrovsky)}$$

or

$$2H^* \rightarrow H_2 \quad \text{(Tafel)}.$$

Sathe *et al* prepared B-doped graphene as an efficient electrocatalyst for HER in 0.5 M H_2SO_4, with a Tafel slope of ~99 mV dec^{-1} and a j_0 of 1.4×10^{-3} mA cm^{-2} [108]. Zheng *et al* demonstrated the coupling of g-C_3N_4 with N-doped graphene as a metal-free hybrid catalyst for the HER that showed a small overpotential of ~240 mV to achieve a current density of 10 mA cm^{-2} and a small Tafel slope of 51.5 mV dec^{-1} [109]. Chen's group demonstrated that S,N-doped graphene exhibited excellent HER activity with an overpotential of −0.28 V at 10 mA cm^{-2} and a Tafel slope of 80.5 mV dec^{-1} (figure 7.10). It was explained that the defects in the graphene surface alone are not an active catalyst for the HER, but the defects created by the coupling of the S and N dopants in the graphene lattice creates a synergistic effect that will alter the ΔGH^* to obtain an excellent catalytic activity for the HER [110]. Zheng's group designed N- and P-doped graphene and activated the adjacent C atom and increased its HER activity. N and P double-doped graphene has the lowest ΔGH^* value of −0.08 eV, suggesting the highest HER activity because of their most appreciable hydrogen adsorption–desorption characteristic. As predicted, the N- and P-doped graphene showed the highest HER activity, with the lowest Tafel slope of 91 mV dec^{-1} and the lower overpotential of ~420 mV for the current density of 10 mA cm^{-2} in 0.5 M H_2SO_4 solution [111]. The N-doped graphene (2 wt.% of N)–CdS nanocomposites revealed a higher activity for H_2 evolution than pure CdS and the other catalysts from water under photo-irradiation.

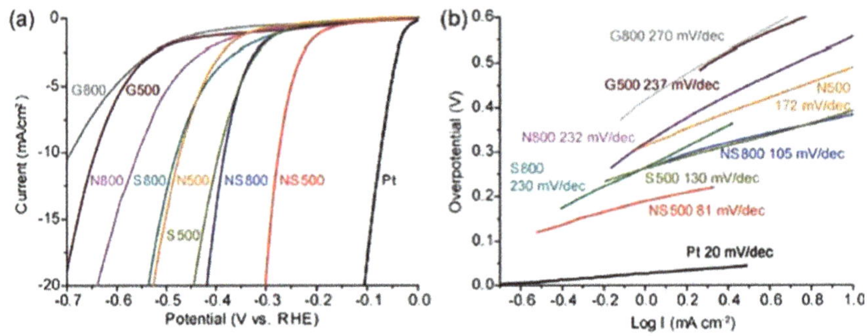

Figure 7.10. HER activity of chemically doped nanoporous graphenes. (a) CV curves of the samples produced at different CVD temperatures and with different dopants in comparison to undoped nanoporous graphene. (b) Tafel plots for the different graphene samples. (Reproduced with permission from [107]. Copyright 2014 Wiley.)

The catalytic activity occurred in the following order: N-graphene–CdS > graphene–CdS > GO–CdS > CdS.N–graphene–CdS. The synthesized catalyst does not lose its activity for the HER even after 30 h of exposure, suggesting that the CdS surface is completely covered by N-doped graphene as a protective layer and this protects the surface from corrosion during irradiation. N-doped graphene/CdS (210 μmol h^{-1}) has five times higher catalytic activity for the HER than pure CdS (40 μmol h^{-1}) [112]. Rao *et al* have synthesized a nanocomposite of N-doped graphene with few-layer 2H-MoS_2 (NEG–MoS_2(2H)) which was identified as an excellent catalyst for the HER. Nitrogenation of graphene enhances the electron donating ability of graphene nanosheets, leading to the enhanced catalytic activity of the N-graphene–2H-MoS_2 composites. The exfoliated form of the composite (EG)–MoS_2(2H) shows higher H_2 evolution (0.54 mmole/gh) than the pure 2H-MoS_2 (0.05 mmole/gh). The rate of H_2 evolution was obtained as even higher (0.83 mmole/gh) after the nitrogenation of the exfoliated graphene layer with the MoS_2 composite NEG–MoS_2(2H). The TOF for HER of NEG–MoS_2(2H) is 2.1 times higher than that of EG–MoS_2(2H) and 57 times higher than with pure 2H-MoS_2. N incorporation makes graphene electron-rich. MoS_2 acted as a p-type and was ready to accept electrons from the electron-rich N-graphene, which resulted in higher catalytic activity. The catalytic activity remained the same even after six cycles of reaction [113].

The HER activities of all catalysts in an acidic medium is higher than those in alkaline solutions. This discrepancy may arise because of the aforementioned different mechanisms for the two electrolytes. For example, the alkaline solution contains an additional water dissociation step, and usually graphene based materials have difficulties in achieving this step and this leads to a relatively high energy barrier throughout the HER process [12].

7.5 Conclusion

In conclusion, this chapter presents an overview of the synthesis of graphene and heteroatom (nitrogen, fluorine, boron, phosphorus etc) incorporated graphene and its metal composites for energy conversion (ORR, OER, and HER) applications. The decreasing amount of fossil fuels around the world and their negative impact on the environment have led to the search for alternative green energy sources. Fuel cells (the ORR) and water-splitting (the OER and HER) are known as the most promising energy conversion methods of recent years. Intense research into graphene and heteroatom doped graphene composites began after the exfoliation of a single layer by Geim and Novoselov. Because of the unique physical and chemical properties, and morphology and structure of graphene, tremendous attention has been paid to this material for the last few decades. In this chapter various synthesis methods of graphene have been explained in detail. Nitrogen and other heteroatom doped graphene composites have also been discussed intensively. All the reported methods have their own advantages and disadvantages. Despite the fact that the identification of the controlled bulk scale synthesis of highly oriented graphene is a challenging task, as more research takes place in the field of graphene synthesis there is hope that the goal can be achieved soon. The incorporation of heteroatoms such as nitrogen, fluorine, or boron on the graphene surface efficiently alters the electronic structure and chemical properties of graphene nanosheets to develop their potential for better electrocatalytic performance by means of creating new catalytic active sites and altering the adsorption–desorption abilities of the reactant molecules and intermediates. The heteroatom doping was achieved via the covalent interaction between the heteroatom and C atom in the graphene, which gives more durability to the catalysts during the catalysis reactions. For example, unlike the Pt catalyst in ORR, the heteroatom doped graphene composites do not have the poisoning effect because of the CO molecule and give more stability to the fuel cell. The heteroatom doping enhances the active sites for the catalytic reaction and also, because of synergistic effects, the catalytic activity and the durability of the catalyst increase drastically and are even better than in many noble metals. Although the graphene composites have admirable durability and excellent catalytic activity, they cannot replace the standard commercial catalysts such as Pt and IrO_2 because of a lack of large scale synthesis, high-cost synthesis processes, and harsh chemical reactions. Moreover, the mechanism of the catalytic reactions with graphene composites are also not well understood because of the complex structure of graphene. As a ray of hope, we believe that all the aforementioned disadvantages will be rectified soon and the graphene based composite materials will play a very important role in the field of green energy conversion in the near future.

Acknowledgements

P G would like to acknowledge ANID-FONDECYT POSTDOCTORADO (Project No 3180176) Government of Chile, for providing financial support.

References

[1] Geim A K 2009 Graphene: status and prospects *Science* **324** 1530–4

[2] Allen M J, Tung V C and Kaner R B 2009 Honeycomb carbon: a review of graphene *Chem. Rev.* **110** 132–45

[3] Rao C N R, Sood A K, Subrahmanyam K S and Govindaraj A 2009 Graphene: the new two-dimensional nanomaterial *Angew. Chem. Int. Ed.* **48** 7752–77

[4] Bolotin K I, Sikes K J, Jiang Z, Klima M, Fudenberg G, Hone J, Kim P and Stormer H L 2008 Ultrahigh electron mobility in suspended graphene *Solid State Commun.* **146** 351–5

[5] Castro Neto A H, Guinea F, Peres N M R, Novoselov K S and Geim A K 2009 The electronic properties of graphene *Rev. Mod. Phys.* **81** 109–62

[6] Lee C, Wei X D, Kysar J W and Hone J 2008 Measurement of the elastic properties and intrinsic strength of monolayer graphene *Science* **321** 385–8

[7] Balandin A A, Ghosh S, Bao W Z, Calizo I, Teweldebrhan D, Miao F and Lau C N 2008 Superior thermal conductivity of single-layer graphene *Nano Lett.* **8** 902–7

[8] Geim A K and Novoselov K S 2007 The rise of graphene *Nat. Mater.* **6** 183–91

[9] Nair R R, Blake P, Grigorenko A N, Novoselov K S, Booth T J, Stauber T, Peres N M R and Geim A K 2008 Fine structure constant defines visual transparency of graphene *Science* **320** 1308

[10] Edwards R S and Coleman K S 2013 Graphene synthesis: relationship to applications *Nanoscale* **5** 38–51

[11] Lightcap I V and Kamat P V 2013 Graphitic design: prospects of graphene-based nanocomposites for solar energy conversion, storage, and sensing *Acc. Chem. Res.* **46** 2235–43

[12] Duan J, Chen S, Jaroniec M and Qiao S Z 2015 Heteroatom-doped graphene-based materials for energy-relevant electrocatalytic processes *ACS Catal.* **5** 5207–34

[13] Chang D W, Choi H J, Filer A and Baek J B 2014 Graphene in photovoltaic applications: organic photovoltaic cells (OPVs) and dye-sensitized solar cells (DSSCs) *J. Mater. Chem.* A **2** 12136–49

[14] Wu J, Pisula W and Müllen K 2007 Graphenes as potential material for electronics *Chem. Rev.* **107** 718–47

[15] Panchakarla L S, Subrahmanyam K S, Saha S K, Govin-daraj A, Krishnamurthy H R, Waghmare U V and Rao C N R 2009 Synthesis, structure, and properties of boron- and nitrogen-doped graphene *Adv. Mater.* **21** 4726–30

[16] Lotya M *et al* 2009 Liquid phase production of graphene by exfoliation of graphite in surfactant/water solutions *J. Am. Chem. Soc.* **131** 3611–20

[17] Reina A, Jia X, Ho J, Nezich D, Son H, Bulovic V, Dresselhaus M S and Kong J 2009 Large area, few-layer graphene films on arbitrary substrates by chemical vapor deposition *Nano Lett.* **9** 30–5

[18] Eda G and Chhowalla M 2010 Chemically derived graphene oxide: towards large-area thin-film electronics and optoelectronics *Adv. Mater.* **22** 2392–415

[19] Novoselov K S, Geim A K, Morozov S V, Jiang D, Zhang Y, Dubonos S V, Grigorieva I V and Firsov A A 2004 Electric field effect in atomically thin carbon films *Science* **306** 666–9

[20] Emtsev K V *et al* 2009 Towards wafer-size graphene layers by atmospheric pressure graphitization of silicon carbide *Nat. Mater.* **8** 203–7

[21] Wang X, Zhi L, Tsao N, Tomović Ž, Li J and Müllen K 2008 Transparent carbon films as electrodes in organic solar cells *Angew. Chem. Int. Ed.* **47** 2990–2

[22] Qu L, Liu Y, Baek J B and Dai L 2010 Nitrogen-doped graphene as efficient metal-free electrocatalyst for oxygen reduction in fuel cells *ACS Nano* **4** 1321–6

[23] Stankovich S, Dikin D A, Dommett G H B, Kohlhaas K M, Zimney E J, Stach E A, Piner R D, Nguyen S T and Ruoff R S 2006 Graphene-based composite materials *Nature* **442** 282–6

[24] Yan X Y, Tong X L, Zhang Y F, Han X D, Wang Y Y, Jin G Q, Qin Y and Guo X Y 2012 Cuprous oxidenanoparticles dispersed on reduced graphene oxide as an efficient electrocatalyst for oxygen reduction reaction *Chem. Commun.* **48** 1892–4
[25] Zhou X, Wang F, Zhu Y and Liu Z 2011 Graphene modified $LiFePO_4$ cathode materials for high power lithium ion batteries *J. Mater. Chem.* **21** 3353–8
[26] Marconcini P, Cresti A, Triozon F, Fiori G, Biel B, Niquet Y M, Macucci M and Roche S 2012 Atomistic boron-doped graphene field-effect transistors: a route toward unipolar characteristics *ACS Nano* **6** 7942–7
[27] Du X A, Guo P, Song H H and Chen X H 2010 Graphene nanosheets as electrode material for electric double-layer capacitors *Electrochim. Acta* **55** 4812–9
[28] Li X, Wang X, Zhang L, Lee S and Dai H 2008 Chemically derived, ultrasmooth graphene nanoribbon semiconductors *Science* **319** 1229–32
[29] Li X, Zhang G, Bai X, Sun X, Wang X, Wang E and Dai H 2008 Highly conducting graphene sheets and Langmuir–Blodgett films *Nat. Nanotechnol.* **3** 538–42
[30] Hernandez Y *et al* 2008 High-yield production of graphene by liquid-phase exfoliation of graphite *Nat. Nanotechnol.* **3** 563–8
[31] Hao R, Qian W, Zhang L and Hou Y 2008 Aqueous dispersions of TCNQ-anion-stabilized graphene sheets *Chem. Commun.* **2008** 6576–8
[32] Sun Z, Kohama S I, Zhang Z, Lomeda J R and Tour J M 2010 Soluble graphene through edge-selective functionalization *Nano Res.* **3** 117–25
[33] Lee J H, Shin D W, Makotchenko V G, Nazarov A S, Fedorov V E, Kim Y H, Choi J Y, Kim J M and Yoo J B 2009 One-step exfoliation synthesis of easily soluble graphite and transparent conducting graphene sheets *Adv. Mater.* **21** 4383–7
[34] Zheng J, Di C, Liu Y, Liu H, Guo Y, Du C, Wu T, Yu G and Zhu D 2010 High quality graphene with large flakes exfoliated by oleyl amine *Chem. Commun.* **46** 5728–30
[35] Qian W, Hao R, Hou Y, Tian Y, Shen C, Gao H and Liang X 2009 Solvothermal-assisted exfoliation process to produce graphene with high yield and high quality *Nano Res.* **9** 706–12
[36] Brodie B C 1859 XIII. On the atomic weight of graphite *Philos. Trans. R. Soc.* **149** 249–59
[37] Staudenmaier L 1898 Verfahren zur Darstellung der Graphitsäure *Ber. Dtsch Chem. Ges.* **31** 1481–7
[38] Hummers W S and Offeman R E 1958 Preparation of graphitic oxide *J. Am. Chem. Soc.* **80** 1339
[39] Huang Y, Liang J and Chen Y 2012 An overview of the applications of graphene-based materials in supercapacitors *Small* **8** 1805–34
[40] Dreyer D R, Park S, Bielawski C W and Ruoff R S 2010 The chemistry of graphene oxide *Chem. Soc. Rev.* **39** 228–40
[41] Stankovich S, Dikin D A, Piner R D, Kohlhaas K A, Kleinhammes A, Jia Y, Wu Y, Nguyen S T and Ruoff R S 2007 Synthesis of graphene-based nanosheets via chemical reduction of exfoliated graphite oxide *Carbon* **45** 1558–65
[42] Kosynkin D V, Higginbotham A L, Sinitskii A, Lomeda J R, Dimiev A, Price B K and Tour J M 2009 Longitudinal unzipping of carbon nanotubes to form graphene nanoribbons *Nature* **458** 872–6
[43] Jiao L, Zhang L, Wang X, Diankov G and Dai H 2009 Narrow graphene nanoribbons from carbon nanotubes *Nature* **458** 877–80
[44] Kumar P, Panchakarla L S and Rao C N R 2011 Laser-induced unzipping of carbon nanotubes to yield graphene nanoribbons *Nanoscale* **3** 2127–9

[45] Xie L, Wang H, Jin C, Wang X, Jiao L, Suenaga K and Dai H 2011 Graphene nanoribbons from unzipped carbon nanotubes: atomic structures, Raman spectroscopy, and electrical properties *J. Am. Chem. Soc.* **27** 10394–7
[46] Cho S, Kikuchi K and Kawasaki A 2011 Radial followed by longitudinal unzipping of multiwalled carbon nanotubes *Carbon* **49** 3865–72
[47] Kang Y R, Li Y L and Deng M Y 2012 Precise unzipping of flattened carbon nanotubes to regular graphene nanoribbons by acid cutting along the folded edges *J. Mater. Chem.* **22** 16283–7
[48] Li Y S, Liao J L, Wang S Y and Chiang W H 2016 Intercalation-assisted longitudinal unzipping of carbon nanotubes for green and scalable synthesis of graphene nanoribbons *Sci. Rep.* **6** 1–12
[49] Wintterlin J and Bocquet M L 2009 Graphene on metal surfaces *Surf. Sci.* **603** 1841–52
[50] Mattevi C, Kim H and Chhowalla M 2011 A review of chemical vapour deposition of graphene on copper *J. Mater. Chem.* **21** 3324–34
[51] Wang S M, Pei Y H, Wang X, Wang H, Meng Q N, Tian H W, Zheng X L, Zheng W T and Liu Y C 2010 Synthesis of graphene on a polycrystalline Co film by radio-frequency plasma-enhanced chemical vapour deposition *J. Phys. D: Appl. Phys.* **43** 455402
[52] An H, Lee W-J and Jung J 2011 Graphene synthesis on Fe foil using thermal CVD *Curr. Appl Phys.* **11** S81–5
[53] Rut'kov E V, Kuz'michev A V and Gall N R 2011 Carbon interaction with rhodium surface: adsorption, dissolution, segregation, growth of graphene layers *Phys. Solid State* **53** 1092–8
[54] Sutter E, Albrecht P and Sutter P 2009 Graphene growth on polycrystalline Ru thin films *Appl. Phys. Lett.* **95** 133109
[55] Gao T, Xie S, Gao Y, Liu M, Chen Y, Zhang Y and Liu Z 2011 Growth and atomic-scale characterizations of graphene on multifaceted textured Pt foils prepared by chemical vapor deposition *ACS Nano* **5** 9194–201
[56] Kwon S-Y, Ciobanu C V, Petrova V, Shenoy V B, Bareno J, Gambin V, Petrov I and Kodambaka S 2009 Growth of semiconducting graphene on palladium *Nano Lett.* **9** 3985–90
[57] Oznuluer T, Pince E, Polat E O, Balci O, Salihoglu O and Kocabas C 2011 Synthesis of graphene on gold *Appl. Phys. Lett.* **98** 183101
[58] Mueller F, Grandthyll S, Zeitz C, Jacobs K, Huefner S, Gsell S and Schreck M 2011 Epitaxial growth of graphene on Ir (111) by liquid precursor deposition *Phys. Rev.* B **84** 075472
[59] Dai B, Fu L, Zou Z, Wang M, Xu H, Wang S and Liu Z 2011 Rational design of a binary metal alloy for chemical vapour deposition growth of uniform single-layer graphene *Nat. Commun.* **2** 1–6
[60] Weatherup R S, Bayer B C, Blume R, Ducati C, Baehtz C, Schlögl R and Hofmann S 2011 *In situ* characterization of alloy catalysts for low-temperature graphene growth *Nano Lett.* **11** 4154–60
[61] Liu N, Fu L, Dai B, Yan K, Liu X, Zhao R, Zhang Y and Liu Z 2011 Universal segregation growth approach to wafer-size graphene from non-noble metals *Nano Lett.* **11** 297–303
[62] Baraton L, He Z, Lee C S, Maurice J L, Cojocaru C S, Lee Y H and Pribat D 2012 GraphITA 2011 *Physical Review* ed L Ottaviano and V Morandi (Berlin: Springer) ch 1 pp 1–7

[63] Kumar S, McEvoy N, Lutz T, Keeley G P, Nicolosi V, Murray C P, Blau W J and Duesberg G S 2010 Gas phase controlled deposition of high quality large-area graphene films *Chem. Commun.* **46** 1422–4 .
[64] Tromp R M and Hannon J B 2009 Thermodynamics and kinetics of graphene growth on SiC (0001) *Phys. Rev. Lett.* **102** 106104
[65] Juang Z Y, Wu C Y, Lo C W, Chen W Y, Huang C F, Hwang J C, Chen F R, Leou K C and Tsai C H 2009 Synthesis of graphene on silicon carbide substrates at low temperature *Carbon* **47** 2026–31
[66] Aristov V Y *et al* 2010 Graphene synthesis on cubic SiC/Si wafers. perspectives for mass production of graphene-based electronic devices *Nano Lett.* **10** 992–5
[67] Sheng Z H, Shao L, Chen J J, Bao W J, Wang F B and Xia X H 2011 Catalyst-free synthesis of nitrogen-doped graphene via thermal annealing graphite oxide with melamine and its excellent electrocatalysis *ACS Nano* **5** 4350–8
[68] Zhang L S, Liang X Q, Song W G and Wu Z Y 2010 Identification of the nitrogen species on N-doped graphene layers and Pt/NG composite catalyst for direct methanol fuel cell *Phys. Chem. Chem. Phys.* **12** 12055–9
[69] Shao Y, Zhang S, Engelhard M H, Li G, Shao G, Wang Y, Liu J, Aksay I A and Lin Y 2010 Graphene based electrochemical sensors and biosensors: a review *Electroanalysis* **22** 1027–36
[70] Luo Z, Lim S, Tian Z, Shang J, Lai L, MacDonald B, Fu C, Shen Z, Yu T and Lin J 2011 Pyridinic N doped graphene: synthesis, electronic structure, and electrocatalytic property *J. Mater. Chem.* **21** 8038–44
[71] Ghosh A, Late D J, Panchakarla L S, Govindaraj A and Rao C N R 2009 NO_2 and humidity sensing characteristics of few-layer graphenes *J. Exp. Nanosci.* **4** 313–22
[72] Deng D H *et al* 2011 Toward N-doped graphene via solvothermal synthesis *Chem. Mater.* **23** 1188–93
[73] Wei D, Liu Y, Wang Y, Zhang H, Huang L and Yu G 2009 Synthesis of N-doped graphene by chemical vapor deposition and its electrical properties *Nano Lett.* **9** 1752–8
[74] Choi E-K, Jeon I-Y, Bae S-Y, Lee H-J, Shin H S, Dai L and Baek J-B 2010 High-yield exfoliation of three-dimensional graphite into two-dimensional graphene-like sheets *Chem. Commun.* **46** 6320–2
[75] Long D, Li W, Ling L, Miyawaki J, Mochida I and Yoon S-H 2010 Preparation of nitrogen-doped graphene sheets by a combined chemical and hydrothermal reduction of graphene oxide *Langmuir* **26** 16096–102
[76] Han J, Zhang L L, Lee S, Oh J, Lee K-S, Potts J R, Ji J, Zhao X, Ruoff R S and Park S 2013 Generation of B-doped graphene nanoplatelets using a solution process and their supercapacitor applications *ACS Nano* **7** 19–26
[77] Khai T V, Na H G, Kwak D S, Kwon Y J, Ham H, Shim K B and Kim H W 2012 Comparison study of structural and optical properties of boron-doped and undoped graphene oxide films *Chem. Eng. J.* **211** 369–77
[78] Sheng Z-H, Gao H-L, Bao W-J, Wang F-B and Xia X-H 2012 Synthesis of boron doped graphene for oxygen reduction reaction in fuel cells *J. Mater. Chem.* **22** 390–5
[79] Kim Y A *et al* 2012 Raman spectroscopy of boron-doped single-layer graphene *ACS Nano* **6** 6293–300
[80] Panchakarla L S, Subrahmanyam K S, Saha S K, Govindaraj A, Krishnamurthy H R, Waghmare U V and Rao C N R 2009 Synthesis, structure, and properties of boron-and nitrogen-doped graphene *Adv. Mater.* **21** 4726–30

[81] Cattelan M, Agnoli S, Favaro M, Garoli D, Romanato F, Meneghetti M, Barinov A, Dudin P and Granozzi G 2013 Microscopic view on a chemical vapor deposition route to boron-doped graphene nanostructures *Chem. Mater.* **25** 1490–5
[82] Wang H, Zhou Y, Wu D, Liao L, Zhao S, Peng H and Liu Z 2013 Synthesis of boron-doped graphene monolayers using the sole solid feedstock by chemical vapor deposition *Small* **9** 1316–20
[83] Gebhardt J, Koch R J, Zhao W, Hoefert O, Gotter-barm K, Mammadov S, Papp C, Goerling A, Steinrueck H P and Seyller T 2013 Growth and electronic structure of boron-doped graphene *Phys. Rev.* B **87** 155437
[84] Tang Y-B *et al* 2012 Tunable band gaps and p-type transport properties of boron-doped graphenes by controllable ion doping using reactive microwave plasma *ACS Nano* **6** 1970–8
[85] Lu X, Wu J, Lin T, Wan D, Huang F, Xie X and Jiang M 2011 Low-temperature rapid synthesis of high-quality pristine or boron-doped graphene via Wurtz-type reductive coupling reaction *J. Mater. Chem.* **21** 10685–9
[86] Ho K-I, Huang C-H, Liao J-H, Zhang W, Li L-J, Lai C-S and Su C-Y 2015 Fluorinated graphene as high performance dielectric materials and the applications for graphene nanoelectronics *Sci. Rep.* **4** 5893
[87] Wang Y, Lee W C, Manga K K, Ang P K, Lu J, Liu Y P, Lim C T and Loh K P 2012 Fluorinated graphene for promoting neuro-induction of stem cells *Adv. Mater.* **24** 4285–90
[88] Nair R R *et al* 2010 Fluorographene: a two-dimensional counterpart of Teflon *Small* **6** 2877–84
[89] Baraket M, Walton S G, Lock E H, Robinson J T and Perkins F K 2010 The functionalization of graphene using electron-beam generated plasmas *Appl. Phys. Lett.* **96** 231501
[90] Wang Z, Wang J, Li Z, Gong P, Liu X, Zhang L, Ren J, Wang H and Yang S 2012 Synthesis of fluorinated graphene with tunable degree of fluorination *Carbon* **50** 5403–10
[91] Gong P, Wang Z, Li Z, Mi Y, Sun J, Niu L, Wang H, Wang J and Yang S 2013 Photochemical synthesis of fluorinated graphene via a simultaneous fluorination and reduction route *RSC Adv.* **3** 6327–30
[92] Bruna M, Massessi B, Cassiago C, Battiato A, Vittone E, Speranza G and Borini S 2011 Synthesis and properties of monolayer graphene oxyfluoride *J. Mater. Chem.* **21** 18730–7
[93] Chronopoulos D D, Bakandritsos A, Pykal M, Zbořil R and Otyepka M 2017 Chemistry, properties, and applications of fluorographene *Appl. Mater. Today* **9** 60–70
[94] Dubois M *et al* 2014 Thermal exfoliation of fluorinated graphite *Carbon* **77** 688–704
[95] Mathkar A, Narayanan T N, Alemany L B, Cox P, Nguyen P, Gao G, Chang P, Romero-Aburto R, Mani S A and Ajayan P M 2013 Synthesis of fluorinated graphene oxide and its amphiphobic properties *Part. Part. Syst. Charact.* **30** 266–72
[96] Wang X, Zhi L and Mullen K 2008 Transparent, conductive graphene electrodes for dye-sensitized solar cells *Nano Lett.* **8** 323–7
[97] Yu D, Nagelli E, Du F and Dai L 2010 Metal-free carbon nanomaterials become more active than metal catalysts and last longer *J. Phys. Chem. Lett.* **1** 2165–73
[98] Gong K, Du F, Xia Z, Durstock M and Dai L 2009 Nitrogen-doped carbon nanotube arrays with high electrocatalytic activity for oxygen reduction *Science* **323** 760–4
[99] Wang S, Zhang L, Xia Z, Roy A, Chang D W, Baek J-B and Dai L 2012 BCN graphene as efficient metal-free electrocatalyst for the oxygen reduction reaction *Angew. Chem. Int. Ed.* **51** 4209–12

[100] Kakaei K and Balavandi A 2017 Hierarchically porous fluorine-doped graphene nanosheets as efficient metal-free electrocatalyst for oxygen reduction in gas diffusion electrode *J. Colloid Interface Sci.* **490** 819–24
[101] Liang Y, Li Y, Wang H, Zhou J, Wang J, Regier T and Dai H 2011 Co_3O_4 nanocrystals on graphene as a synergistic catalyst for oxygen reduction reaction *Nat. Mater.* **10** 780–6
[102] Mirzakulova E, Khatmullin R, Walpita J, Corrigan T, Vargas-Barbosa N M, Vyas S, Oottikkal S, Manzer S F, Hadad C M and Glusac K D 2012 Electrode-assisted catalytic water oxidation by a flavin derivative *Nat. Chem.* **4** 794
[103] Suntivich J, May K J, Gasteiger H A, Goodenough J B and Shao-Horn Y 2011 A perovskite oxide optimized for oxygen evolution catalysis from molecular orbital principles *Science* **334** 1383–5
[104] Zhao Y, Nakamura R, Kamiya K, Nakanishi S and Hashimoto K 2013 Nitrogen-doped carbon nanomaterials as non-metal electrocatalysts for water oxidation *Nat. Commun.* **4** 1–7
[105] Chen S and Qiao S-Z 2013 Hierarchically porous nitrogen-doped graphene–$NiCo_2O_4$ hybrid paper as an advanced electrocatalytic water-splitting material *Acs Nano* **7** 10190–6
[106] Chen S, Duan J, Ran J and Qiao S Z 2015 Paper-based N-doped carbon films for enhanced oxygen evolution electrocatalysis *Adv. Sci.* **2** 1400015
[107] Ma T Y, Dai S, Jaroniec M and Qiao S Z 2014 Graphitic carbon nitride nanosheet–carbon nanotube three-dimensional porous composites as high-performance oxygen evolution electrocatalysts *Angew. Chem.* **126** 7409–13
[108] Sathe B R, Zou X and Asefa T 2014 Metal-free B-doped graphene with efficient electrocatalytic activity for hydrogen evolution reaction *Catal. Sci. Technol.* **4** 2023–30
[109] Zheng Y, Jiao Y, Zhu Y, Li L H, Han Y, Chen Y, Du A, Jaroniec M and Qiao S Z 2014 Hydrogen evolution by a metal-free electrocatalyst *Nat. Commun.* **5** 1–8
[110] Ito Y, Cong W, Fujita T, Tang Z and Chen M 2015 High catalytic activity of nitrogen and sulfur co-doped nanoporous graphene in the hydrogen evolution reaction *Angew. Chem.* **127** 2159–64
[111] Zheng Y, Jiao Y, Li L H, Xing T, Chen Y, Jaroniec M and Qiao S Z 2014 Toward design of synergistically active carbon-based catalysts for electrocatalytic hydrogen evolution *ACS Nano* **8** 5290–6
[112] Jia L, Wang D H, Huang Y X, Xu A W and Yu H Q 2011 Highly durable N-doped graphene/CdS nanocomposites with enhanced photocatalytic hydrogen evolution from water under visible light irradiation *J. Phys. Chem.* C **115** 11466–73
[113] Maitra U, Gupta U, De M, Datta R, Govindaraj A and Rao C N R 2013 Highly effective visible-light-induced H_2 generation by single-layer 1T-MoS_2 and a nanocomposite of few-layer 2H-MoS_2 with heavily nitrogenated graphene *Angew. Chem. Int. Ed.* **52** 13057–61

IOP Publishing

Nanostructured Materials for Sustainable Energy and Environmental Remediation

R V Mangalaraja, R Udayabhaskar, P Sathishkumar and Joydeep Dutta

Chapter 8

Materials science of advanced carbon nanomaterials for photovoltaic and photothermal devices

Thangaraj Pandiyarajan, R V Mangalaraja, Balasubramanian Karthikeyan and M A Gracia-Pinilla

The rapid increase in industrialization has led to global warming and solar energy is one of the best possible solutions to combat climate change as it can significantly reduce greenhouse gas emissions. In this century, the fundamental aspects, fabrication, and production of green and renewable energy sources is one of the most exciting areas of semiconductor science. Since the discovery of graphene in 2014, which offers some outstanding physico-chemical properties, great efforts have been devoted to exploring potential applications of graphene in various technological fields including photovoltaic devices and biomedical applications. This chapter is divided into two parts. The first deals with the fundamental aspects and fabrication of solar cells through the utilization of the fascinating material graphene as the active layer, electron/hole collection layer, counter electrode, transparent conducting electrode, and Schottky junction solar cell. Various configurations with the inclusion of graphene material in solar cells and its photovoltaic parameters with overall photoconversion efficiency are summarized in tables. The second part of the chapter presents the recent development of multifunctional graphene based materials for photothermal therapeutic cancer treatment, where the conversion of light energy into thermal energy using graphene as the active material for biomedical applications is discussed comprehensively.

8.1 Introduction

In the past few decades, the exponential growth of technological industries and a population explosion have led to an increase in global energy demands. Fossil fuel

doi:10.1088/978-0-7503-5143-0ch8

reserves and non-renewable energy sources have been used to meet this energy demand, which has consequently increased the greenhouse effect. In recent years, severe droughts, flooding, intense storms, and an increase of global mean surface temperature have been seen across the globe and are caused by climate change [1]. There is an immediate need for the reduction of carbon emissions and less use of fossil fuels and non-renewable energy sources. Therefore, it is necessary to find an alternative eco-friendly and renewable energy source for the future world. It is expected that the renewable energy source contribution across the globe in 2040 will be 47.7% and increase significantly up to 80% in 2100 [2]. Solar energy is one of the best possible alternative renewable energy sources among all other renewable sources, such as wind energy, hydropower, biomass, marine energies, and geothermal energy. The Sun emits radiation at a rate of 3.8×10^{23} kW of which approximately 1.8×10^{14} kW power reaches the Earth's surface [3]. The effective utilization of available solar energy in energy production depends on a number of factors such as the choice of materials, fabrication process, cost of production, long-term stability, and photoconversion efficiency of the cell, which all need to be considered. The choice of materials is one of the most important parameters. The search for new materials has moved to low-dimensional materials, in accordance with the famous words of Richard Feynman: 'There is plenty of room at the bottom'. The fundamental aspects and various fabrication techniques of carbon based nanostructured materials (fullerenes, carbon nanotubes (CNTs), graphene, and nanodiamonds) have been studied extensively. The allotropes of carbon materials have several advantages, such as being the most abundant available resource, cheap, and eco-friendly, and having extraordinary desirable physical and chemical properties [4]. Among the various carbon allotropes, 2D graphene has received great attention due to its unique optoelectronic properties. Monolayer graphene (MLG) exhibits 97.7% transparency and absorbs only 2.3% of visible light, and has a tunable bandgap in bilayer graphene upon electrical gating. The electrical conduction can also be modified as a metal and semiconductor by tuning the configuration of the zigzag and armchair edges, respectively. Graphene films 10 nm thick exhibit a high electrical conductivity of 550 S cm^{-1}, whereas polycrystalline graphite shows the highest conductivity of 1250 S cm^{-1} [5]. These remarkable optoelectronic properties make graphene an ideal candidate for future eco-friendly photovoltaic applications.

According to a recent WHO report, cancer is one of the most dangerous diseases and the second leading cause of death globally, with about one in six deaths being related to cancer. Also, it was estimated that the over 9.6 million deaths in 2018 will reach to 29 million by 2040 [6]. The mortality rate can be reduced by early detection, monitoring, and diagnosis. The photothermal (PT) effect, i.e. conversion of light to heat energy, has become one of the hot areas of research for invasive treatments in cancer therapy. The PT effect has several advantages, such as the selective thermal ablation of cancer cells, as conventional hyperthermia (laser, ultrasound, microwave, or radio frequency) treatment does not possess such selectivity and will affect normal cells as well, which causes side effects to living organisms [7]. Carbon based nanostructured materials have exceptional properties, such as biocompatibility, high specific surface area, and easy functionalization with biomolecules for targeted drug delivery. Monolayer or multilayer graphene structures with sp^2 bonded carbon atoms have shown excellent absorption of near-infrared (NIR) photons and can convert them into thermal energy due to the band structure [8]. In addition, graphene oxide (GO) enriched with a number of functional groups and having good thermal conductivity, makes graphene based materials versatile candidates for photothermal therapy cancer treatment.

8.2 Classifications of carbon allotropes

Over the last century, a variety of carbon allotropes has been discovered and can be classified based on the dimensionality of the system. A schematic representation of the classification of carbon based materials along with their length scale is depicted in figure 8.1.

8.2.1 Three-dimensional materials

Graphite is the main carbon material found on Earth, consisting of many stacked sp^2 hybridized layers of six carbon atoms arranged in a hexagonal lattice structure [9]. It is a three-dimensional material due to its delocalized charge carriers in all three directions and no confinement in any direction. The structure

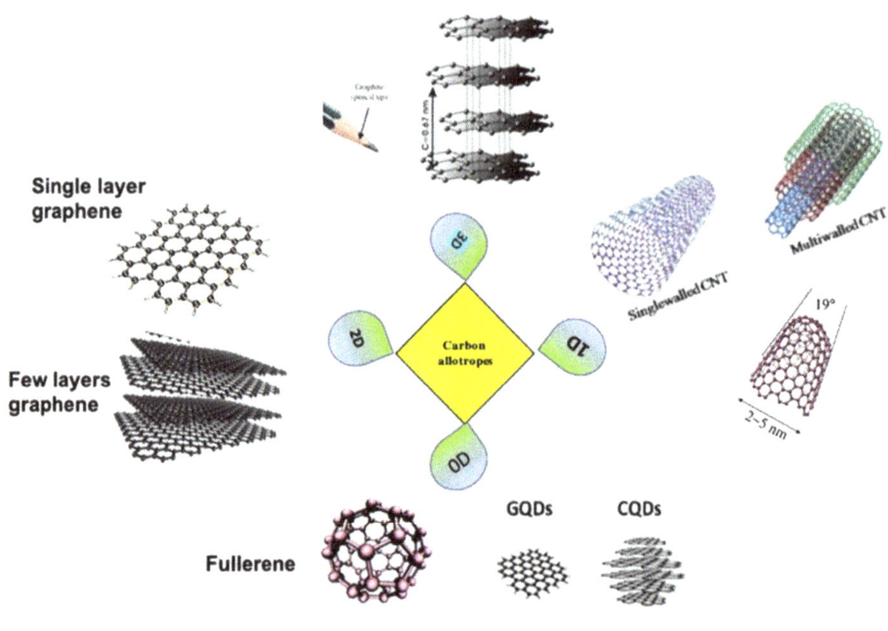

Figure 8.1. Schematic representation of different carbon based nanostructures. Multilayers stacked into three-dimensional graphite (e.g. in pencils), single and few layers of graphene as a two-dimensional material, single and multilayers of carbon atoms rolled up into one-dimensional single-walled and multi-walled carbon nanotubes, zero-dimensional fullerenes, graphene, and carbon quantum dots. (Reproduced with permission from [212]. Copyright 2019 *Frontiers*.)

of graphite was determined initially by J D Bernal in the year 1924 [10]. The proposed structure was a hexagonal net of carbon atoms stacked in layers in which half of the atoms in one layer is are directly above those in the layer below. The bonds between the carbon atoms are very strong due to the sigma bonds, whereas the interconnections between layers are rather weak due to the van der Walls force acting between them.

8.2.2 Two-dimensional materials

In the twentieth century and later, graphene based materials have attracted the attention of the scientific community due to their extraordinary properties. Graphene is a two-dimensional (2D) and sp^2 bonded carbon based material which has been studied extensively since the first single layer of graphene was exfoliated in 2004 [11, 12]. Graphene exhibits remarkable properties, such as like high mechanical strength [13], good flexibility, excellent conducting properties with high transparency [14], and chemical and thermal stability, which make the material useful in various technological industries.

8.2.3 One-dimensional materials

CNTs belong to the class of one-dimensional (1D) carbon allotropes due to the spatial confinement in two directions and the delocalization of carriers along one direction, i.e the carriers can move freely only along the axis of the cylinder. Iijima [15] in 1991 reported new types of finite carbon structures consisting of needle-like structures with outer diameters of 4–30 nm and a length of up to 1 mm, which were prepared through an arc-discharge evaporation route and later named carbon nanotubes (CNTs). Since then research has been accelerated for the fabrication of high purity single-walled and multi-walled CNTs due to their unique structural properties, such as a mechanical strength greater than that of Fe, a density lower than that of Al, and a thermal stability of 1400 °C in a vacuum [16], which makes these materials suitable for energy production and storage applications. Also, they are widely used as probe tips in scanning tunneling microscopy [17] and atomic force microscopy equipment [18]. Another subclass of 1D carbon structures is carbon nanohorns (sometimes also known as nanocones), which are conical carbon nano-structures constructed from an sp^2 carbon sheet. Usually, carbon nanohorns have advantages from the synthesis point of view, such as mass production at room temperature and production without a toxic metal catalyst, which are not possible in the case of carbon nanotubes [24]. Terrones [19] first proposed the transformation of ordinary graphene into graphitic cone-like structures. A schematic representation of the construction of carbon nanocones is shown in figure 8.2(c), where a maximum of six pentagons can be introduced since six pentagons would result in a closed nanotube tip structure. Since then, several research groups across the globe have been working on the production of nanohorns for industrial applications, such as drug carriers [20], biofuel cells [21], highly sensitive biosensors [22], and also in photothermal cancer treatment [23]

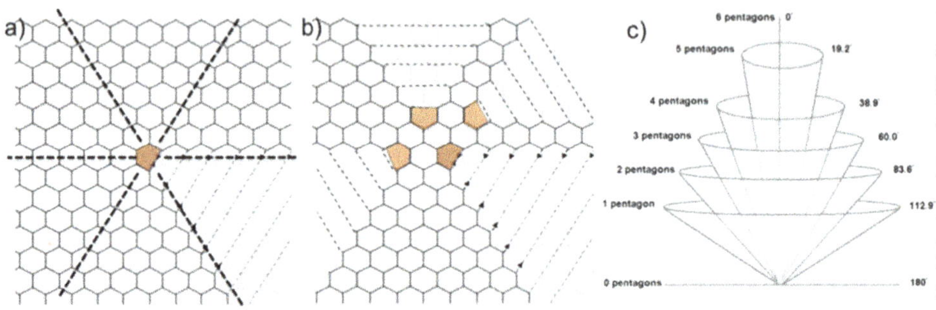

Figure 8.2. Schematic representation of the construction of carbon nanocones by cutting a wedge (disclination) from graphene and reconnecting the resultant dangling bonds (dotted arrows). The colored one is the created pentagons. (a) Single pentagon cone. (b) Four pentagon cone. (c) Possible cones with their cone tip angle and number of pentagons. (Reproduced with permission from [24]. Copyright 2016 American Chemical Society.)

8.2.4 Zero-dimensional materials

Fullerenes belongs to new class of allotropes which are made of hollow carbon cages and they come under the umbrella of zero-dimensional materials due to their spatial confinement in all three directions. The number of carbon atoms in each cage can vary and is generally represented as C_n, where n represents the number of carbon atoms present in the cages. Fullerene was discovered by Richard Smalley's team in 1985 and they were awarded the Nobel Prize in Chemistry in 1996 [25]. The most widely studied fullerenes are C_{60}, in which the structure consists of 60 carbon atoms making up 20 six-membered rings and 12 five-membered rings, composed of 60 vertices; all carbon atoms are equivalent.

8.3 Graphene synthesis route

8.3.1 Exfoliation

The quality of graphene materials and the types of defects depend strongly on the synthesis method. To date, several techniques have been developed for the mass production of graphene, which allow a wide choice in terms of size, quality, and price, and are shown in figure 8.3. Bulk graphite is made up of stacked parallel layers separated by 3.41 Å, which can be separated for promising applications in various technological fields. The first experiment on the exfoliation of graphite was done by Bordie by a using mixture of concentrated nitric and sulfuric acids. Thereafter, several experiments were performed to try to exfoliate a single layer of graphene until Novoselov and Geim achieved single-layer graphene through a simple method using common adhesive tape.

8.3.1.1 Mechanical exfoliation

Mechanical exfoliation is one of the simplest and most successful top-down approaches for obtaining high quality single-layer graphene for fundamental research. Looking back into its history, the properties of graphene were introduced

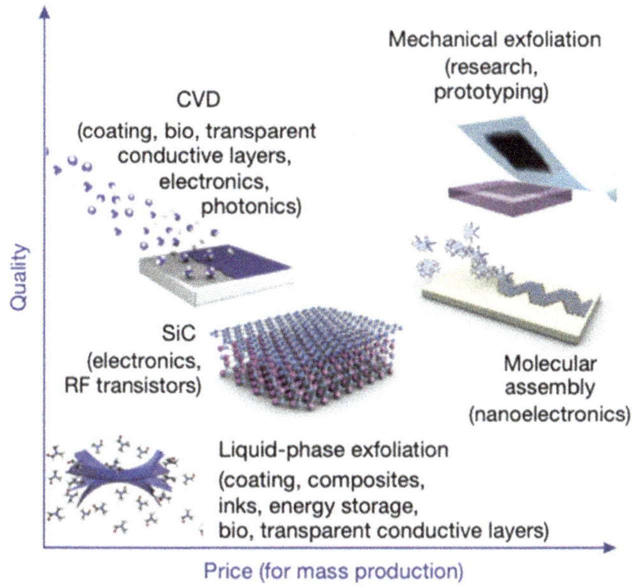

Figure 8.3. There are several methods for the mass production of graphene, which allow a wide choice in terms of size, quality, and price for any particular application. (Reproduced with permission from [26]. Copyright 2012 Nature.)

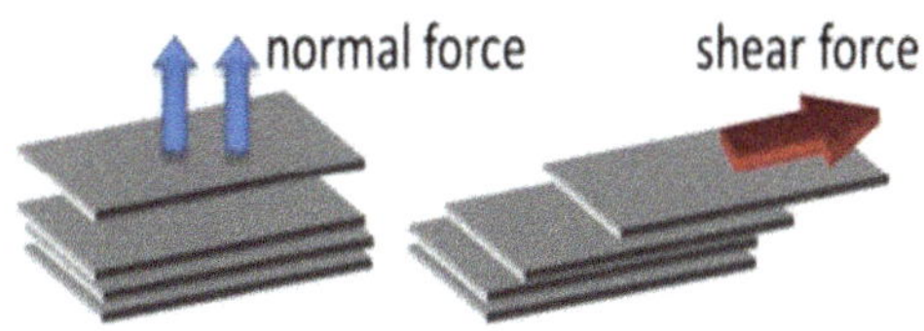

Figure 8.4. Different types of mechanism involved in exfoliating graphite into graphene layers. (Reproduced with permission from [30]. Copyright 2015 The Royal Society of Chemistry.)

by Wallace in 1947, where he identified the linear $E(k)$ dispersion for the electronic structure of graphene around the K point of the Brillouin zone [27]. Later, in the 1960s, Boehm and co-workers synthesized mono- and few-layer graphene successfully through reduction techniques, reported in a highly appreciated and cited article [28]. In 2004 Novoselov and Geim used adhesive tape repeatedly to separate graphite layers into small flakes and then transferred them to a suitable SiO_2 substrate [29]. In general, the graphite layers are attracted by van der Waals attraction force between the layers and this needs to be overcome to make the single-layer graphene [30]. This can be achieved through repeated peeling of graphite layers. In mechanical exfoliation carried out through the 'Scotch tape' route, one can expect a normal force to be exerted on the graphite layers in order to overcome the van der Waals attractive forces (figure 8.4) [30, 31]. The very important drawback of this route is that the normal force applied to the graphite layers to overcome the van der Waals force will be applied in an uncontrollable destructive manner.

A schematic representation of the Scotch tape method, including the exfoliation of graphite into graphene and the thin layer of graphene being transferred onto the SiO_2 substrate, is shown in figure 8.5. However, this particular route has certain limitations for large-scale production and potential applications, because the productivity rate is very low and it is a time consuming process, since the number of peelings required to achieve single-layer graphene is highly unpredictable. Thus it is necessary to find reliable and adequate methods to prepare large-scale graphene materials to realize the potential applications of graphene in various technological fields of interest.

Over the years, several approaches have been identified for the large-scale production of single- and few-layer graphene. In the last few decades, the simple ball milling route has been appreciated for the synthesis of graphene structures due its advantages, such as its non-destructive nature, low cost, high quality product, and scalability. In general, ball milling uses shear force to peel off graphene layers (depicted schematically in figures 8.6(a)–(c)) and sometimes it can reduce the size of the graphite flakes. The advantage of this route is that the shear force is applied to the graphite layers to overcome the van der Waals force in a controllable and non-destructive way. In addition, to provide support for the non-destructive exfoliation of graphene layers, a milling assisting agent such as oxalic acid can be used [33]. Usually, balling milling of graphite can be carried out in two different states, either dry or wet. In the beginning ball milling research was focused on the synthesis of semiconductor nanoparticles and ceramics, and later Knieke modified the route with the mixing of the liquid phase and ball milling [34]. Several solvents, such as sodium borohydride, dimethylformamide, N-methyl-2-pyrrolidone and hydroquinone, have been used to weaken the van der Waals force between the graphite layers by matching the surface energy [35, 36]. In addition, ball milling can be used in eco-friendly sustainable methodologies. Recently, the large-scale production of graphene was reported using a green route through the ball milling of graphite with sucrose as an auxiliary milling agent, with fewer defects [37]. Xing *et al* combined the annealing process with ball milling to produce nitrogen doped carbon nanoparticles at a large scale [38]. In a similar way, several researchers tried various bio-reducers obtained from plants such as *Artemisia vulgaris* [39] and fruit extracts such as from Cannonball tree fruit (*Couroupita guianensis*, Aubl.) [40]. Ayala *et al* reported the preparation of few-layer reduced graphene oxide via ball milling with *Opuntia ficus-*

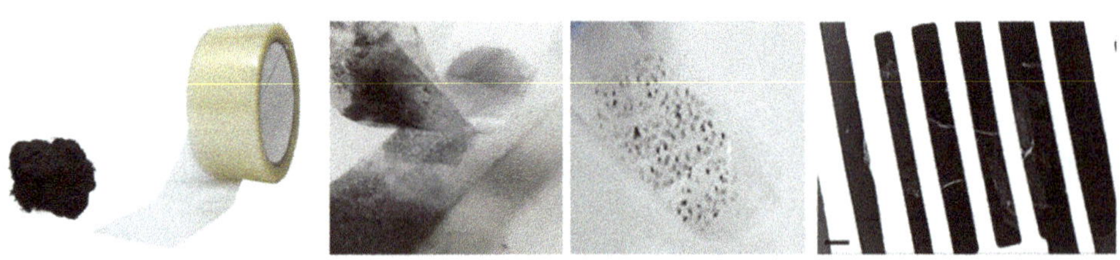

Figure 8.5. (a)–(c) The schematic procedure of single-layer graphene synthesis through the Scotch tape route. (d) Transmission electron microscopy image of a single-layer graphene sheet freely suspended on a micrometer-sized metallic scaffold. (Reproduced with permission from [31, 32]. Copyright 2012 and 2007 Nature.)

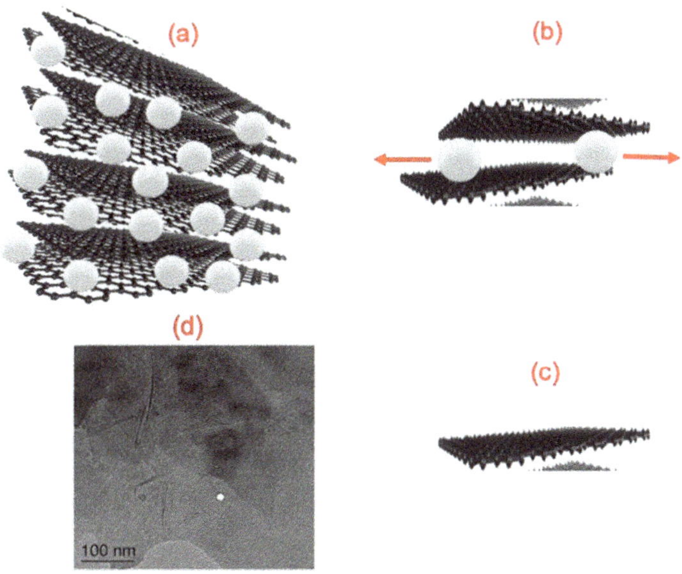

Figure 8.6. (a)–(c) Schematic representation of single-layer graphene sheets by the ball milling route. (d) TEM image of ball milled graphene sheets deposited onto a grid from the supernatant. (Reproduced with permission from [42]. Copyright 2010 The Royal Society of Chemistry.)

indica plant extract [42]. Also, González *et al* explored mechanochemical treatment with various carbohydrates such as glucose, fructose, and saccharose, which was successfully applied to exfoliate few-layer graphene materials. Among these carbohydrates, glucose showed the best behavior in terms of exfoliation to achieve few-layer graphene with a relatively low number of defects [41] (figure 8.7).

8.3.1.2 Thermal exfoliation

Thermal exfoliation is another simple and important environmentally friendly and sustainable route for the exfoliation of graphite into single- and few-layer graphene sheets. The mechanism behind this route is the pressure generated during the heating process due to decomposition of the graphitic layers which produce gases. This pressure exceeds the van der Waals force between the interactive graphitic layers. To evaluate the pressure required to exfoliate the graphene layers, Lifshitz's formulation of van der Waals forces could be applied to calculate the binding energy between the two adjacent layers. The pressure needed to exceeds the van der Waals force is given by [44, 45]

$$P = \frac{\partial G}{\partial l} = \frac{A_{\text{Ham}}}{6\pi l^3},$$

where G is the interaction free energy per unit area between two semi-infinite slabs, A_{Ham} is the Hamaker coefficient, and l is the interlayer distance. McAlister *et al* calculated that parameters of interlayer separation of $l = 0.27$ nm and the Hamaker coefficient could yield pressure of $P = 2.49 \times 10^6$ Pa. Another important parameter

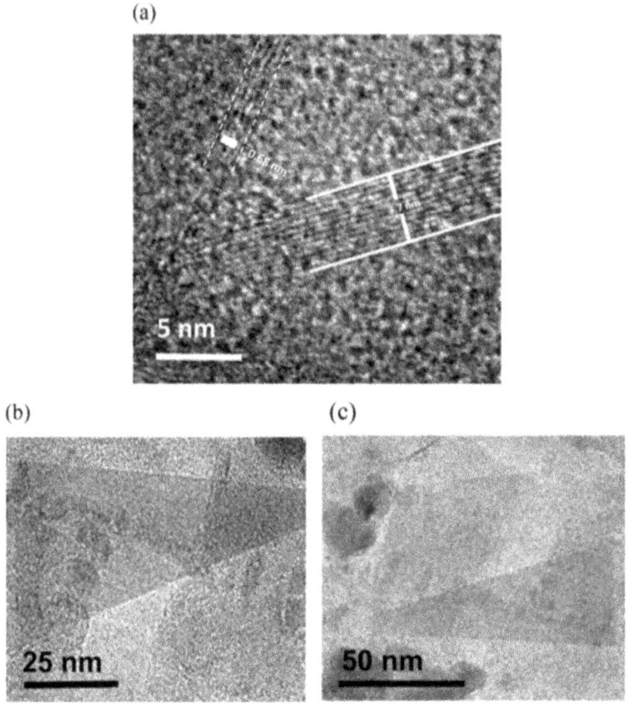

Figure 8.7. Transmission electron microscopy image of some reduced graphene oxide sheets prepared by a ball milling: a green approach. (Reproduced with permission from [41]. Copyright 2013 The Royal Society of Chemistry.)

is the diffusion time scale as a function of temperature since the lateral diffusion can relieve the generated pressure and the expansion will occur slowly. The diffusion can be calculated by [45]

$$\frac{\partial C_i}{\partial t} = D_i \frac{1}{r}\frac{\partial}{\partial r}\left(r\frac{\partial C_i}{\partial r}\right) + R,$$

where C_i is the concentration of species i, t is time, D_i is the diffusivity of species i, r is the radial position, and R is the rate of formation of the gaseous species. The relation between the reaction and diffusion time scales with respect to the temperature is depicted in figure 8.8 and the graph reveals that at a temperature of 550 °C the reaction time scale is shorter than the diffusion time scale. Thus a temperature above 550 °C is needed to exfoliate graphene quickly. Furthermore, samples prepared at a preheated temperature of 1050 °C for 30 s showed that more than 80% of the observed flakes were single-layer sheets and this was further confirmed with statistical analysis.

Thermal exfoliation has many advantages over other routes, such as less processing time with a very fast reaction rate (high temperature thermal exfoliation would produce the graphene layers in few seconds to minutes) and it usually occurs in a gaseous environment, thus reducing the use of hazardous chemicals. It has been

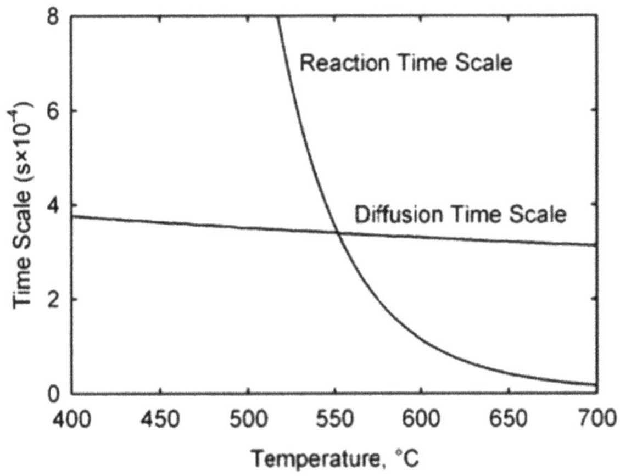

Figure 8.8. The relation between the reaction time scale and diffusion time scale as a function of temperature. (Reproduced with permission from [45]. Copyright 2007 American Chemical Society.)

predicted theoretically [45] that the temperature needed for the complete exfoliation of graphite is above 550 °C, however, experimentally it is observed to be more than 1000 °C. This approach requires higher energy consumption, and the generated carbon oxides and water vapor may react with carbon atoms at the high temperature, which causes more defects (vacancies) [45]. Recently, Ahmeda *et al* [47] reported defect-free exfoliation of graphene at an ultra-high temperature, and their heat-treatment method at an ultra-high temperature provided a physical procedure for defect-free graphene exfoliation which exhibited enhanced capacitance behaviors.

It can be observed from SEM images of the powder samples that the high temperature (2000 °C) heat-treatment process could cause a remarkable morphological expansion of the graphene layers and that the interlayer expansion increases with increasing processing temperature from 90 °C to 900 °C to 2000 °C. Among these temperatures, a high degree of homogeneous expansion is achieved at 2000 °C. SEM images of graphene samples exfoliated at different temperatures are shown in figure 8.9.

In addition to high temperature exfoliation, there are other techniques available for the exfoliation of graphitic layers in a few seconds with a fast reaction rate, without water washing or effluent treatment, such as the microwave heating and arc-discharge routes. Very recently, Wu *et al* [48] reported a two-step microwave assisted process for the fast exfoliation of graphite to few-layer graphene. The first step of the process involved using an ordinary house-hold microwave oven and graphite intercalation compound (GIA) was formed with microwave irradiation of 30 s at 700 W. Then the prepared GIC was mixed with 2 ml deionized water for 1 min and exposed to another microwave irradiation for 80 s to produce graphene aggregates. In the second step, the obtained GAs were completely exfoliated into ~100% few-layer graphene under 100 W low-power ultrasonication for only 5 min (figure 8.10).

The key advantage of this route is the exfoliation yield of graphene is about 100%, including 66% of graphene sheets with $\leqslant$ 5 layers with the average area of the

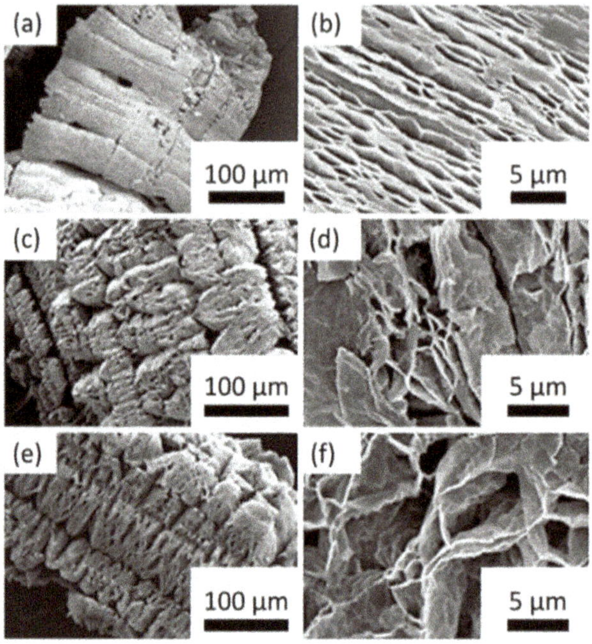

Figure 8.9. SEM images of graphene exfoliated at (a) and (b) 90 °C, (c) and (d) 900 °C, and (e) and (f) 2000 °C. (Reproduced with permission from [47]. Copyright 2018 Elsevier.)

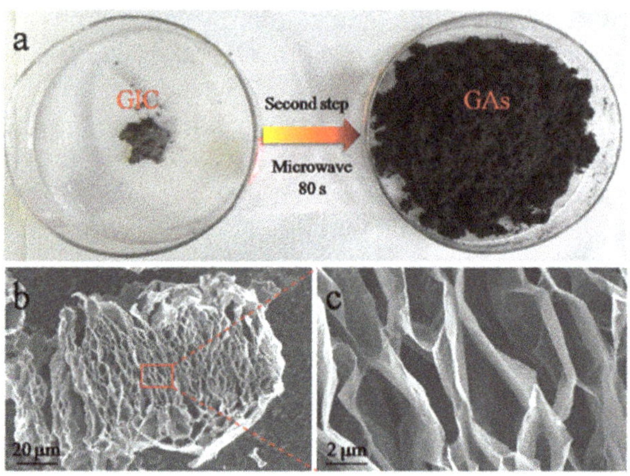

Figure 8.10. (a) Photograph showing the evolution process from graphite intercalation compound (GIA) (left) to graphene aggregate GAs (right) powders during the second microwave irradiation step. (b) and (c) SEM images of GAs. (Reproduced with permission from [48]. Copyright 2020 Elsevier.)

graphene sheets being up to 256.8 μm^2. The produced graphene film had a conductivity up to 2.03×10^5 S m^{-1} which is highest value for graphene reported so far. In addition to these advantages, exfoliated graphene sheets do not require water washing or effluent treatment during any part of the preparation process, which is normally done with KOH for chemical activation [49]. The monolayer to few-layer graphene sheets were identified through high resolution transmission electron microscopy (HR-TEM) and atomic force microscopy (AFM) studies, as depicted in figure 8.11.

In contrast to the advantages of high temperature thermal exfoliation, it requires sufficiently high energy consumption and creates more defects (vacancies) due to the fact that the generated carbon oxides and water vapor may react with the carbon atoms at high temperature. In addition, the fast heating rates may also cause more topological defects due to the rapid decomposition of functional groups, such as more ripples [50]. Therefore, it is necessary to perform thermal exfoliation under mild temperatures the industrial applications. Lv *et al* [51] proposed a vacuum-assisted low-temperature exfoliation of graphene for mass production with exfoliation temperatures of less than 200 °C, which is a breakthrough for low-cost and sustainable energy consumption for the mass production of graphene. The mechanism behind the vacuum-assisted thermal exfoliation is the strong pulling force from the vacuum environment, whereas conventional thermal exfoliation is based on the pressure generated from the decomposition of functional groups which is caused by rapid heating to a very high temperature. Similarly, Zhang *et al* [52] prepared few-layer graphene sheets with an average thickness of 0.9 nm through vacuum-assisted thermal exfoliation and *in situ* reduction of graphite oxide in a large quantity at a temperature as low as 135 °C. Also, the prepared sheets exhibited a high specific surface area (758 $m^2\ g^{-1}$) which is comparable to that of conventional graphene generated at 1050 °C at atmospheric pressure (700 $m^2\ g^{-1}$).

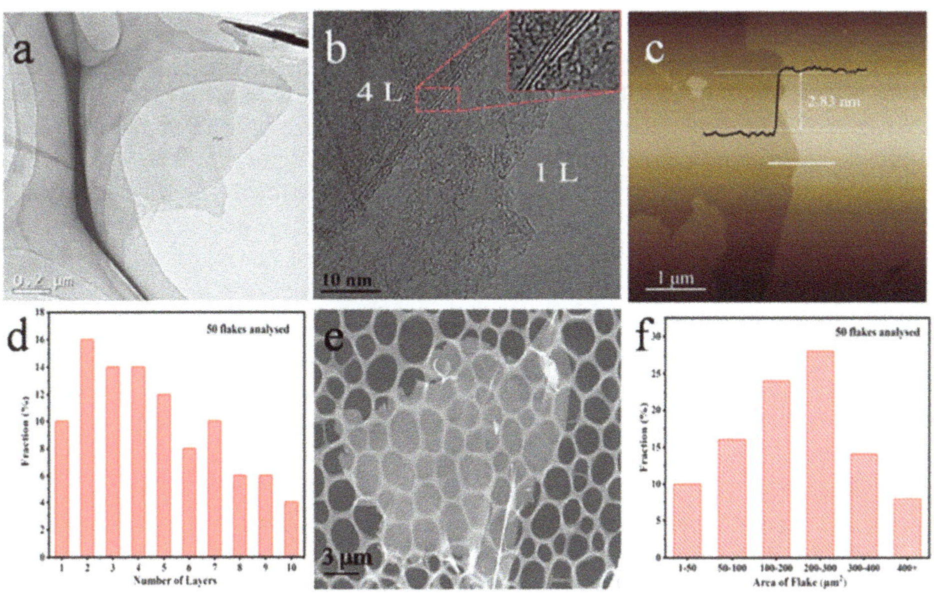

Figure 8.11. Characterization of the obtained graphene. (a) Typical TEM image. (b) HR-TEM image. (c) Typical AFM image. (d) Layer distribution analysed by HR-TEM statistics. (e) Typical SEM image of graphene dispersed on the micro-grid. (f) Areal distribution of graphene analysed from 50 flakes. (Reproduced with permission from [48]. Copyright 2020 Elsevier.)

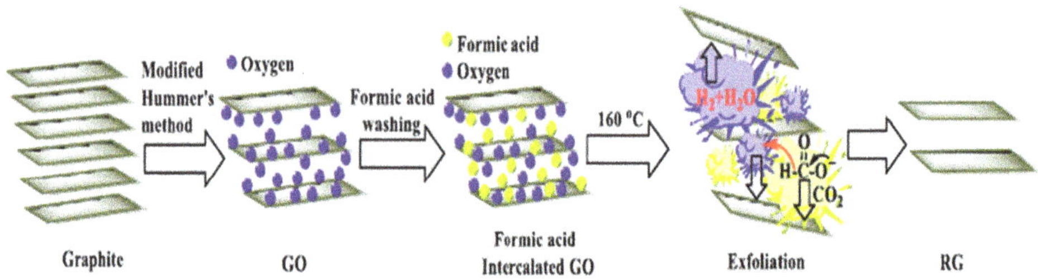

Figure 8.12. Schematic representation of formic acid assisted low-temperature exfoliation of GO. (Reproduced with permission from [55]. Copyright 2015 The Royal Society of Chemistry.)

Another way to realize low-temperature exfoliation is H_2-induced exfoliation [53], HCl assisted exfoliation [54], and formic acid assisted exfoliation under ambient atmosphere. Singh *et al* [55] prepared few-layer reduced graphene at a low temperature (160 °C) through exfoliation of graphite oxide under ambient atmospheric conditions with the assistance of formic acid in a short time (14 min). The prepared reduced graphene showed very high bulk electrical conductivity (1.6×10^3 S cm^{-1}) at room temperature due to restoration of the extended conjugation of the sp^2 network during rapid exfoliation. The external chemical assistance generates enough pressure from evolved gases that it can overcome the van der Waals forces holding the graphite layers (a schematic representation is provided in figure 8.12).

8.3.2 Bottom-up approaches

Although the top-down approaches have several advantages, the production of large-scale and cost-effective high quality graphene to meet demand in the near future is still a challenging task. Over the years many researchers across the globe have explored several techniques, including chemical vapor deposition (CVD), epitaxial growth, liquid exfoliation, chemical reduction of graphene oxide, and organic synthesis. The CVD route showed the best performance among them in terms of resolving the critical issues such as controlling grain size and layer numbers, and other issues that existed in other techniques. One of the major components in CVD is the substrate, since only it is exposed to the precursor materials and these react/decompose on the substrate surface to grow graphene. To date, Cu, Ni and Pt have been widely used substrate materials. Another advantage of CVD is that it is compatible with gas, liquid, and solid precursors [56]. A schematic illustration of typical CVD equipment is provided in figures 8.13(a)–(c), representing the basic steps involved in the growth of graphene using the CVD technique. The growth of graphene can be classified into two different routes. Route 1 contains a carbon-dissolvable substrate and follows four basic steps: (i) adsorption and decomposition of precursors; (ii) diffusion and dissolution of carbon species; (iii) segregation of dissolved carbon species; and (iv) nucleation and growth of graphene. Route 2 contains a carbon non-dissolvable substrate and involves only two basic steps: (i) adsorption and decomposition of precursors; and (ii) nucleation and growth of graphene [57].

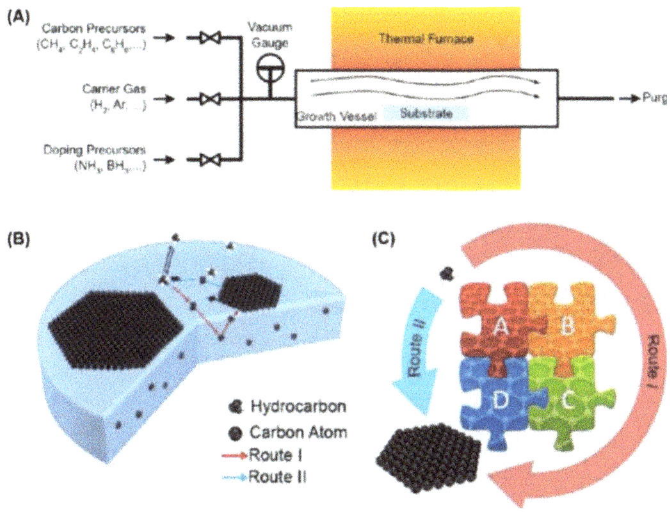

Figure 8.13. (a) Schematic illustration of a typical CVD system for the fabrication of graphene. (b) Elementary steps involved in the CVD process. The red arrows represent the case for a carbon-dissolvable substrate (route 1), and the blue arrows demonstrate the case for a carbon non-dissolvable substrate (route 2). (c) Schematic illustration of process engineering. Four elementary steps are connected for the synthesis of graphene. Two distinct routes exist from hydrocarbon precursors to graphene. (Reproduced with permission from [57]. Copyright 2013 American Chemical Society.)

Li *et al* introduced the CVD growth of the first large-area (centimeters) reproducible high quality and uniform monolayer graphene films on copper foils. The obtained films are predominantly single-layer graphene, with a small percentage (less than 5%) of the area having a few layers, and are continuous across copper surface steps and grain boundaries [58]. A successful demonstration of the fabrication of continuous films with monolayer to 12 layers of graphene was grown in ambient pressure using CVD on polycrystalline Ni and was transferred to a large variety of substrates. The films exhibited a large fraction of single- and bilayer graphene regions up to ~20 μm in lateral size. This technique avoids the use of bulk single crystal substrates or complex processing conditions for the scalable production of graphene for technological applications [59].

A high quality monolayer graphene film was grown on a roll of copper foil with dimensions of 30 inches in the diagonal direction using a CVD system (figure 8.14) and the graphene films fabricated on to a flexible Cu substrate sheet had resistances as low as 125 Ω sq^{-1} with an optical transmittance of 97.4%. The combination of roll-to-roll and CVD methods allows the commercial production of large-scale graphene flexible transparent electrodes which could replace the existing use of indium doped tin oxide (ITO) in the near future [60].

The homogeneity and layer distribution of graphene films depends highly on various parameters, such as the growth pressure, temperature, rate of cooling, and concentration of carbon sources. Cooling leads to the segregation of carbon species which will precipitate on the surface yielding the inhomogeneous graphene growth. An alternative strategy has been demonstrated through the utilization of a Mo

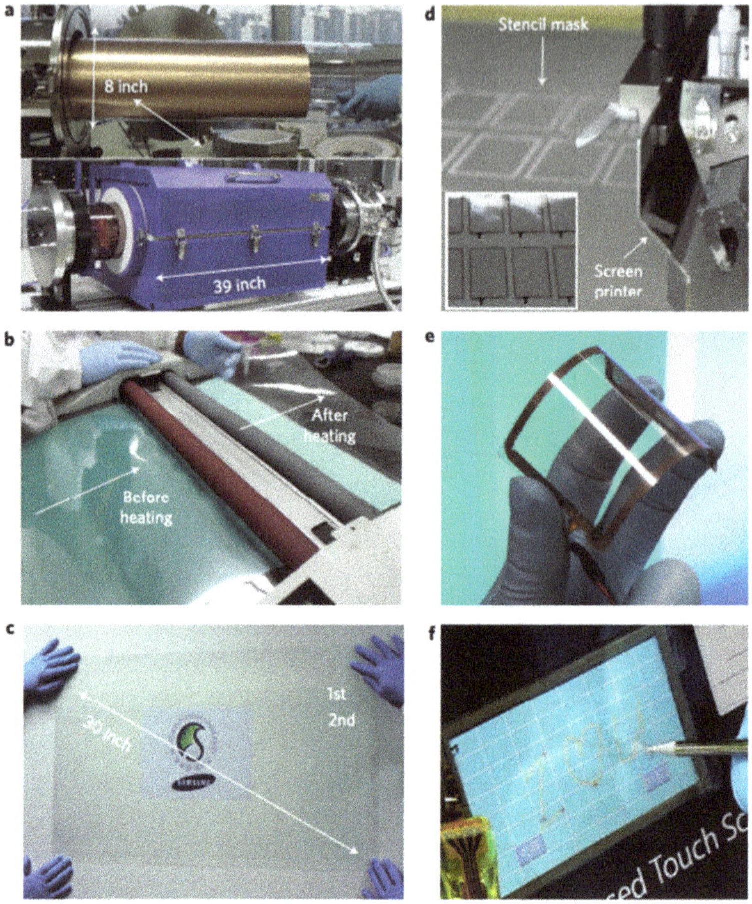

Figure 8.14. Photographs of the roll-based production of graphene films. (a) Copper foil wrapping around a 7.5 inch quartz tube to be inserted into an 8 inch quartz reactor. The lower image shows the stage in which the copper foil reacts with the CH_4 and H_2 gases at high temperatures. (b) Roll-to-roll transfer of graphene films from a thermal release tape to a PET film at 120 °C. (c) A transparent ultralarge-area graphene film transferred onto a 35 inch PET sheet. (d) Screen printing process of silver paste electrodes onto graphene–PET film. The inset shows the 3.1 inch graphene–PET panels patterned with silver electrodes before assembly. (e) An assembled graphene–PET touch panel showing outstanding flexibility. (f) A graphene based touch-screen panel connected to a computer with control software. (Reproduced with permission from [59]. Copyright 2009 American Chemical Society.)

component in a binary alloy for trapping all of the excess dissolve carbon species in the form of molybdenum carbids, which ultimately reduce the carbon precipitation process (figure 8.15). The formed molybdenum carbids are chemically stable during high temperature growth. In addition, the growth of uniform monolayer graphene with 100% surface coverage was achieved successfully using Ni–Mo alloy [61].

Although monolayer graphene has several advantages for the electronic industries, it is difficult to utilize monolayer graphene in transistor device applications due to its lack of bandgap. The alternative way is to prepare bilayer graphene, specifically

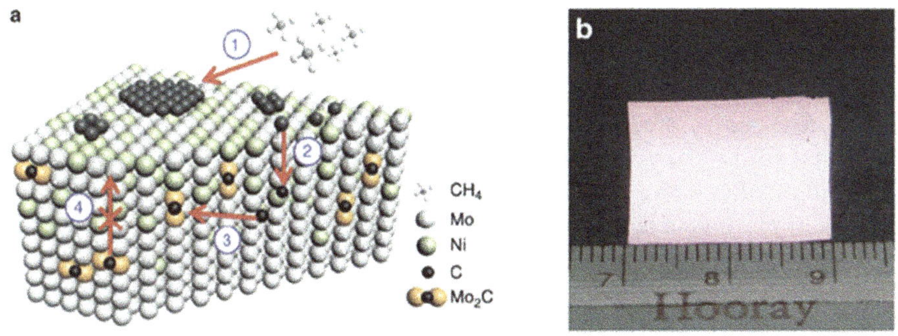

Figure 8.15. (a) Schematic drawing of a designed binary alloy for CVD growth of single-layer graphene by suppressing carbon precipitation, which was demonstrated by an Ni–Mo model system. (b) Photograph of a 2 × 2 cm single-layer graphene grown on a Ni (200 nm)–Mo (25 μm) foil. (Reproduced with permission from [61]. Copyright 2011 The Royal Society of Chemistry.)

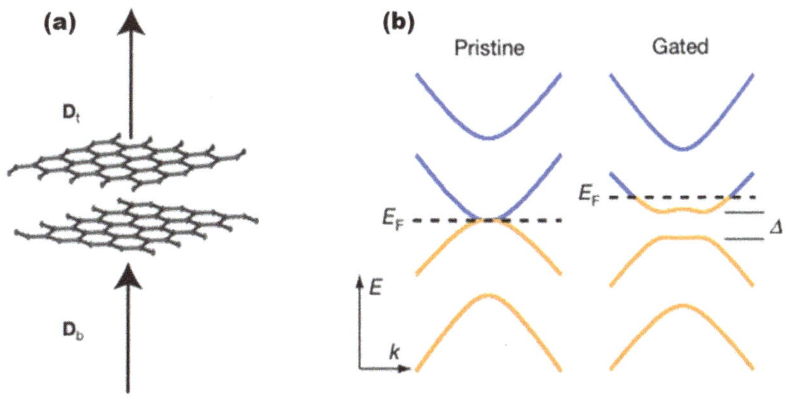

Figure 8.16. (a) Sketch showing how gating of the bilayer induces top (D_t) and bottom (D_b) electrical displacement fields. (b) Left, the electronic structure of a pristine bilayer has zero bandgap (k denotes the wavevector). Right, upon gating, the displacement field induces a non-zero bandgap and a shift of the Fermi energy E_F. (Reproduced with permission from [62]. Copyright 2009 Nature.)

AB-stacked bilayer graphene, which shows zero band and can have a continuously tunable bandgap upon applying a gated electric field (figure 8.16). The zero bandgap has the lowest conduction band edge (D_t) and highest valence band edge (D_b) which touch each other without an gate electric field and are displaced upon electrical gating. There is a shift of Fermi energy (E_F) due to carrier doping because of the difference in electrical displacement fields D_t and D_b, where, D_t and D_b are top and bottom of electric field displacement, respectively. The non-zero bandgap could be generated due to the breaking of the inversion symmetry of the bilayer [62].

Li *et al* demonstrated that the self-limiting effect of graphene growth on Cu foil can be broken by using a high H_2/CH_4 ratio in a low-pressure CVD process to enable the continued growth of bilayer graphene, and a rational design of a two-step CVD process was developed for the growth of bilayer graphene with a high AB stacking ratio (up to 90%) and high coverage (up to 99%) [63]. Similarly, synchronous growth of AB-stacked bilayer graphene on Cu was achieved by simply controlling hydrogen pressure in the CVD process and produced bilayer graphene with a high coverage of over 95% and a high AB stacking ratio of up to 90%. The bilayer structure was obtained within a short growth time of 30 min [64]. The thermal CVD growth techniques require multiple steps and a high temperature, which makes them relatively time consuming and expensive. To overcome these issues, plasma enhanced chemical vapor deposition (PECVD) techniques have been explored and showed the synthesis of high quality monolayer graphene without the need for high temperatures. Chen *et al* utilized a direct PECVD method to achieve efficient growth of single crystalline bilayer graphene (BLG) hexagons and large BLG films (millimeter size) with small twist angles through precise control of the CH_4/H_2 pressure ratio [65]. The resulting gas and substrate temperatures under continuous plasma exposure were kept at ~425 °C and ~400 °C for the growth of BLG and multilayer graphene (MLG), respectively. However, the temperature used for the fabrication process remains high and to realize its full potential for industrial applications the temperature needs to be reduced, and techniques for transfer-free growth on desired substrates are needed. An efficient alternative route is microwave plasma chemical vapor deposition (MPCVD). Vishwakarma and co-workers also demonstrated the growth of large-area graphene directly on 2 × 2 cm^2 insulating substrates (quartz and glass) through magnetron-generated microwave plasma chemical vapor deposition at a substrate temperature of 300 °C. The fabricated graphene films exhibited 97.5 $cm^2\ V^{-1}\ s^{-1}$, $5.09 \times 10^{13}\ cm^{-2}$, and 1259 $\Omega\ sq^{-1}$ for carrier mobility, sheet carrier concentration, and sheet resistance, respectively [66].

Graphene nanoribbons (GNRs) are another class of materials exhibiting a non-zero bandgap [67] and the bandgap can be tuned simply by adjusting their width and edge structures, which makes this material more suitable for future nanoelectronics industries [68]. Over the years, several techniques, such as lithographic patterning and unzipping of carbon nanotubes [69], have been explored. Unfortunately these routes showed inherent structural disorder at the edges, inhomogeneity in the width, and minimum of 10 nm has been achieved. Theoretical calculations revealed that arrow graphene nanoribbon widths of <2 nm have a bandgap comparable to that of silicon (1.1 eV) which makes the materials suitable for technological applications.

Bottom-up approaches have been explored to overcome the issues of top-down approaches. Large-scale GNRs with dimensions of 1 nm width and >100 nm length have been synthesized using a novel solution synthesis route [70]. The produced GNRs exhibited a large electronic bandgap of 1.3 eV, which is significantly higher than any value prepared by top-down approaches (figure 8.17). The polymerization of pre-synthesized molecular precursors by a Ni^0-mediated Yamamoto coupling followed by cyclodehydrogenation via a Scholl reaction using iron (III) chloride was used to form GNRs. Similarly, extended GNRs with lateral dimensions of ~2 nm and longitudinal dimensions of >100 nm were achieved based on the AB-type Diels–Alder polymerization process and exhibited a low optical bandgap of approximately 1.24 eV [71].

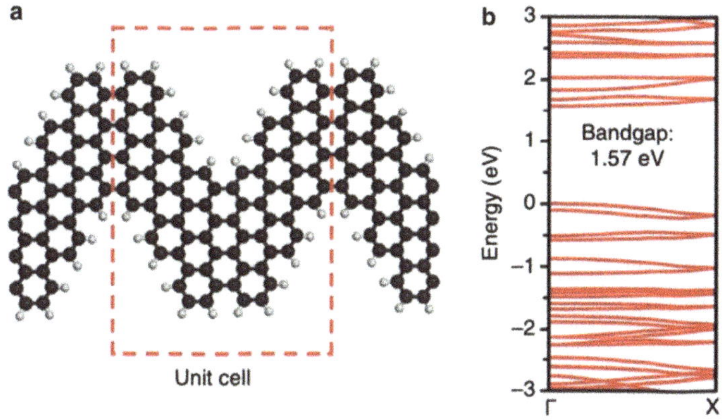

Figure 8.17. (a) Schematic of GNRs synthesized through a bottom-up approach and (b) the corresponding calculated band structure. (Reproduced with permission from [70]. Copyright 2014 Nature.)

8.4 Photovoltaic devices

Over the decades, the rapid growth of population, urbanization, and industrialization has caused energy crises around the world. To date, worldwide energy consumption is around 12 TW and fossil fuels are a major source of energy production. Moreover, the International Energy Agency (IEA) reports that in developing countries around the world nearly 770 million people are living without electricity [72].

The demand for energy is expected to increase up to 24 TW by 2050 [73] and fossil fuels are inadequate to meet the requirements since these resources are non-renewable and very limited. In addition, these kinds of resources produce severe environmental hazards such as global warming, which ultimately causes climate change. In order to reduce the risks of climate change, it is necessary to adopt renewable energy sources, such as wind, tidal, geothermal, biomass, solar, etc, which are purely eco-friendly and clean energy sources. Among the available renewable energy sources, solar is the most abundant (figure 8.18) and renewable environmentally friendly energy source. Using solar thermal energy provides savings of over 118 million tons of CO_2 annually.

8.4.1 Fundamental processes and important parameters

A photovoltaic (PV) cell is a device which generates power through harvesting light energy from the Sun. Any PV cell comprises three main components, the photoactive layer, anode, and cathode, and each component has its own characteristics and will determine the efficiency of a solar cell (figure 8.19).

In any solar cell, the characteristics of the cell can be extracted from four important parameters: (i) short-circuit current, (ii) open-circuit voltage, (iii) fill factor, and (iv) photoconversion efficiency

Short-circuit current. The short-circuit current (I_{sc}) is the current through the solar cell when the voltage across the solar cell is zero (i.e. when the solar cell is short

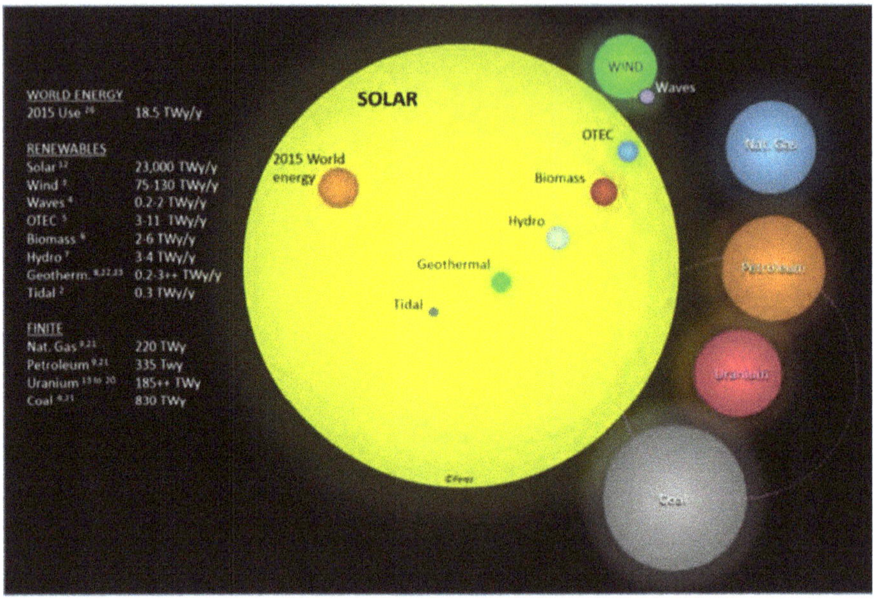

Figure 8.18. 2015 estimated finite and renewable planetary energy reserves (terawatt-years). Total recoverable reserves are shown for the finite resources. Yearly potential is shown for the renewables. (Reproduced with permission from [213]. Copyright 2013 The Royal Society of Chemistry.)

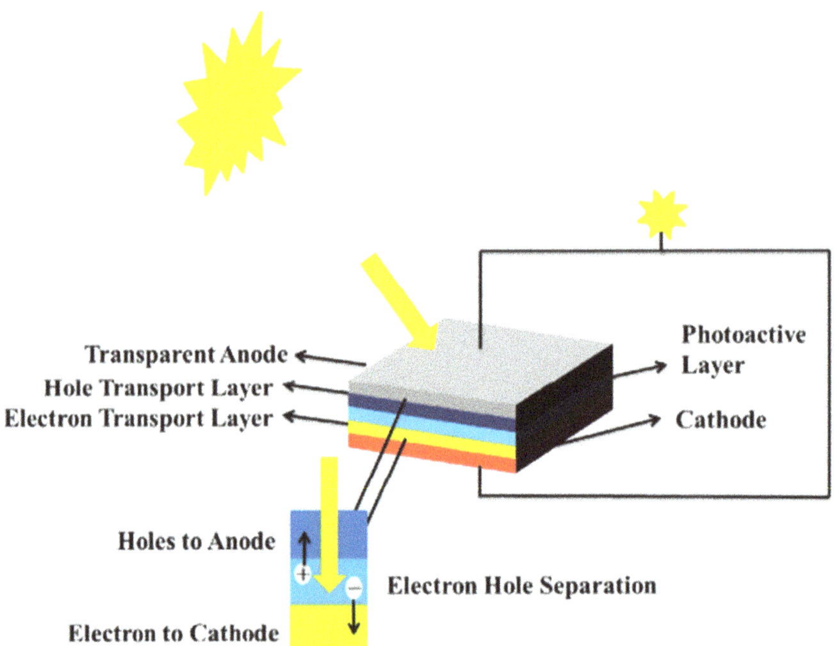

Figure 8.19. Schematic illustration of a basic photovoltaic cell. (Reproduced with permission from [75]. Copyright 2018 Elsevier.)

circuited). It is also referred to as the measure of the total current produced from the cells and it can be deduced through the relation [76]

$$I_{sc} = I_{ph} - I_0\left(e^{\frac{I_{sc}R_s}{V_T}} - 1\right) - \frac{I_{sc}R_s}{R_{sh}},$$

where I_{ph} is the photogenerated current, and R_s and R_{sh} are the series resistance and shunt resistance of the solar cell, respectively. I_0 is the reverse saturation current and it can be calculated by

$$I_0 = AA^*T^2e^{-\frac{\phi_b}{kT}},$$

where A is the area of the graphene/silicon junction, A^* is the Richardson constant ($\approx$ 112 A cm^{-2} K^{-2} for n-Si), T is the absolute temperature, φ_b is the Schottky barrier height, and k is the Boltzmann constant.

Open-circuit voltage. The open-circuit voltage (V_{oc}) is the maximum voltage drawn from a solar cell, when there is no current in the circuit. The open-circuit voltage corresponds to the amount of forward bias on the solar cell due to the bias of the solar cell junction with the light generated current and can calculated through the relation [77]

$$V_{oc} = \eta V_T \ln\left(1 + \frac{I_{ph}}{I_0}\right).$$

Here η is the diode ideality factor and V_T is the thermal voltage.

Fill factor. The fill factor (FF) is the parameter of conjunction of V_{oc} and I_{sc}, which determines the deviation of the actual power from the ideal cell and the measurement of maximum power from a solar cell. The FF is defined as the ratio of the maximum power extracted from the solar cell and V_{mp} and I_{mp} are the voltage and current at the maximum power point, respectively. The FF can be calculated through relation [78]

$$FF = \frac{P_{max}}{V_{oc}I_{sc}} = \frac{V_{mp}I_{mp}}{V_{oc}I_{sc}},$$

where P_{max} is the maximum power obtained from the cell, and V_{mp} and I_{mp} are the voltage and current at the maximum power point, respectively.

Solar cell efficiency. The solar cell efficiency (also known as the power conversion efficiency (PCE)) represents the ratio of maximum power obtained from the solar cell over the incident power of solar light:

$$PCE = \frac{P_{max}}{P_{inc}} = \frac{V_{mp}I_{mp}}{P_{inc}} = \frac{V_{oc}I_{sc}FF}{P_{inc}}.$$

8.4.2 Types of solar cells

Figure 8.20 presents the overall power conversion efficiencies for different type of solar cells. Crystalline silicon is the most mature technology and has reached record

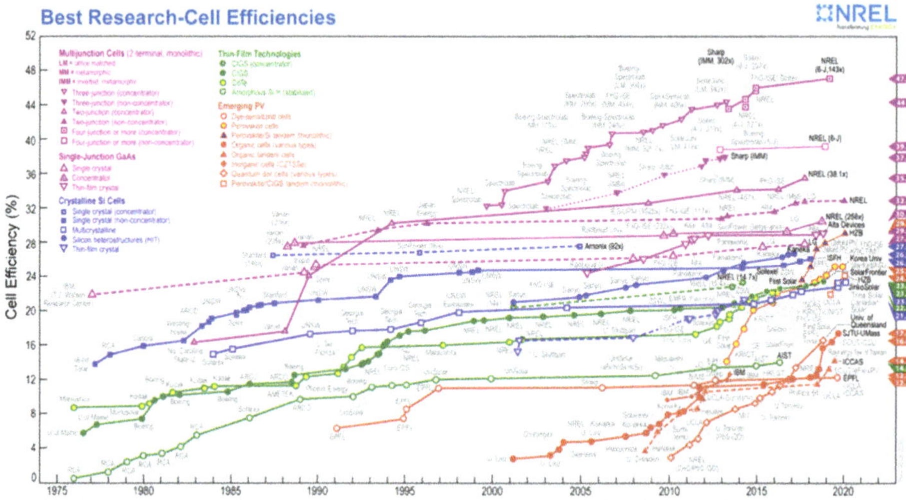

Figure 8.20. Conversion efficiencies of the best researched solar cells worldwide for various photovoltaic technologies since 1976, by the National Renewable Energy Laboratory (NREL). (Reproduced with permission from [79]. Copyright NREL.)

efficiencies of ~27.6%, whereas the highest efficiency of up to 47.1% has been achieved by using multijunction solar cells, and single junction based GaAs solar cells have achieved to 29.1%. The thin film based third-generation solar cells using $Cu(In,Ga)Se_2$ have reached a power conversion efficiency of up to 20.4%. The emerging photovoltaic solar cells have achieved the least efficiency. For example, 14.1% and 7.0% efficiency were reached by dye-sensitized solar cells and quantum dot sensitized solar cells, respectively. From the efficiency chart in figure 8.20, it can be concluded that the quantum dot sensitized solar cells achieved lower efficiency, but at the same they have great potential to improve their efficiency considerably due to their extraordinary properties compared to other-generation solar cells. In addition, they can provide cost effectiveness and environmental friendliness.

8.4.3 Graphene based photovoltaic cells

Since the discovery of graphene in 2004, research focused on graphene based materials has accelerated tremendously toward many technological fields due to its excellent optical, optoelectronic, and mechanical properties. More specifically, graphene and its derivatives have been investigated extensively in the field of photovoltaic cells because of their attractive properties, such as high optical transparency and highly electrical conduction [80]. Graphene has an optical transmission rate of over 98% in the visible bandwidth region which is an improvement on its closest competitor ITO, which makes graphene a highly viable option in optoelectronic devices [81].

Graphene and its derivatives can be incorporated in to various photovoltaic devices. Graphene based materials could be used as transparent cathodes and anodes

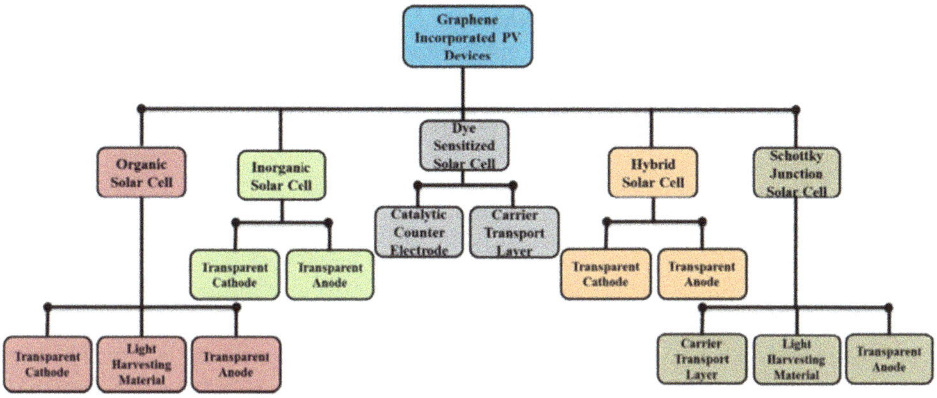

Figure 8.21. Schematic illustration of different photovoltaic device by using graphene as a source materials. (Reproduce with permission from reference [75]. Copyright 2018 Elsevier.)

for hybrid, organic, and inorganic solar cells. Similarly, it can be utilized as a light harvesting materials in organic and Schottky junction solar cells. Graphene can also be used as the electron acceptor or the hole transport layer in photovoltaic devices [83]. Furthermore, it can be used as catalytic counter electrodes and carrier transport layers in dye-sensitized solar cells (DSSCs). Figure 8.21 summarizes the reported works on using graphene as a source material in various photovoltaic systems.

8.4.3.1 Graphene as transparent conductive electrodes (TCEs)

Solar cell efficiency depends strongly on three parameters: (i) absorption, i.e. the number of photons absorbed; (ii) dissociation efficiency, which determines the fraction of electrons and holes that are dissociated; and (iii) collection efficiency, which determines the fraction of dissociated charges that reach the electrodes. Identifying the proper network structure for TCEs is essential for improving solar cell efficiency since it determines the absorption as well as collection efficiencies [84]. To date, a widely used TCE material has been ITO for various application such as touch screens, flexible devices, and photovoltaic devices, since it has optical transparencies of >90% at a wavelength of 550 nm and low sheet resistances of 10–30 $\Omega\ sq^{-1}$. Although it has several advantages, there are some inherent disadvantages to ITO, such as a high fabrication cost because it requires sputtering, evaporation, pulsed laser deposition, and electroplating techniques. Also, it is brittle when there is applied stress, and and there is also a scarcity of indium compounds [85]. These limitations force us to explore new alternative materials to meet the demand of the near future. One group of such alternative materials is carbonaceous materials. For example, in addition to its extraordinary physical and chemical properties, single-layer graphene is zero bandgap semiconductor which has low sheet resistance and could absorb 2.3% of white light, leading to the better alternative materials for transparent conducting electrodes for photovoltaic cells [86]. The optical transmittance of free-standing graphene can be obtained through the Fresnel equations for a thin film [87],

$$T = (1 + 0.5n\alpha) - 2 \approx 1 - n\alpha,\ \text{where}\alpha = e^2/(4\pi\varepsilon_0 hc) = G_0/(\pi\varepsilon_0 c) \approx 1/137.$$

Substituting all values yields that single-layer graphene should transmit 97.7% of the incident light and have 2.3% visible light absorption.

The transmittance (T), sheet resistance (R_S), visible absorbance coefficient (α), and electrical conductivity (σ) of the material can be related through the relation [88]

$$\frac{\sigma}{a} = \frac{188.5}{R_s(T^{-1/2} - 1)}.$$

In addition, the transparency of graphene films also depends on the number of layers. Similarly, sheet resistance (R_S) also depends on the number of layers in accordance with the equation $R_{sn} = R_{s1}/n$, where n is the number of layers. The sheet resistances of single-layer graphene films are 2.1 kΩ sq^{-1}, whereas four-layer graphene has a value of 350 Ω sq^{-1}, while still maintaining a relatively high transmittance of about 90% [89]. The relationship between number of layers, transparency, and sheet resistance is shown in figure 8.22.

Although graphene possesses several advantages, the major problem associated with this material is its large-scale fabrication since it has high sheet resistance, which is much higher than the predicted theoretical value for TCE materials. Kim *et al* [90] fabricated large-scale graphene films through CVD on thin nickel layers and the transferred graphene films showed very low sheet resistance of 280 Ω sq^{-1}, with an optical transparency of 80%. The techniques involved for in the growth, etching, and transferring processes of large-scale graphene films are depicted in figure 8.23.

The prepared graphene films exhibited less than a few layers of graphene. Furthermore, it was found that there was an increase in sheet resistance upon decreasing the number of graphene layers as a function of ultraviolet/ozone treatment time (figure 8.24).

In a similar way, a solution-based vacuum filtration method was adopted to fabricate single monolayers to several layers of reduced graphene oxide over large areas (figure 8.25). This controlled method showed nearly 100% transfer of reduced graphene oxide thin films independent of the substrate. This is due to van der Waals interactions giving rise to sufficiently strong cohesive forces within the film [92].

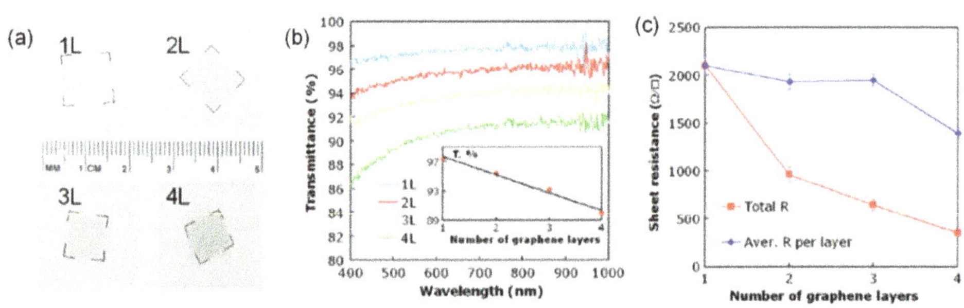

Figure 8.22. Photographs of 1 cm^2 films with from one up to four layers of stacked graphene films on cover glass slips. (b) Transmittance of n-layer graphene films shown in (a). The inset shows the relationship between the transmittance, T (%), at λ = 550 nm as a function of the number of stacked graphene layers. (c) Sheet resistance of n-layer graphene films as a function of the number of stacked graphene layers. (Reproduced with permission from [59]. Copyright 2009 American Chemical Society.)

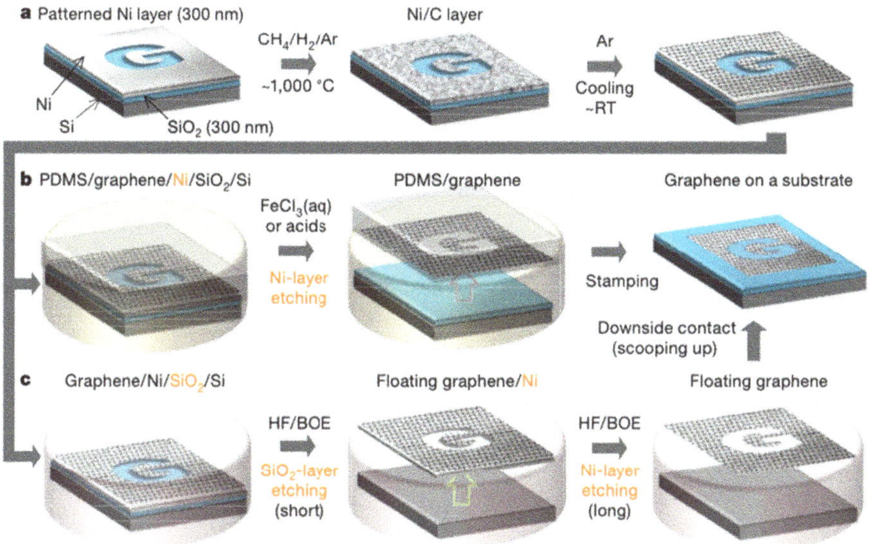

Figure 8.23. (a) Synthesis of patterned graphene films on a thin nickel layer. (b) Etching using $FeCl_3$ (or acids) and transfer of graphene films using a polydimethylsiloxane stamp. (c) Etching using buffered oxide etchant or hydrogen fluoride (HF) solution and transfer of graphene films. (Reproduced with permission from [89]. Copyright 2009 Nature.)

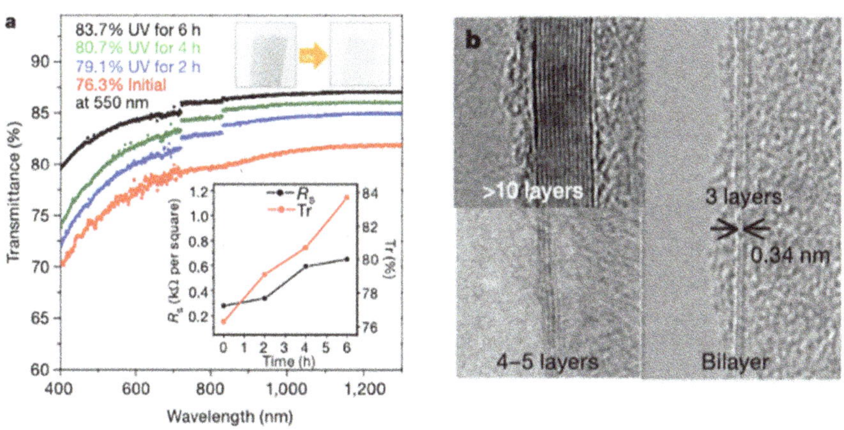

Figure 8.24. (a) Transmittance of graphene films on a quartz plate. The upper inset shows the ultraviolet (UV)-induced thinning and the consequent enhancement of transparency. The lower inset shows the changes in transmittance and sheet resistance as functions of ultraviolet illumination time. (b) TEM images of graphene films of different thicknesses. (Reproduced with permission from [89]. Copyright 2009 Nature.)

Recently, the shear force of slot-die printing during the roll-to-roll process has been used to fabricate a thin layer of reduced graphene oxide (RGO) in poly(3,4-ethylenedioxythiophene)–polystyrene sulfonate–RGO composite and the fabricated flexible transparent conducting electrodes (FTEs) exhibited a low sheet resistance of

Figure 8.25. Solution-processed GO thin films. Photographs of GO thin films on (a) filtration membrane, (b) glass, and (c) plastic substrates. (Reproduced with permission from [91]. Copyright 2008 Nature.)

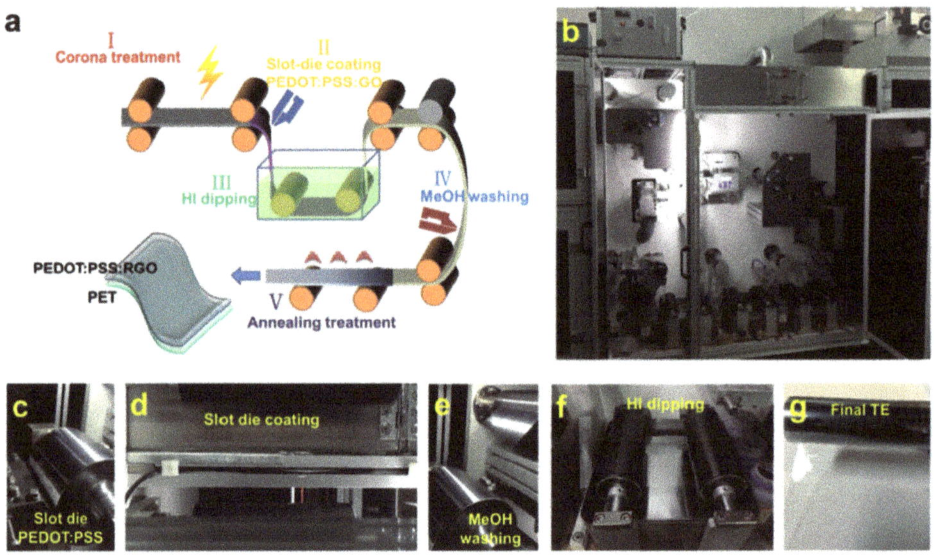

Figure 8.26. Fabrication of PEDOT:PSS–RGO transparent electrodes. (a) A schematic of the five essential steps in the roll-to-roll processes for PEDOT:PSS–RGO films. (b) A photograph of the roll-to-roll equipment. (c)–(e) Photographs of the slot-die coating process. (f) A photograph of the HI dipping process. (g) Large-area flexible PEDOT:PSS–RGO transparent electrodes. (Reproduced with permission from [92]. Copyright 2017 Wiley.)

51 Ω sq^{-1} and 82% transmittance (figure 8.26). Moreover, the prepared FTEs showed highly flexural strength even under an extreme bending situation. An added advantage of this route is that the total cost is only around \$2.8 [93]. It is reported that the transmittance of the film could be controlled by the coating speed which controls the film thickness. The transmission changes from 87% to 69% as the PEDOT:PSS–RGO film thickness increases from 50 to 170 nm at a wavelength 550 nm. The transparency and sheet resistance of graphene based TCEs are listed in table 8.1.

Structural changes have been observed during slot-die printing, which is depicted in figure 8.27(a). The GO can be uniformly distributed and oriented in the PEDOT: PSS matrix upon the impact of shear force and oriental structure, and treatment with HI can improve the π–π effect between the PEDOT chains and RGO which

Table 8.1. Sheet resistance and transparency of graphene based TCEs. (Reproduced with permission from [125]. Copyright 2014 Elsevier.)

TCE	Sheet resistance	Transparency %	Reference
Reduced graphene oxide	70 kΩ sq^{-1}	65	[93]
Graphene films	5 kΩ sq^{-1} to 1 MΩ sq^{-1}	>80	[94]
Graphene	450 Ω sq^{-1}	82	[95]
Single-layer graphite oxide	1.9×10^7 Ω sq^{-1}	95.4	[96]
Multilayer graphene	119 Ω sq^{-1}	92.1	[97]
Graphene/polyethylene terephthalate (PET)	3 Ω sq^{-1}	80	[98]

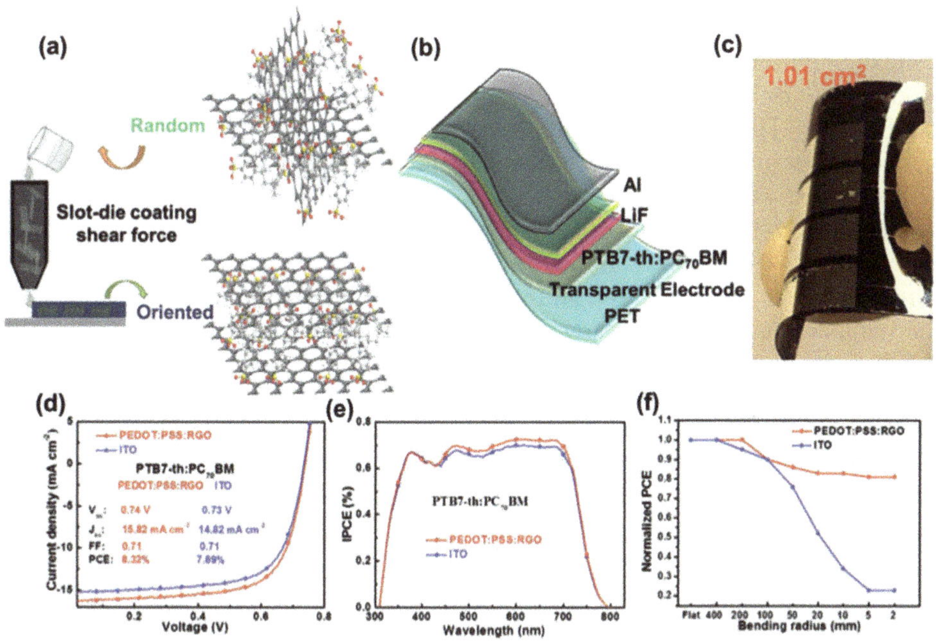

Figure 8.27. (a) Schematic diagram of the structure change of the PEDOT:PSS composite after slot-die coating. (b) Device architecture of the flexible polymer solar cell (PSCs). (c) Photograph of the flexible PSCs. (d) Comparison of the *J–V* characteristics of flexible PSCs with PEDOT:PSS–RGO and ITO transparent electrodes. (e) IPCE spectra of flexible PSCs with PEDOT:PSS–RGO and ITO transparent electrodes. (f) Normalized PCE of flexible PSCs with PEDOT:PSS–RGO and ITO transparent electrodes under bending tests. (Reproduced with permission from [92]. Copyright 2017 Wiley.)

leads to the improvement of the FTE. The photovoltaic cell incorporated with PEDOT:PSS–RGO FTE exhibited a photoconversion efficiency of 8.32% over the area of 1 cm^2. Moreover, the PEDOT:PSS–RGO FTE exhibited 80% of its initial photoconversion efficiency even after a 2 mm curvature radius bending test (figure 8.27).

8.4.3.2 Graphene as photoanode material

The choice of photoanode material is a crucial factor in the design of DSSCs since it determines the light scattering/absorption properties which ultimately lead to enhancement of the PCE. Over the decades, anatese TiO_2 has been used widely as a photoanode material due to its cost effectiveness, good stability, easy availability, better optical and electronic characteristics, and non-toxicity. To date anatese TiO_2 based materials have yielded a photoconversion efficiency of approximately 12%–14% [100, 101]. The PCE can also be improved through hybrid structures of TiO_2 with carbonaceous materials such as graphene. The pathway for the electron transfer mechanism of bare TiO_2 and graphene coated TiO_2 photoanode materials is shown in figure 8.28.

Covalently functionalized amino acid (alanine, isoleucine, leucine, methionine, phenylalanine, tyrosine, and valine) graphene nanosheets were incorporated TiO_2 nanoparticle thin film, and it was found that electron transport behavior was enhanced and there was an improvement in the performance of the photoanode in the DSSC. Furthermore, aromatic amino acid functionalized graphene–TiO_2 photoanodes showed that the adsorbed dye was increased due to the higher roughness of the graphene–TiO_2 composite nanoparticles, or anchoring the dye to the active functional groups of the GO sheets [103]. The results revealed that the GO functionalized with aliphatic amino acids resulted in a reduced PCE, whereas the photoanodes containing GO functionalized with aromatic amino acids had better cell efficiencies (figure 8.29(a)). GO functionalized with tyrosine and phenylalanine exhibited better power conversion efficiency than the other amino acids. The photovoltaic parameters, such as V_{oc} (V), J_{sc} (mA cm^{-2}), FF (%), and PCE (η), and the flat band potential obtained from the Mott–Schottky analysis of the assembled solar cells with amino acid functionalized graphene incorporated TiO_2 photoanodes are listed in table 8.2. A theoretical approach has also been investigated to identify the linkage between GO and amine compounds. It was found that the aromatic rings of tyrosine and phenylalanine exhibited interesting geometry in relation to the GO sheet compared to the other. Two rings of tyrosine are aligned in parallel to the GO sheet, whereas one of the rings of phenylalanine that binds to GO

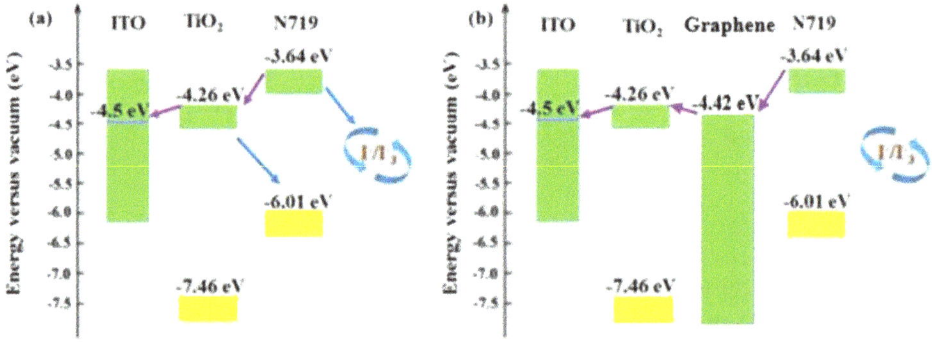

Figure 8.28. Schematic representation of the electron transfer mechanism in (a) a bare TiO_2 photoanode and (b) a graphene coated TiO_2 photoanode. (Reproduced with permission from [101]. Copyright 2019 Elsevier.)

Table 8.2. The photovoltaic parameters of the assembled solar cells with the amino acid functionalized graphene incorporated TiO_2 photoanodes and the flat band potential obtained from the Mott–Schottky analysis [104]. (Reproduced with permission from [126]. Copyright 2014 Elsevier.)

GO amino acids	V_{oc} (V)	J_{sc} (mA cm^{-2})	FF (%)	η (%)	E_{fb} (V versus NHE)
None	0.771 ± 0.001	2.112 ± 0.066	61.0 ± 1.6	0.99 ± 0.04	−0.624
Alanine	0.664 ± 0.003	1.234 ± 0.060	50.4 ± 0.5	0.41 ± 0.01	−1.262
Isoleucine	0.583 ± 0.003	3.146 ± 0.012	57.1 ± 0.5	1.05 ± 0.01	−1.094
Leucine	0.456 ± 0.011	2.509 ± 0.060	36.4 ± 2.3	0.42 ± 0.02	−5.015
Methionine	0.699 ± 0.011	2.718 ± 0.009	67.0 ± 1.1	1.27 ± 0.06	−1.470
Phenylalanine	0.762 ± 0.009	6.638 ± 0.008	65.8 ± 1.0	3.33 ± 0.07	−0.670
Tyrosine	0.694 ± 0.003	8.190 ± 0.007	72.6 ± 0.3	4.13 ± 0.02	−1.459
Valine	0.662 ± 0.004	3.461 ± 0.071	48.1 ± 4.1	1.10 ± 0.01	−0.788

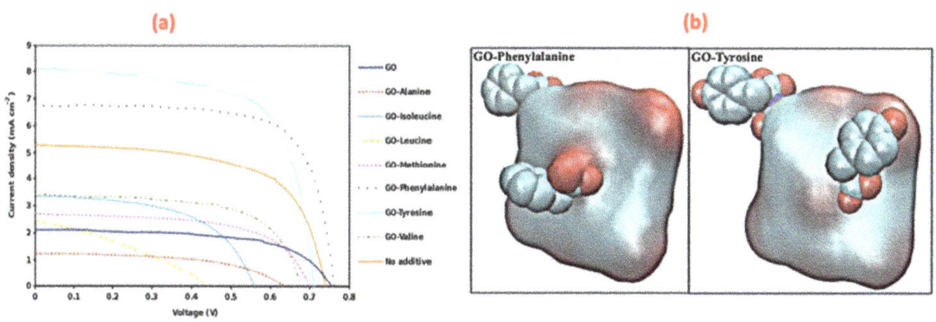

Figure 8.29. (a) Photovoltaic performance of the fabricated DSSCs for the photoanodes with amino acid functionalized graphene. (b) The calculated mesh-surface illustrations of the aromatic amino acid (phenylalanine and tyrosine) functionalized GO. (Reproduced with permission from [103]. Copyright 2017 Elsevier.)

is placed perpendicular to the GO sheet [104]. Calculated mesh-surface illustrations of the aromatic amino acid functionalized GO are shown in figure 8.29.

Graphene and its composite material can also be used as a photoanode for organic photovoltaic (OPV) applications. Chemically reduced GO deposited on PET substrate showed a transparency of above 65% (figure 8.30). Studies were also conducted on the effect of RGO thickness on the photovoltaic performance with different thicknesses, such as 4 nm, 10 nm, 16 nm, and 21 nm. It was found that a film thickness of 4 nm had a sheet resistance of 16 kΩ sq^{-1} with 88% of transmittance, and exhibited a PCE of 0.28%. Similarly, the values were 6.6 kΩ sq^{-1} and 78% for 10 nm, 3.2 kΩ sq^{-1} and 65% for 16 nm, and 1.6 kΩ sq^{-1} and 55% for 21 nm film thickness yields PCEs of 0.61%, 0.78%, and 0.77%, respectively [105].

Long-term stability is a crucial parameter in any DSSCs, and must be addressed for commercialization of PV cells. One of the major problems associated with the utilization of TiO_2 anodes in DSSCs is that the oxidative holes in the valence band promote dye degradation and hinders the long-term stability of photovoltaic cells.

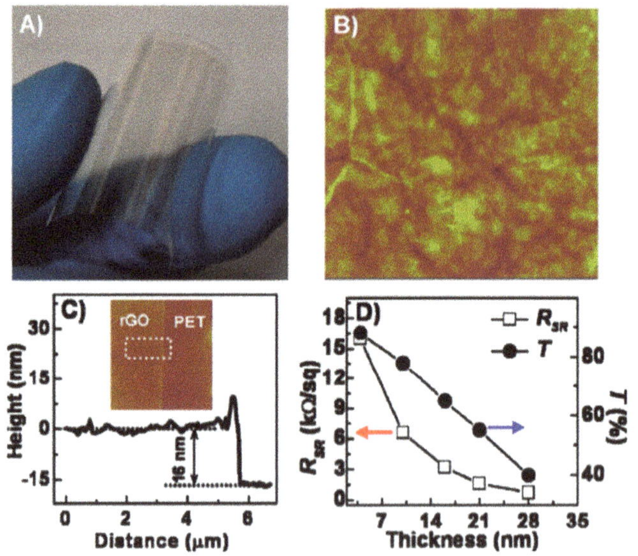

Figure 8.30. (a) Photograph of RGO–PET. (b) AFM topographic image of RGO–PET. (c) AFM image and section analysis of 16 nm thick RGO film on PET, i.e. RGO–PET in panel B. (d) Sheet resistance (R_{sr}) and transmittance (T) of RGO film as a function of its thickness. (Reproduced with permission from [104]. Copyright 2010 American Chemical Society.)

To overcome this issue a hybrid structure of graphene with various semiconducting materials has been explored. Incorporating graphene microplatelets with different concentrations into SnO_2–TiO_2 mesoporous heterostructured anode materials achieved improvements in the cell efficiency and long-term stability of DSSCs (figure 8.31). The graphene concentrations increased from 0.01 wt% to 0.04 wt% and J_{sc} was increased from 8.59 mA cm^{-2} to 9.06 mA cm^{-2}, which led to an enhancement in the PCE of 3.14% to 3.22%. The addition of graphene microplatelets in a controlled way could provide charge transport which considerably increases J_{sc}, specifically, graphene improves the effective surface area for dye absorption [106]. The photovoltaic parameters of DSSCs contains various graphene composite photoanodes are listed in table 8.3.

8.4.3.3 Graphene as counter electrodes

The counter electrode is one of the important components for dye sensitized solar cells, and Pt is one of the widely used materials due to the fact that it has high electrical conductivity and good catalytic behavior [117]. Although it has several advantages, Pt is one of the most expensive metals on Earth and it is not suitable for the large-scale production of photovoltaic cells. Therefore, there is a pressing demand to find alternative materials for counter electrodes with low cost and without sacrificing solar cell efficiency. Over the decades many carbonaceous materials (carbon black, graphene, and CNTs) have been identified due to their high electrical conductivity, and conducting polymers (polyaniline (PANI), polypyrrole, poly(3,4-ethylenedioxythiophene (PEDOT)) are commonly used since they

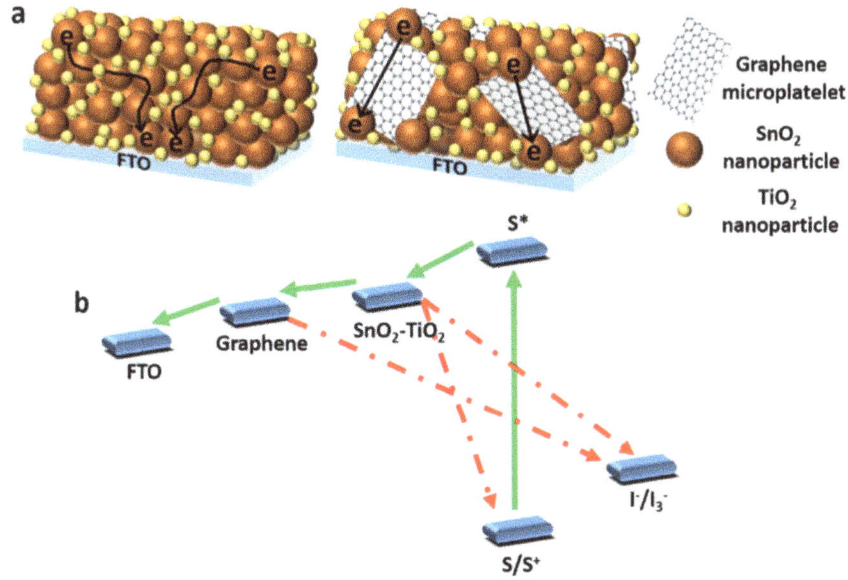

Figure 8.31. (a) Schematic illustration of improved charge transport in the anode and charge collection in the FTO with the introduction of graphene microplatelets. (b) Electronic band diagram of the DSSC incorporated with graphene microplatelets. (Reproduced with permission from [105]. Copyright 2020 Elsevier.)

Table 8.3. The photovoltaic parameters of DSSCs contains graphene composite photoanodes.

	Photovoltaic parameters				
Graphene composite anodes	J_{sc} (mA cm^{-2})	V_{oc} (mV)	FF	η (%)	Reference
TiO_2–SLG–I	0.66	14.64	73.01	7.06	[106]
TiO_2–GQDs	15.49	0.75	60	6.97	[107]
RGO–TiO_2	12.72	0.70	52	4.54	[108]
NFS–GQDs modified TiO_2	22.6	0.79	70	11.7	[109]
RGO–TiO_2 nanocrystals	12.16	668	67.7	5.5	[110]
RGO–TTRGO–TW	10.7	0.78	63.9	5.33	[111]
		0.75	54.3	4.36	
TiO_2 microsphere–RGO	16.75	0.74	65	7.57	[112]
TiO_2–graphene nanoribbon	14.22	0.724	70	7.18	[113]
RGO–TiO_2 composite nanofibers	10.92	0.651	62.21	4.43	[114]
RGO–TiO_2 blocking layer	15.29	0.74	66	7.48	[115]

have excellent catalytic performance [118–121]. Natural graphene has a large surface area, high optical transmittance, and excellent electrocatalytic properties and could be considered as a better alternative for the counter electrodes in photovoltaic devices. The photoconversion efficiency can be changed with an appropriate annealing temperature [123]. Figure 8.32 depicts the resistance (R_{ct1}) between the electrolyte and as-deposited graphene counter electrode and it is found that the value

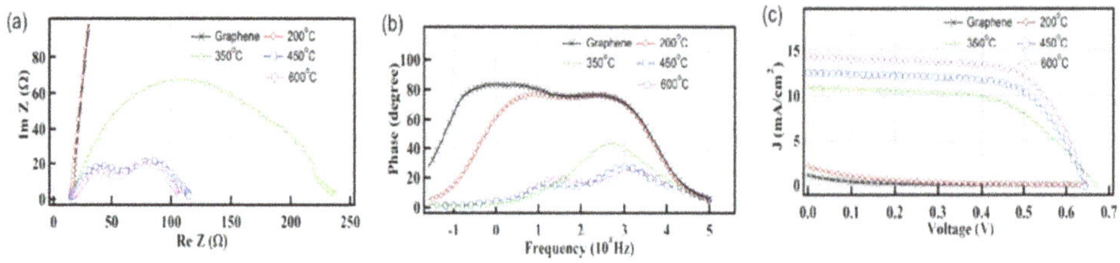

Figure 8.32. (a) Nyquist plots and (b) Bode phase plots of DSSCs with graphene counter electrodes annealed at different temperatures. (c) The current density–voltage characteristics of DSSCs with graphene counter electrodes annealed at different temperatures. (Reproduced with permission from [122]. Copyright 2011 The Royal Society of Chemistry.)

of R_{ct1} is drastically decreased from 75000 to 38 Ω when the annealing temperature is increased from 200 °C to 600 °C. This factor influences the photoconversion efficiency, for example 0.06% was achieved when the samples were as grown and this is enhanced enormously to 5.69% for a sample annealed at 600 °C. Similarly, thermal treatment at 1000 °C and 1500 °C on graphene sheets could decrease the number of oxygen containing functional groups and heal of lattice defects, respectively. A photovoltaic cell including functionalized graphene (FGS) with a C/O ratio of 13 yields a cell efficiency of 4.99% [122].

In recent years, the combination of graphene and polymers has been identified as an ideal candidate for alternative low-cost counter electrodes which are comparable in efficiency to Pt counter electrodes. Among the available polymers, polyaniline (PANI) is considered to be an important electrode material due to its extraordinary properties, such as low cost, environmental stability, high degree of processability, and interesting redox properties. To date there has been plenty of research carried on the utilization of G–PANI composites as Pt-free counter electrodes. GN–PANI was coated on the FTO glass by electro-polymerization, using an aqueous solution containing GN and aniline (GN–ANI) particles, which were synthesized via a reflux process. The short-circuit current density J_{sc} = 16.28 mA cm^2, FF = 0.67, and cell efficiency = 7.17% for the DSSC with a catalytic film of GN–PANI are much higher than those for the DSSCs with catalytic films of pristine PANI (J_{SC} = 14.68 mA cm^{-2}, FF = 0.63, and cell efficiency = 6.27%) and for a graphene sheet alone (J_{SC} = 14.79 mA cm^{-2}, FF = 0.50, and cell efficiency = 4.54%). This result clearly indicates that the enhancement of the cell parameter is due to the presence of graphene, which gives the large surface area for the reduction of I_3^- ions to provide the higher catalytic properties [124]. Interestingly, cell efficiency can be increased just by increasing the weight percentage of PANI–graphene composites. The cell efficiency was increased up to 7.70% using PANI 8 wt%–graphene complex counter electrodes which were fabricated using a reflux technique. It was found that the enhancement of the photocurrent density of the PANI 8 wt%–graphene is because the complex counter electrodes can accelerate the recovery of triiodides which promote electron transfer kinetics [125] (figure 8.33).

Another way of improving cell performance and enhancing the photoconversion efficiency of counter electrodes with a multilayer technique has been adopted in the

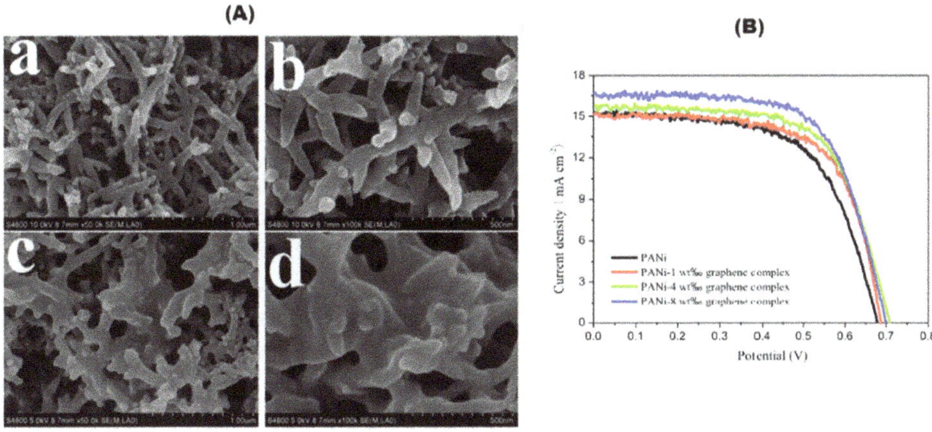

Figure 8.33. (A) SEM images of (a) and (b) PANI and (c) and (d) PANI 8 wt%–graphene complex counter electrodes. (B) *J–V* characteristics of DSSCs from pure PANI and PANI–graphene complex counter electrodes (CEs). (Reproduced with permission from [124]. Copyright 2014 Elsevier.)

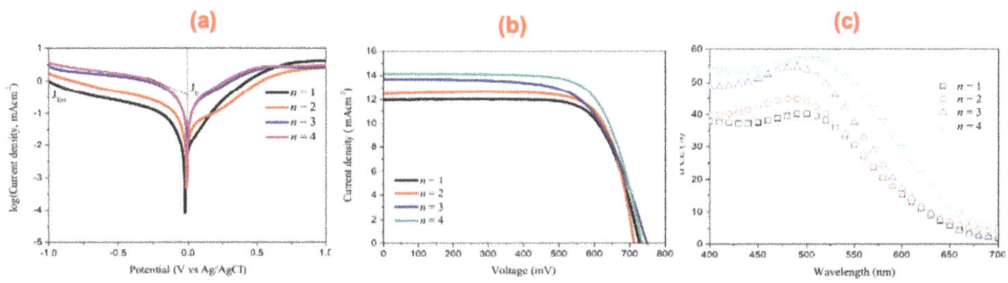

Figure 8.34. (a) Tafel polarization curves of symmetrical cells fabricated with PANI–(GO–PANI)$_n$ multilayer counter electrodes, (b) *J–V* characteristics of DSSCs from PANI–(GO–PANI)$_n$ multilayer counter electrodes, and (c) IPCE curves of DSSCs from PANI–(GO–PANI)$_n$ multilayer counter electrodes. (Reproduced with permission from [125]. Copyright 2014 Elsevier.)

last few years. The layer-by-layer (LbL) self-assembly approach has been utilized because it can provide well-defined nanoarchitectures with improved electrical and electrochemical properties [126]. The polyaniline–(graphene oxide–polyaniline)$_n$ (PANI–(GO–PANI)$_n$; where *n* represents the deposition cycle) multilayer counter electrodes were fabricated through LbL self-assembly. The interfacial charge transfer characteristics at the counter electrode–electrolyte interface can be obtained from the Tafel polarization curve (figure 8.34) and the results reveal a higher exchange current density (J_0) on the electrode and better catalytic activity toward triiodide reduction, in the following order: PANI–(GO–PANI)$_4$ > PANI–(GO–PANI)$_3$ > PANI–(GO–PANI)$_2$ > PANI–(GO–PANi)$_1$.

Similarly, a polyaniline–graphene (PANi–G) complex was prepared through a refluxed route and an LbL assembly method for multilayer counter electrodes with the introduction of graphene oxide (GO) [127]. The results demonstrate that the (PANI–10 wt% G/GO)$_{10}$ multilayer counter electrodes produces a power conversion efficiency of 7.88%. The comparison chart of photovoltaic parameters for multilayer counter electrodes is presented in table 8.4.

Furthermore, the efficiency of the cell has been increased to 7.1% under the irradiation of 100 mW cm^{-2} AM 1.5 simulated solar light through the composite form of RGO–PPy:PEDOT counter electrodes deposited on the FTO substrate using *in situ* polymerization and elctrodeposition routes. Here, PPy acts as a connecting backbone which provides uniform film formation, whereas RG and PEDOT introduce a multitude of catalytic sites which enhance the catalytic activities [128].

The efficiency of the photovoltaic cell could also be increased further to 7.36% through a composite of graphene dots (GDs)–poly(3,4-ethylene-dioxythiophene)–polystyrene sulfonate (PEDOT:PSS) deposited on a flexible paper substrate using the cost-effective 'soak and dry' method that could be easily scaled up for mass production. The various photovoltaic parameters of DSSCs based on carbonaceous and carbonaceous hybrid counter electrodes is listed in table 8.5. Furthermore, it was found that the flexible paper counter electrodes consisting of GD–PEDOT:PSS exhibited a cell efficiency of (4.91%), which was almost three times higher than that of the cell made with the sputtered Pt paper counter electrode (1.70%). This reverse behavior is due to the fact that GD based ink was better able to fill the pores in the paper substrate, whereas the sputter coating of Pt was unable to sufficiently reach all the surfaces and coat the metal inside all the voids [129] (figure 8.35).

8.4.3.4 Active layers

In recent years, bulk heterojunction polymer solar cells (PSCs) have been studied widely due to their low cost and easy fabrication. In general a PSC consists of four major components, (i) an ITO anode, (ii) a hole transport layer (HTL), (iii) a photoactive layer; and (iv) a top cathode, and each component is important for enhancing the performance of the solar cell. Among them, the HTL and electron transporting layer (ETL) are two major components of a PSC because they directly control the charge carrier recombination process at the interfaces to enhance the

Table 8.4. Comparison of short-circuit current density (J_{sc}), open-circuit voltage (V_{oc}), fill factor (FF), and power conversion efficiency (η (%)) in the resultant DSSCs with respect to number of layers. (Data reproduced from [126, 127].)

	Photovoltaic parameters			
Counter electrodes	J_{sc} (mA cm^{-2})	V_{oc} (mV)	FF	η (%)
PANI–(GO–PANi)$_1$	11.92	728	0.73	6.63
PANI–(GO–PANi)$_2$	12.47	712	0.75	6.69
PANI–(GO–PANi)$_3$	13.64	748	0.70	7.12
PANI–(GO–PANi)$_4$	14.14	734	0.71	7.41
(PANI–10 wt%G–GO)$_{n=1}$	8.61	657	0.615	3.48
(PANI–10 wt%G–GO)$_{n=3}$	8.48	687	0.667	3.88
(PANI–10 wt%G–GO)$_{n=5}$	13.46	735	0.668	6.61
(PANI–10 wt%G–GO)$_{n=7}$	13.26	711	0.704	6.66
(PANI–10 wt%G–GO)$_{n=10}$	16.79	727	0.646	7.88

Table 8.5. Photovoltaic parameters of DSSCs based on carbonaceous and carbonaceous hybrid counter electrodes.

Counter electrodes	J_{sc} (mA cm^{-2})	V_{oc} (mV)	FF	η (%)	Reference
Solar reduced graphene oxide (SRGO)	12.20	0.72	0.44	3.96	[129]
Thermally reduced graphene sheets + hydriodic acid	14.352	0.693	78.374	7.8	[130]
GO	11.83	0.66	71.5	5.6	[131]
PANI–/RGO	12.58	0.63	0.55	3.98	[132]
PEDOT–RGO	15.82	0.73	0.67	7.79	[133]
NiS–RGO	7.39	0.74	0.19	1.04	[134]
$Mo_{0.7}Fe_{0.3}$–RGO	13.67	635	62.69	5.44	[135]
GO/(Cu–Co) 1/9	24.10	0.69	0.46	6.31	[136]
$CoFe_2O_4$–graphene	22.08	0.771	53.1	9.04	[137]
$Co_{0.5}Ni_{0.5}Se$–$GN_{0.50}$	18.33	0.75	0.685	9.42	[138]
Polyimide–graphene	25.8	0.84	70.2	15.2	[139]
MoS_2–graphene aerogel	17.24	0.37	63.83	7.86	[140]
Polyisothianaphene (PITN)/graphene 1/4	14.51	660	0.55	5.27	[141]

PCE of the cell. Also, the HTL and ETL determine the generation, separation, and transport of charge carriers in the PSC and the choice of the material is a crucial factor. To date, several materials have been used, such as polymers and metal oxides (NiO, MoO_3, V_2O_5, and WO_3), and the most commonly used HTL to improve the PCE is poly(3,4-ethylenedioxythiophene):poly(styrenesulfonate) (PEDOT:PSS) because of its high work function value of 5.2 eV, which will match the highest occupied molecular orbital (HOMO) level of most donor polymers, and excellent hole injection properties. However, it has inherent disadvantages, such as high acidity and a hygroscopic nature, which limit the stability and life time [143]. Therefore, it is necessary to find an alternative material which exhibits a long life time, is stable in air, and has inert HTL materials. A schematic diagram of a GO modified HTL and photoactive layer device and their corresponding energy level diagrams are depicted in figure 8.36. The photovoltaic performance of a device consisting of PEDOT:PSS–GO as the HTL layer and Poly(3-hexylthiophene): Phenyl-C61-butyric acid methyl ester (P3HT:PCBM) as the active layer was a PCE of 2.8% ± 0.12%. The PCE of the cell was increased further to 4.4% ± 0.08% for P3HT:PCBM–GO-1 as the active layer and PEDOT:PSS as the HTL layer. Introduction of GO in the PEDOT:PSS HTL would yield simultaneous selective electron blocking and hole injection at the HTL, which impairs charge carrier recombination to improve the PCE of the cell. Similarly, insertion of GO in the active layer could enhance the light absorption efficiency, which ultimately enhances the photogenerated charge carriers [144].

A perovskite solar cell (PeSC) with composite HTL was fabricated through simple sequential spin coating techniques [145]. The device fabricated without GO and containing only conventional PEDOT:PSS showed a PCE of 8.2%, whereas the device

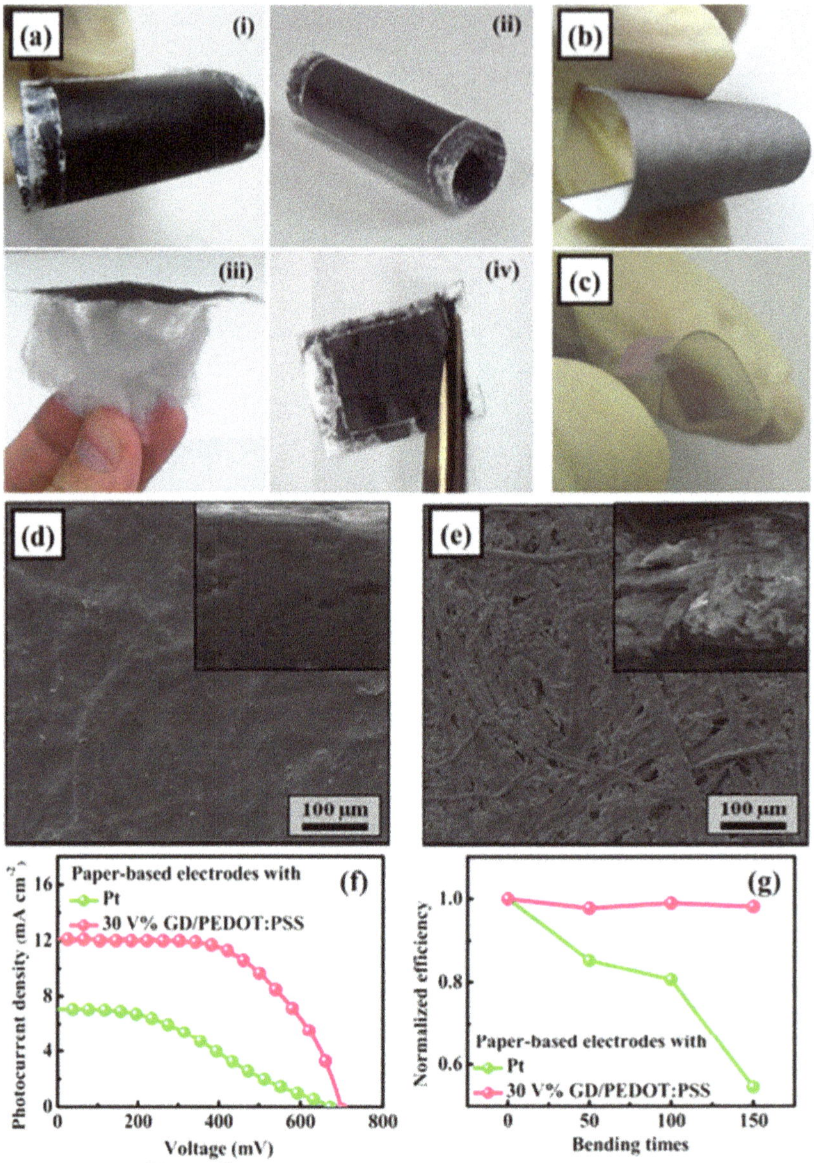

Figures 8.35. (a) Images of the paper-based counter electrodes made with the 30 vol% GD–PEDOT:PSS composite materials. (b) Image of the paper-based counter electrode made with sputtered Pt. (c) A flexible dye-sensitized TiO_2 photoanode using the ITO–PEN substrate. (d) Top view and side view (inset) SEM images of the paper-based counter electrode made with the 30 vol% GD–PEDOT:PSS composite. (e) Top view and side view (inset) SEM images of the paper-based counter electrode made with sputtered Pt. (f) The *J–V* curves of the flexible DSSCs using the two types of paper-based counter electrodes. (g) Normalized efficiency of the flexible DSSCs using the two types of paper-based counter electrodes after being bent up to 150 times. Both of the flexible DSSCs used the dye–TiO_2–ITO–PEN photoanode substrate. (Reproduced with permission from [128]. Copyright 2017 Elsevier.)

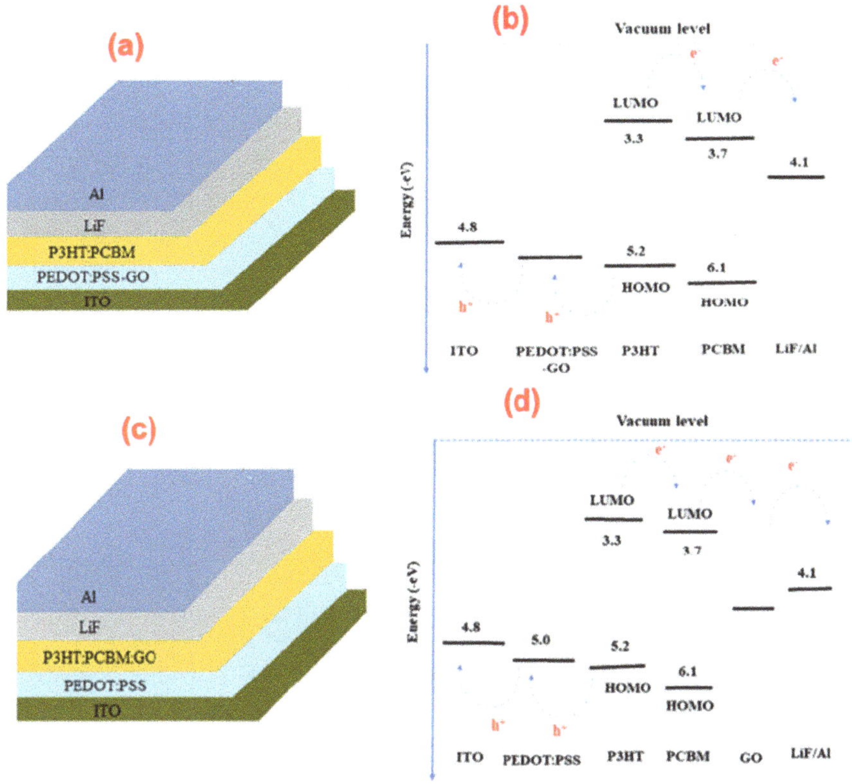

Figure 8.36. (a) Schematic diagram of the GO modified HTL device and (b) energy level diagram of the device's constituent materials. (c) Schematic diagram of the GO modified active layer devices and an energy level diagram of the device's constituent materials. (d) (Reproduced with permission from [143]. Copyright 2018 Elsevier.)

with GO–PEDOT:PSS exhibited a higher PCE of 9.7%. The increase of PCE is due to the decrease of series resistance (R_S) and increased shunt resistance (R_{Sh}). This is because the work-functions between GO (4.9 eV) and PEDOT:PSS (5.1 eV) are matched perfectly which can result in more efficient charge transport and an overall decrease in R_s.

The charge extraction property in bulk heterojunction (BHJ) solar cells can be altered by simple charge neutralization of the –COOH groups in GO with a cesium derivative and this material can utilized as both the ETL and HTL in BHJ solar cells [146]. GO natural acts as a HTL layer due to its functionalized –COOH group and the replacement of –COOH with –COOCs could act as an electron extraction layer (figure 8.37).

The work function of GO is 4.7 eV and it matches the HOMO level of P3HT which facilitates hole extraction. When the –COOH functional group in GO is charge neutralized with Cs derivative the work function is reduced to 4 eV and thus matches the lowest unoccupied molecular orbit (LUMO) level of PCBM which could provide efficient electron extraction properties. Though simply control of the functionalization of GO derivatives, they can be utilized as single electron/hole extraction material systems (figure 8.38).

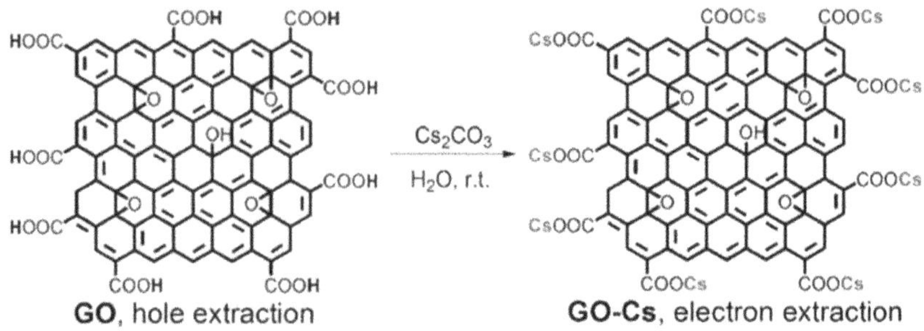

Figure 8.37. Chemical structures and synthesis route of GO and charge neutralization of GO with Cs derivatives. (Reproduced with permission from [214]. Copyright 2018 Elsevier.)

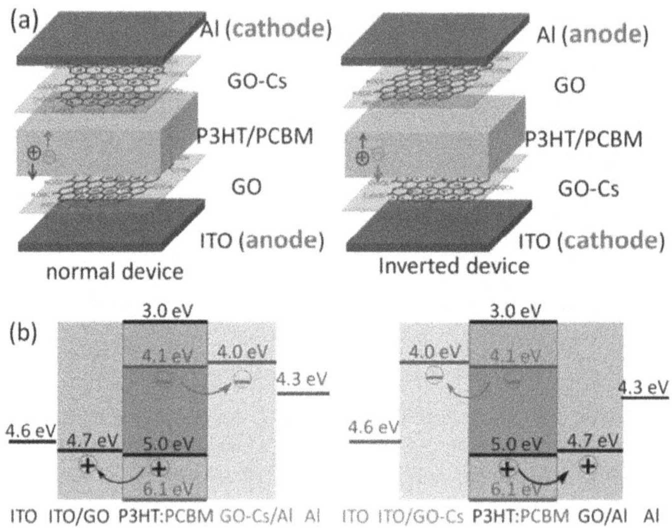

Figure 8.38. (a) Device structures and (b) energy level diagrams of the normal device and the inverted device with GO as the hole extraction layer and GO–Cs as the electron extraction layer. (Reproduced with permission from [145]. Copyright 2012 Wiley.)

Poly[*N*-9′-heptadecanyl-2,7-carbazole-alt-5,5-(4′,7′-di-2-thienyl-2′,1′,3′-benzothiadiazole)] (PCDTBT):[6,6]-phenyl C71 butyric acid methyl ester (PC71BM) BHJ solar cells with stretchable GO as the ETL were fabricated through a stamping process. A schematic of the device and the steps involved in the fabrication are shown in figure 8.39. The *J–V* characteristics studies confirmed that GO played the important role of the ETL to improve the PCE of the cells. The device fabricated with GO has a higher J_{sc} of 12.36 mA cm^{-2} than that without GO (10.66 mA cm^{-2}) and consequently enhanced the PCE of the cells. The photovoltaic parameters of graphene based solar cells with different structures are listed in table 8.6. Furthermore, device stability has been investigated and it was found that the cell fabricated with the GO ETL layer exhibited a reduction of the PCE of the cell by only 3%–4%, whereas the cell without the GO ETL layer showed a sharp PCE decay of 56% [148].

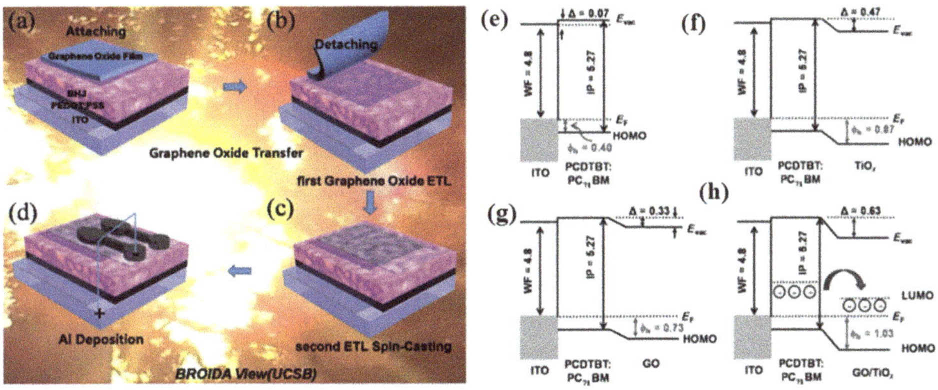

Figure 8.39. Schematic illustration and fabrication steps of BHJ solar cells with an ETL of graphene oxide (GO) applied by stamping transfer. (a) Attachment of the transfer film on top of the BHJ active layer. (b) After detachment of the film, the first ETL of the GO is uniformly transferred and coated onto the BHJ layer. (c) Spin-casting of a second ETL of TiO_x on top of the GO. (d) Completed device structure after Al deposition. Energy level diagrams of BHJ films (e) without ETL, and with ETLs of (f) TiO_x, (g) GO, and (h) GO/TiO_x. (Reproduced with permission from [146]. Copyright 2013 Wiley.)

Table 8.6. Photovoltaic parameters of graphene based solar cells with different structures.

	Photovoltaic parameters				
Structure	J_{sc} (mA cm^{-2})	V_{oc} (mV)	FF	η (%)	Reference
ITO–GO–PCDTBT:PC70BM–TiO_2–Al OPV	11.04	0.83	0.56	5.1	[147]
ITO–F-RGO–PTB7-Th:PC71BM–PFN–Al	16.89	0.786	0.648	8.6	[148]
ITO–ZnO–PC70BM–P3HT–GO–PEDOT:PSS–Au	7.75	0.58	0.625	2.75	[149]
(ITO)–RGO–PEDOT:PSS–$CH_3NH_3PbI_3$–PC61BM–BCP–Ag	16.75	0.87	0.75	10.75	[150]
ITO–GO–PEDOT:PSS–$CH_3NH_3PbI_3$–PCBM	15.72	0.84	0.738	9.74	[151]
ITO–F_5-RGO–PEDOT:PSS–PTB7–PC71BM–PFN–FM	15.31	0.78	0.63	7.52	[152]
Glass–ITO–ERGO–P3HT:PCBM–Ca–Al	9.35	0.60	0.626	3.52	[153]
ITO–GO–PEDOT:PSS–PCDTBT–PCBM–Al	10.44	0.82	0.50	4.28	[154]
ITO–GO (3000 rpm)/PEDOT:PSS–$CH_3NH_3PbI_3$–PCBM–Al	17.14	0.90	0.69	10.65	[155]
mTiO_2 + G–perovskite–GO–spiro-OMeTAD	22.48	1.08	0.751	18.19	[156]

8.4.3.5 Graphene Schottky junction

Metal brought into contact with semiconductors can produced two different types of junctions: (i) an ohmic junction, which is produced using highly doped semiconductors, and (ii) the Schottky junction, which is produced from lightly doped semiconductors. The junction could function as a diode in which there is high current and low resistance in one direction and very low/negligible current and high resistance in the other directions [159]. Thanks to its extraordinary optical properties, such as a near zero bandgap, graphene can act like metals and when semiconductors are brought into contact with graphene it can produce a Schottky junction. There is built-in potential near the semiconductor surface because of a difference in the work function (figure 8.40). The photon carries energy above the bandgap and will separate the photogenerated holes (h^+) and electrons (e^-) and accelerate towards Schottky electrodes. The separated carriers produce an open-circuit voltage if the cell is open and yield current when there is a short circuit in the cell [160].

The first graphene/silicon Schottky junction solar cell was successfully fabricated by Li *et al.* The results revealed that the photogenerated carriers are increased linearly with incident light intensity, which leads to an increase in photovoltaic parameters (figure 8.41). The built-in potential was generated due to separation of the charge carriers and holes are diffused to the junction where they are swept to the graphene sheets side. The fabricated PV device showed a very low efficiency of 1.7% and this was because of the low work function, and the reduction of the absorbed number of photons by 30%–40% for 350–800 nm on planar Si wafers [161]. Various researchers across the globe have introduced different techniques such as *p*-type doping, silicon-pillar-arrays, monolayer graphene (MLG), different numbers of graphene layers, and interfacial layers to improve cell efficiency.

Introduction of monolayer graphene into the Schottky junction could yield better performance since it has a transparency of >97% leading to extraction of more photon absorption in photovoltaic devices. Moreover, a *p*-doped graphene–Si Schottky junction will reduce the series resistance which increases the built-in potential. The PV cell consists of *p*-MLG–*n*-SiNWs and *n*-MLG–*p*-SiNWs structures fabricated through a CVD technique [163]. The identification of the generation and separation of electron–hole

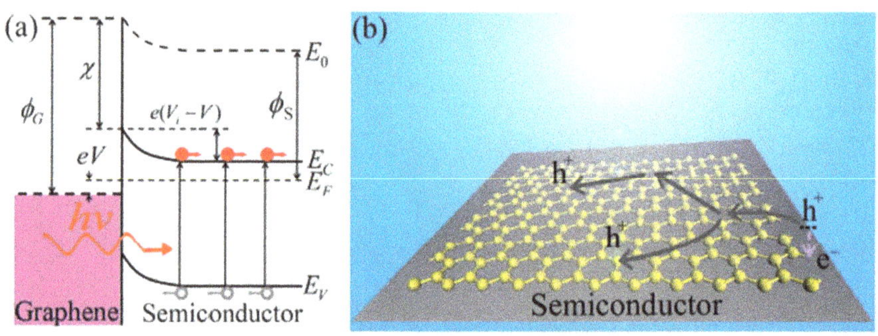

Figure 8.40. (a) Energy band diagram of a semiconductor/graphene Schottky junction solar cell under illumination. (b) Schematic illustration of graphene based Schottky junction solar cell. (Reproduced with permission from [158]. Copyright 2012 The Royal Society of Chemistry.)

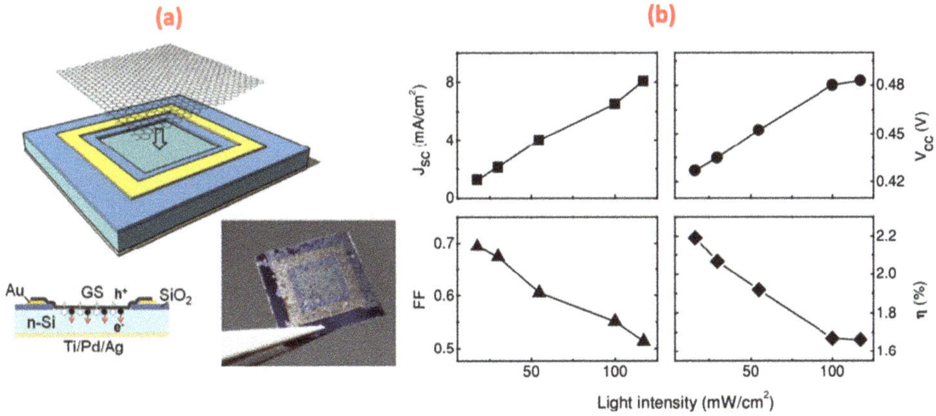

Figure 8.41. (a) Schematic illustration of a GS–*n*-Si Schottky junction. (b) Light-intensity-dependent J_{sc}, V_{oc}, FF, and η plots of a 0.1 cm^2 GS–*n*-Si cell. (Reproduced with permission from [160]. Copyright 2010 The Royal Society of Chemistry.)

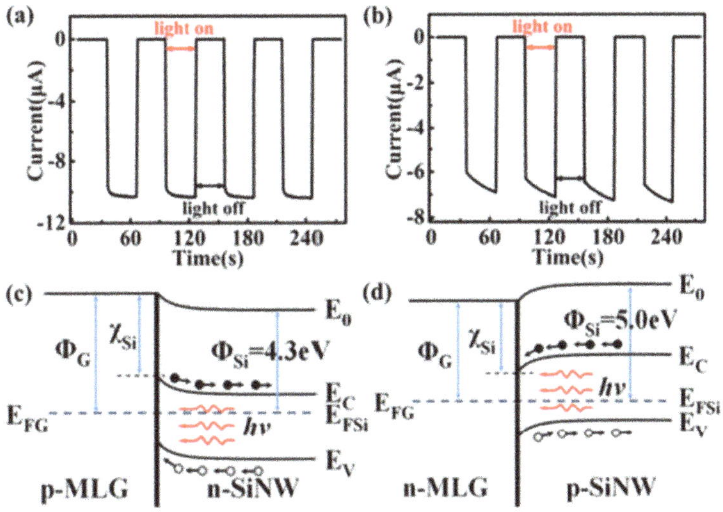

Figure 8.42. The photoresponse of PV devices (a) and (b) at zero bias. (c) and (d) are energy diagrams of *p*-MLG–*n*-SiNW and *n*-MLG–*p*-SiNW Schottky junctions upon light illumination, respectively. (Reproduced with permission from [161]. Copyright 2011 The Royal Society of Chemistry.)

pairs is direct evidence for the efficiency of Schottky junction solar cells and they were characterized by photoresponse studies. The obtained results showed the highest sensitivity to light illumination of $I_{on}/I_{off} > 10^3$ and a fast response speed (figure 8.42). Also, the PCE of the device is directly related to the etching time of the SiNWs and PCEs of 0.10%, 0.40%, 0.05% and 2.15% for 10 min *n*-SiNW pristine MLG, 10 min *n*-SiNW pristine MLG, 20 min *n*-SiNW pristine MLG, and 40 min *n*-SiNWs pristine MLG and 20 min n-SiNWs pristine MLG (graphene suspension filled), respectively.

The pillar-array patterned Si substrate showed better light absorption and anti-reflective properties which yield an enhanced PCE of PV cells. The pillar arrays of Si were realized through a photolithography and reactive ion etching (RIE) route whereas the graphene layers were grown using CVD and then were transferred onto the ten pillar-array-patterned areas. The fabricated pillars had dimensions of ~2 μm in diameter and ~2 μm in spacing, with different heights ranging from 200 nm to 1 μm (figures 8.43(a)–(e)). Antireflection studies revealed a reduction of reflectance from 45% to 25% by introducing a pillar array in the structure (figure 8.43(f)). Also, a pillar structure array showed the larger J_{sc} which more effectively enhanced the PCE (η) than a cell structure with planar Si is because of the anti-reflectance behavior of the structure (figure 8.43(g)). Similarly, Feng *et al* realized a G–SiPA Schottky junction solar cells and found a photoconversion efficiency of 3.55% [164].

The number of layers also plays a vital role in enhancing the performance of solar cell because the number of layers is related to the work function and transmittance [165]. Studies have been carried out on the performance of Schottky junction solar cells based on silicon/graphene (1–6 layers) [166]. The photovoltaic parameters are found to increase monotonically up to four layers, due to adjusting the work function of graphene and improving the built-in potential in the graphene–n-Si Schottky junction, and then decreases because of a reduction in the transmittance. The photovoltaic parameters of graphene based various Schottky junction solar cells are listed in table 8.7.

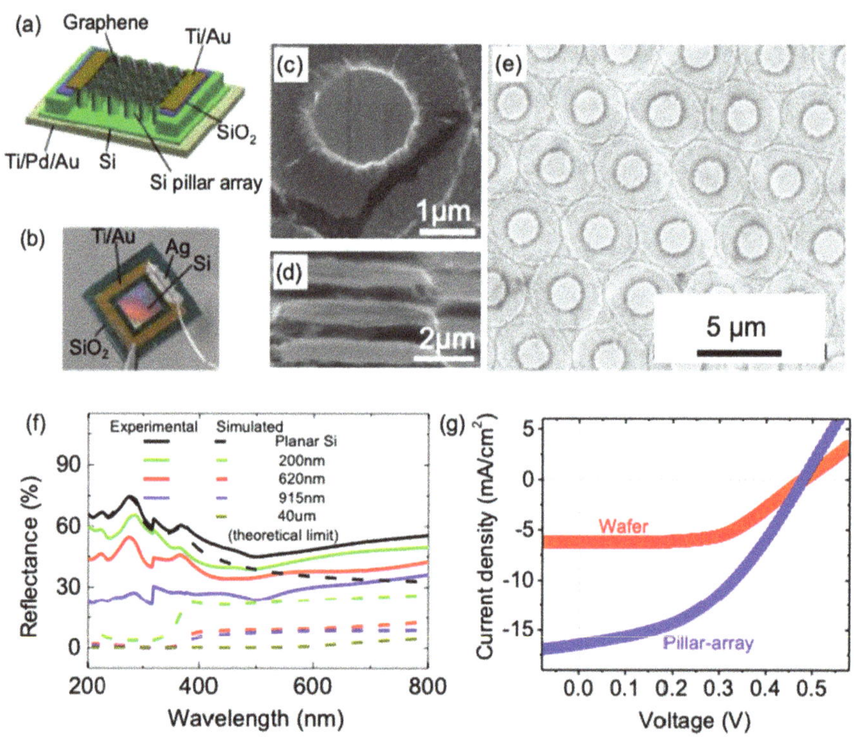

Figure 8.43. (a) A schematic diagram of a G–SiPA solar cell. (b) Photograph of the G–SiPA solar cell. (c) Top view, (d) side view, and (e) low-magnification view of SEM images of the pillar array. (f) Calculated (dashed line) and measured (solid line) reflection spectra of planar Si and Si pillar arrays with different etching depths. (g) Experimental *J–V* curves of solar cells with (blue) and without (red) a pillar-array pattern. (Reproduced with permission from [163]. Copyright 2013 The Royal Society of Chemistry.)

Table 8.7. Photovoltaic parameters of graphene based Schottky junction solar cells.

	Photovoltaic parameters				
Structure	J_{sc} (mA cm^{-2})	V_{oc} (mV)	FF	η (%)	Reference
MLG–H-SiNW array	1.38	0.16	0.358	0.079	[165]
MLG–H-SiNH array	1.42	0.17	0.343	0.083	[165]
G–SPA (HNO_3)	17.22	495	0.51	4.35	[166]
Gr-Al_2O_3 (0.2 nm)–Si	33.4	0.556	0.452	8.4	[167]
Graphene nanoribbon–multiple silicon nanowires (SiNWs)	11.3	0.59	0.221	1.47	[168]
Graphene–GaAs	10.03	0.65		1.95	[169]
GaAs–SAM–graphene	19.3	0.77	0.745	11.1	[170]
Graphene–single GaAs nanowire	425	75	0.276	8.8	[171]
Gr–Si co doped with Au NPs and HNO_3	32.8	0.51	0.61	10.20	[172]
n-graphene–*p*-SiPMMA–TiO_x–Gr	34.3	0.48	0.636	10.5	[173]
Graphene–Si hole array (12.8 μm depth)	31.56	0.52	0.634	10.4	[174]
CdSe nanobelt (NB)–graphene	4.73	0.31	0.361	0.53	[175]
CdSe nanobelt (NB)–graphene	5.75	0.51	0.427	1.25	[176]
Graphene (Gr)–silicon (Si) with HfO_2 interfacial layer	33.4	0.49	0.56	9.1	[177]

8.5 Photothermal therapy

Light interacts with matter in different ways, such as absorption, reflection, transmission, refraction, and scattering, producing interesting physical phenomena such as photoelectric, photoluminescence, photoacoustic, and photothermal effects (figure 8.44(a)) [180]. The photothermal (PT) process is the absorption of light in a wide frequency and converting it to thermal energy. PT methods are widely used in the

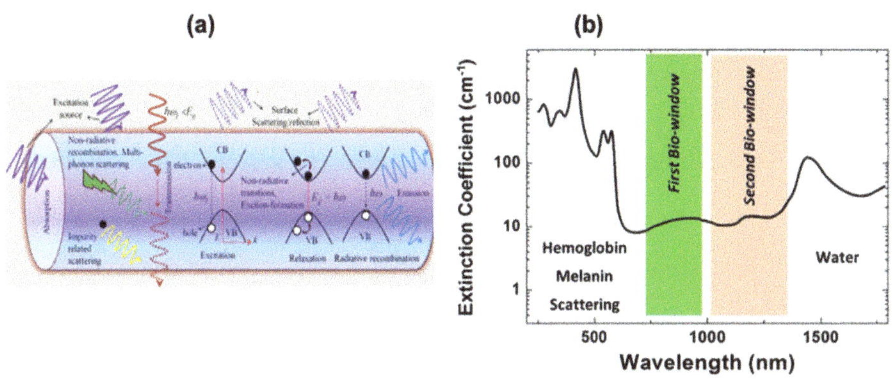

Figure 8.44. (a) Schematic illustration of light-matter interaction. (b) Extinction coefficient of a representative tissue. The different effects leading to light attenuation (such as the presence of hemoglobin, water, and optical scattering). (Reproduced with permission from [180]. Copyright 2014 The Royal Society of Chemistry.)

medical field for the selective destruction of targeted cancer cells, as one branch of photothermal therapy (PTT) [181]. Of the entire electromagnetic spectral range, wavelengths between 700 and 1400 nm show advantages for photothermal therapy since in these ranges the intrinsic tissue absorption and scattering of light are lower and light has a relatively deep penetration depth in tissue. The broad wavelength region can be further classified into two parts. The first biological window (I-BW) spans from 700 nm to 980 nm, which corresponds to the spectral range between the visible absorption band of hemoglobin and the characteristic absorption band (980 nm) of water. Light absorption vanishes strongly but optical extinction residual scattering will be present in this region. (ii) The second biological window spans from 1000 nm to 1400 nm and optical absorption does not vanish completely, whereas optical scattering is minimized due to the longer wavelengths (figure 8.44(b)) [182].

The photothermal transduction efficiency can be obtained from the total energy balance equation [183]

$$\sum_i m_i C_{p,i} \frac{dT}{dt} = Q_{\text{in,np}} + Q_{\text{in,surr}} - Q_{\text{out}},$$

where m and C_p are the mass and heat capacity of the solvent, T is the solution temperature, $Q_{\text{in,np}}$ is the photothermal energy input, $Q_{\text{in,surr}}$ is the heat input which is due to light absorption by the solvent, and Q_{out} is the heat lost to the surroundings. $Q_{\text{in,np}}$ could be calculated through relation $Q_{\text{in,np}} = I(1-10^{(-A\lambda)})\eta$, where I is the laser power, A_λ is the absorbance at the excitation wavelength of 800 nm, and η is the photothermal efficiency. Similarly, Q_{out} can be obtained by this relation, in which h is the heat transfer coefficient, A is the surface area of the container, and T_{surr} is the ambient surrounding temperature. By substituting all those parameters, the photothermal transduction efficiency can be calculated directly from the steady-state temperature increase

$$\eta = \frac{hA(T_{\max} - T_{\text{surr}}) - Q_{\text{in,surr}}}{I(1 - 10_{(}^{-A^{\lambda}}))},$$

where $T_{\max}$ is the maximum steady-state temperature. The value of (hS) could be calculated through the relation [184]

$$hS = \frac{m_D C_D}{\tau_s},$$

where m_{D} is the combined mass of the nanoparticles and solvent, C_{D} is the combined heat capacity, and τ_{s} is the system time constant and could be obtained from the relation

$$\tau_s = -\frac{t}{\ln(\theta)},$$

where θ is a dimensionless parameter which can be calculated from the cooling cure after irradiation and is related with the formula

$$\theta = \frac{T - T_{\text{surr}}}{T_{\text{Max}} - T_{\text{surr}}}.$$

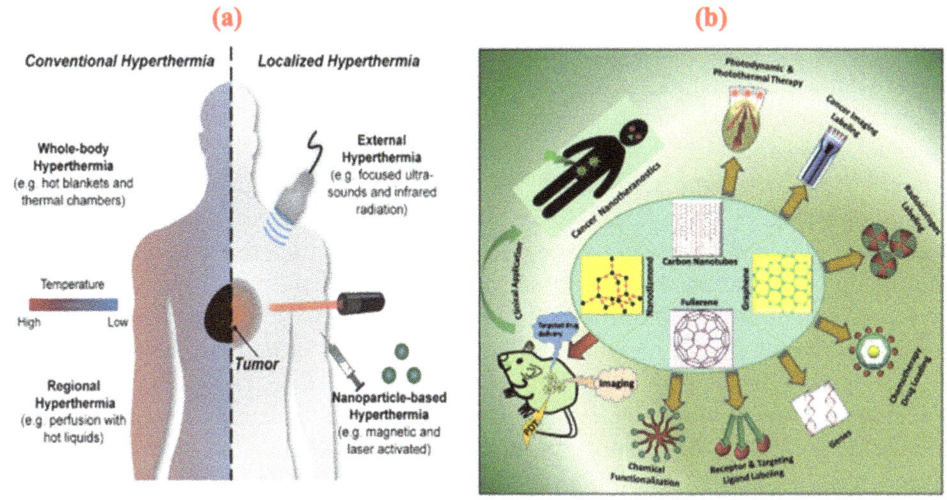

Figure 8.45. (a) An illustration of conventional and localized hyperthermia. Conventional hyperthermia creates a heat gradient from the surface region of the body to the interior. The localized hyperthermia approaches promotes specific heating of the tumor tissue, avoiding damage in the surrounding areas. (b) Schematic illustration of cancer theranostics using carbon allotropes following various approaches. (Reproduced with permission from [187, 188]. Copyright 2020 and 2017 The Royal Society of Chemistry.)

Conventionally, many thermal ablative strategies have been developed, such as laser, ultrasound, microwave, and radio frequency treatments, but unfortunately all these techniques do not specifically target malignant cells and there will always be destruction of normal cells, which causes undesirable side effects (figure 8.45). Therefore, an efficient and better alternative technique is needed which meets specific requirements, such as the selective sensitization of malignant cells to irradiation and reducing the toxicity of thermal ablation. Over the last decade, the development of PTT has been accelerated by advancements in nanomaterials research and as it promotes the selective death of cancer cells by irradiating the target area with a laser light. To date, there have been many inorganic nanomaterials, such as gold [185], tungsten, copper [186], molybdenum [187], and iron oxide [188], and carbon nanomaterials (graphene, CNT, fullerenes, nanodiamonds, and carbon dots) that have been explored for localized PTT treatment.

8.5.1 Graphene based PTT

Among the various nanomaterials, carbon based nanostructures are better alternatives for photothermal reagents for the following reasons: (i) relatively higher absorbance in the wavelength region between 700 and 1400 nm which covers the two bio-windows; (ii) good biocompatibility; (iii) easily modified and functionalized surfaces; and (iv) good dispersibility in bio-liquids [191]. Among the various allotropes of carbon, pristine graphene exhibited strong broad band absorption from the far-IR to UV range, and weak radiative emission, and mostly emission is hindered by carrier relaxation because of the zero bandgap. The moment photons interact with graphene, charge carrying hot electrons and holes are created and it will undergo an independent relaxation process which gives different momenta to prevent recombination. Thus pristine graphene demonstrates the weak/negligible fluorescence under continuous wave excitation. These properties make graphene an excellent candidate for realizing light-to-heat transduction, in which all photo-generated electrons contribute to the heat generation through non-radiative transitions [181]. When NIR photons are irradiated on a graphene surface plasmons are excited and induced, followed by transmitting random dipoles and resonance, which can finally be converted to thermal photon energy output. This makes these materials potential candidates for photothermal therapy [192]. In addition, photothermal materials should possess excellent optical absorption, thermal stability, and thermal conductivity, and have poor emissive properties. Detailed investigation revealed that the graphene materials showed an extraordinary thermal conductivity and stability [193, 194]. In addition, graphene consists of π-conjugated six carbon atom rings, like a giant aromatic poly-molecule, and offers a high specific surface area which further makes the material an ideal candidate for biomedical applications.

Two fundamental mechanisms have been proposed as the cause of cell death which triggered by photothermal treatment, necrosis and apoptosis (figure 8.46(a)) [195]. When NIR photons interact with cells, necrosis can be produced, which is characterized by loss of plasma membrane integrity. This will release the intracellular contents, including damage-associated molecular patterns (DAMPs), into the extracellular milieu, and thus the ruptured membranes are pro-inflammatory in nature, whereas in apoptosis the membrane remains intact with an anti-inflammatory response. A schematic illustration of the apoptosis mechanism resulting from gold nanoprism-mediated photothermal therapy is shown in figure 8.46(b).

A magnetic reduced graphene oxide functionalized with glutaraldehyde (MRGOGA) photothermal agent was synthesized for efficient capture and effective killing of both gram-positive *Staphylococcus aureus* (*S. aureus*) and gram-negative *Escherichia coli* (*E. coli*) bacteria under NIR laser irradiation. Furthermore, it was identified that 80 ppm MRGOGA solution provided rapid and effective killing of up to 99% of both gram-positive and gram-negative bacteria in 10 min upon NIR laser irradiation under the batch operation mode, whereas under dark conditions the survival rates were all above 100, revealing that both bacterial strains are not affected by MRGOGA alone [196].

Yang *et al* prepared a reduced graphene oxide (RGO)–iron oxide nanoparticle (IONP) nanocomposite functionalized with PEG for photothermal therapy of 4T1 murine breast tumor cancer with a laser wavelength of 808 nm, a power density of 0.5 W cm, and 5 min irradiation. Four different groups were used to carried out the PTT: (i) untreated mice (control); (ii) laser treated mice without RGO–IONP–PEG injection (laser only); (iii) RGO–IONP–PEG injected mice without laser irradiation; and (iv) RGO–IONP–PEG injected mice with laser irradiation. The results revealed that the surface temperature of tumors in the RGO–IONP–PEG injected mice was increased to 48 °C upon exposure to the laser whereas the uninjected mice exhibited an increase in temperature to 39 °C under the same conditions. Furthermore, it was

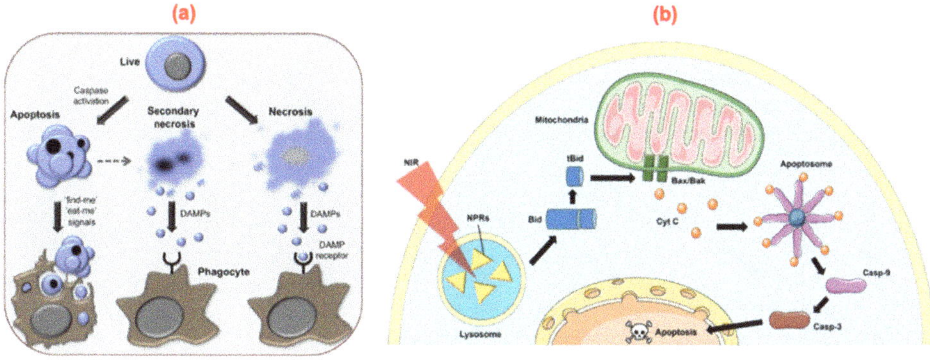

Figure 8.46. (a) Apoptosis and necrosis are the two mechanisms of cell death induced by photothermal therapy. (b) The proposed apoptosis mechanism resulting from gold nanoprism-mediated photothermal therapy. Lysosome rupture following laser irradiation activates Bid, inducing Bax/Bak oligomerization and pore formation in the mitochondrial outer membrane. Cytochrome c is released, leading to apoptosome assembly facilitated by Apaf-1. The apoptosome then activates caspase-9, which subsequently activates caspase-3. Caspase-3 then acts as the executioner initiating apoptotic cell death. (Reproduced with permission from [193]. Copyright 2015 The American Chemical Society.)

identified that post-PTT treatment, the RGO–IONP–PEG injected mice showed almost nil re-growth over 40 days, whereas the uninjected mice showed rapid abnormal tumor growth which led to animal death in less than 20 days [197] (figure 8.47).

The bystander effect is basically a cell-to-cell signal transfer where individual damaged cells transfer their toxic signals to their neighboring cells mediated through gap-junction intercellular communication (GJIC). The bystander effect enhances the sensitivity of the cells to chemotherapy due to diffusion of cytotoxic signals [198]. To utilize the bystander effect to treat cancer cells through PTT activated under NIR photons, antiarrhythmic peptide 10 (AAP10) on the surface of dopamine-modified RGO (AAP10-pDA–RGO) nanocomposite has been successfully synthesized and schematic illustrations of the preparation route and bystander effect are provided in figure 8.48.

It has been identified that the temperature of the pDA–RGO- and AAP10-pDA–RGO-containing solution was increased by 40 °C min^{-1} under the exposure of an NIR 808 nm laser at 1.5 W cm^{-2} and reaches a maximum temperature of 55 °C while the surrounding temperature is 25 °C. The photothermal efficiency was calculated and the AAP10-pDA–RGO nanocomposite value was 49.1%, which suggests that the prepared composites could serve as ideal biocompatible candidates for NIR photothermal treatment [199].

Alzheimer's disease is a neurodegenerative disorder (continuous decline of thinking, reduction of independent behavior) which causes brain cells to die slowly and it has been estimated that more than 100 million people will be affected around the globe by 2050 [200]. With this in mind, researchers are exploring several techniques, such as chemically exfoliated WS_2 nanosheets [201], penetratin peptide-functionalized gold nanostars [202], polyoxometalates (POMs) [203], gold

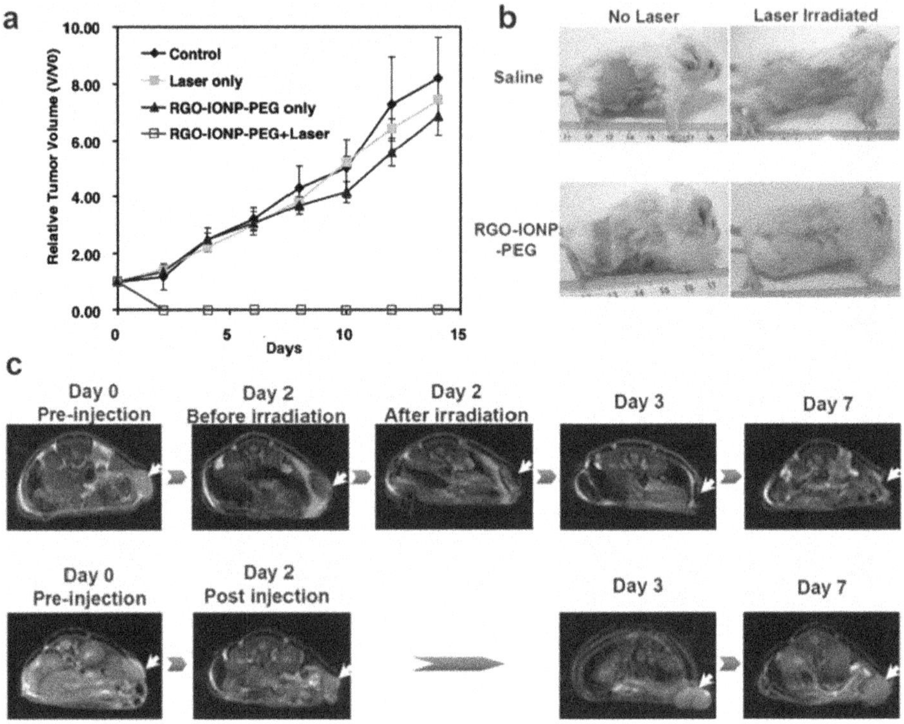

Figure 8.47. *In vivo* photothermal therapy using RGO–IONP–PEG. (a) The tumor growth curves of different groups of mice after various treatments. Four groups, including untreated mice (control), laser treated mice without RGO–IONP–PEG injection (laser only), RGO–IONP–PEG injected mice without laser irradiation (RGO–IONP–PEG only), and RGO–IONP–PEG injected mice with laser irradiation (RGO–IONP–PEG + laser), were used in this experiment (seven mice for each group). The tumor volumes were normalized to their initial sizes. Error bars were based on standard deviations. (b) Representative photos of 4T1 tumor-bearing mice after the various treatments indicated. Laser wavelength = 808 nm. Power density = 0.5 W cm. Irradiation time = 5 min. (c) MR images of RGO–IONP–PEG injected mice with (top row) and without (bottom row) laser irradiation. Arrows point to the tumor sites. Images were collected at day 0 (pre-injection), day 2, day 3 and day 7 post-infection Both tumor growth curves and MR imaging data showed that the tumors on mice injected with RGO–IONP–PEG and irradiated with the NIR laser were completely ablated, while the tumor growth of control mice was not affected. (Reproduced with permission from [195]. Copyright 2012 The Royal Society of Chemistry.)

nanorods (GNRs) loaded with single chain variable fragment (scFv) 12B4 and thermophilic acylpeptide hydrolase (APH) [204] to cure this disease in its early stages. Among these, GO showed excellent performance in treating Alzheimer's disease using the photothermal technique. Thioflavin-S (ThS)-modified GO [205] was synthesized to promote a localized heating environment and dissolve the amyloid aggregation upon NIR laser irradiation; a schematic illustration is provided in figure 8.49(a). Furthermore, a cell viability test was conducted using control, Aβ fibrils, Aβ fibrils with GO–Ths in the absence and presence of NIR radiation, Aβ fibrils under NIR laser radiation, and NIR laser radiation with a control sample (figure 8.49(b)). The results demonstrate that fibrils lead to a decrease of 48% in

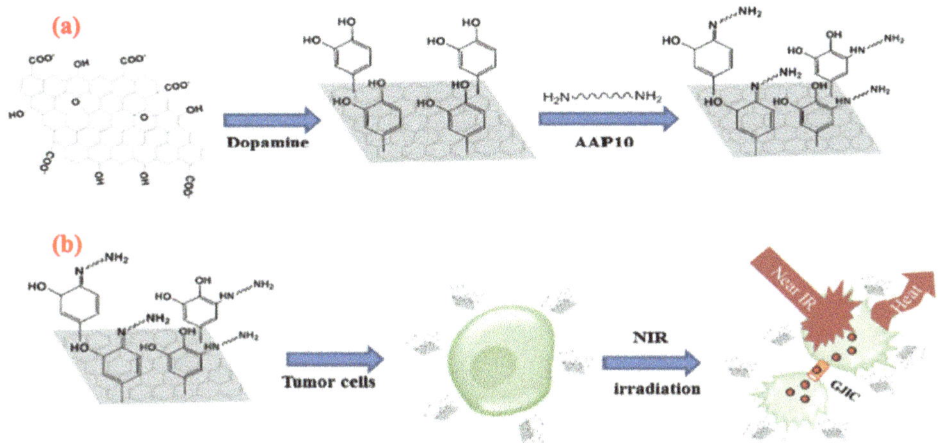

Figure 8.48. (a) Schematic illustration of the route for preparing AAP10-pDA–RGO nanocomposites. (b) Destruction of cancer cells with the aid of the bystander effect upon exposure to NIR irradiation during PTT. (Reproduced with permission from [197]. Copyright 2016 The Royal Society of Chemistry.)

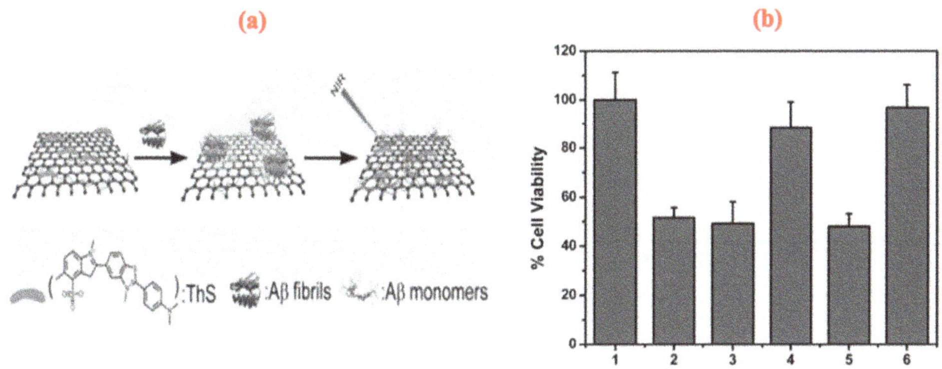

Figure 8.49. (a) Schematic illustration of ThS-modified GO with high NIR absorbance used for Alzheimer's disease treatment. GO–ThS can effectively dissolve amyloid deposits of Aβ1–40 upon NIR laser irradiation. (b) Effect of GO–ThS on the cell toxicity of Aβ. Cell viability was determined using the MTT method and the data points shown are the mean values ± standard error of the mean (SEM) from three independent experiments. (1) Control, (2) Aβ fibrils, (3) Aβ fibrils with GO–ThS in the absence of NIR laser irradiation, (4) Aβ fibrils with GO–ThS under laser irradiation, (5) Aβ fibrils under NIR laser irradiation, (6) NIR laser irradiation with control sample 1. Cell viability was determined using the MTT method and the data points shown are the mean values ± SEM from three independent experiments. Control: Aβ untreated cells; Aβ = 5 x 10^{-6} M; and GO–ThS = 12.5 μg mL^{-1}. (Reproduced with permission from [203]. Copyright 2012 The Royal Society of Chemistry.) In a similar way, the destruction of amyloid fibrils through pristine graphene and graphene oxide has been demonstrated by Yang *et al* [204].

cellular reduction. Treatment of the cells with Aβ in the presence of GO–ThS (12.5 μg mL^{-1}) under NIR laser irradiation for 5 min increased the survival of the cells to about 88% within the experimental error.

8.5.2 Chemo/photothermal synergistic therapy

In recent years, researchers have explored the possibilities of a controlled route of drug delivery to cancer cells by using graphene based materials, since most of doxorubicin (DOX), camptothecin (CPT), and resveratrol (Rves) aromatic molecules are adsorbed on the graphene surface through hydrophobic–hydrophobic interactions and π–π stacking. Figures 8.50(a) and (b) illustrate the synthetic route of Rves adsorption of NRGO–PEG and the photothermally controlled action mechanism of NRGO–PEG-RV in inducing cell death.

Detailed photothermal investigations have been done with different samples such as nanographene–PEG (NGO–PEG), NGO–PEG-RV, nano reduced graphene oxide–PEG (NRGO–PEG), NRGO–PEG-RV, and water only as a control (figure 8.51). The results demonstrate that the water-alone samples do not respond to NIR irradiation for 5 min. The solution containing the sample of NRGO–PEG and NRGO–PEG-RV with the addition of 20 mg ml^{-1} of GO exhibited a temperature increase to 60 °C, whereas the samples containing NGO–PEG and NGO–PEG-RV showed the temperature rise of 39 °C. Furthermore, it was identified that the release rate of Rves was highly dependent on the pH values and it showed the fastest release rate at pH = 5.0, while the slowest release rate was at pH = 7.4. The overall results *in*

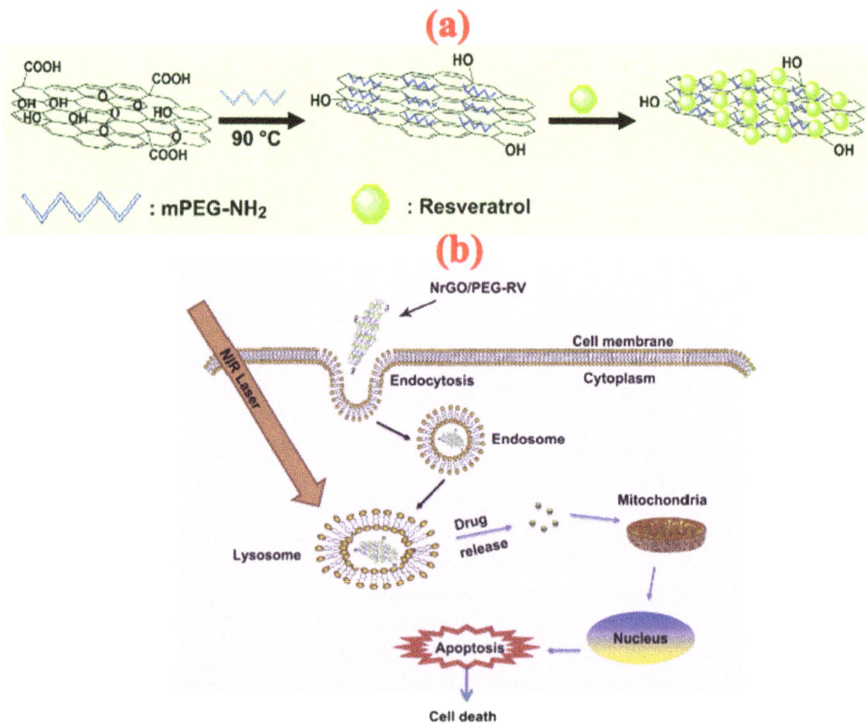

Figure 8.50. (a) Schematic illustration of the synthetic procedure of NRGO–PEG and NRGO–PEG-RV. (b) Photothermally controlled action mechanism of NRGO–PEG-RV in inducing cell death. (Reproduced with permission from [205]. Copyright 2014 Elsevier.)

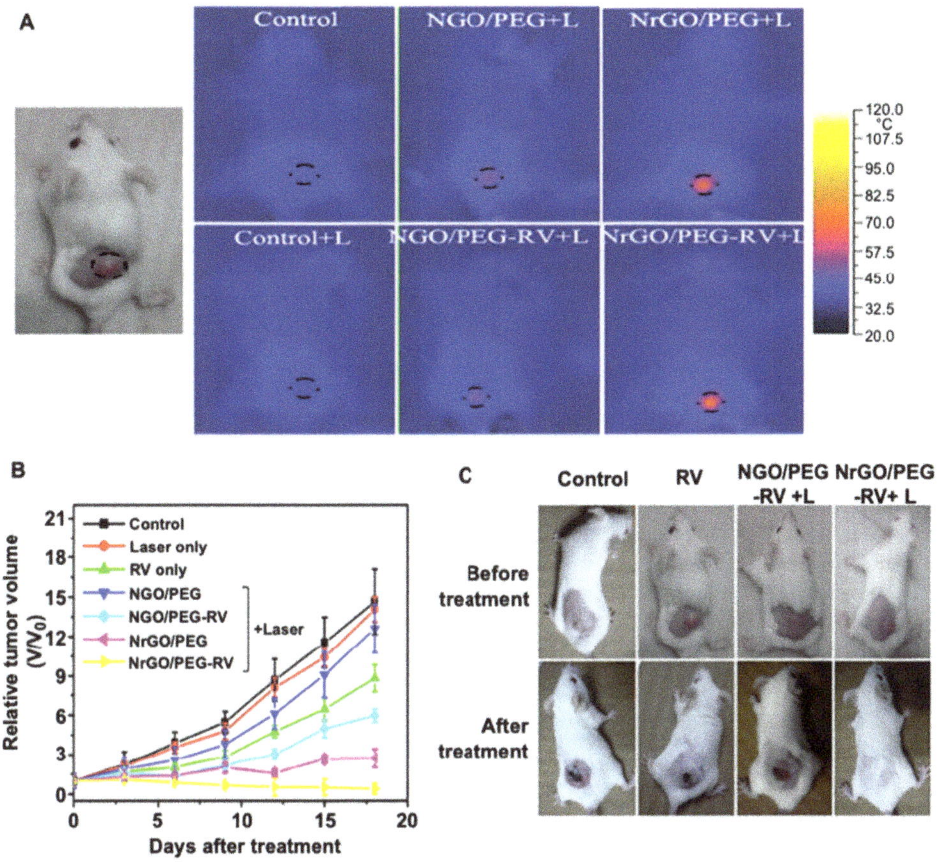

Figure 8.51. (a) Infrared thermal graphic maps of tumor-bearing mice injected with PBS, NGO–PEG, NGO–PEG-Rves, NRGO–PEG, or NRGO–PEG-Rves after NIR irradiation (L) for 5 min. (b) Quantitative analysis of the effects of various treatments on tumor size. (c) Photographs of representative mice before and after different treatment (30 days). (Reproduced with permission from [205]. Copyright 2014 Elsevier.)

vitro and *in vivo* revealed that NRGO–PEG can deliver drug molecules in a remotely and photothermally more controlled way [207]. In a similar way, a facile synthesis of Rves-loaded folate-terminated PEG-phospholipid coated reduced graphene oxide nanoassembly (FA-PEGLip@RGO–Res) was proposed as a potential candidate for photothermal treatment to ablate tumor cells by Hai *et al* [208]. The prepared FA-PEG-Lip@RGO solutions were exposed to an NIR laser wavelength of 780 nm with 0.6 W cm^{-2} power and showed a rapid increase of temperature to 45.4 °C, 51.6 °C, 56.6 °C, and 59.6 °C for FA-PEGLip@RGO concentrations of 0.07 mg mL^{-1}, 0.14 mg mL^{-1}, 0.28 mg mL^{-1}, and 0.56 mg mL^{-1}, respectively.

The major challenge associated with chemotherapy treatment is the ability of transport an appropriate amount of drugs with fewer acute and long-term side effects to normal cells. Over the decades, controlled targeted drug delivery systems (TDDSs) have become an important field in the biomedical and pharmaceutical industries. In the advancement of nanomaterials research, mesoporous silica nanoparticles have

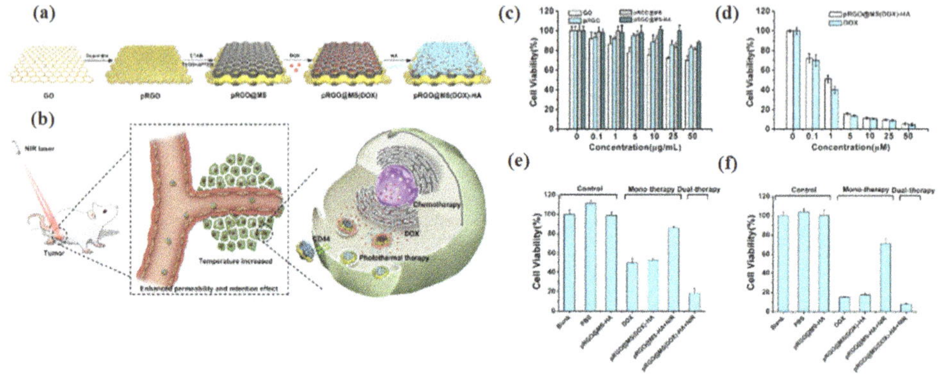

Figure 8.52. Schematic illustration of (a) the synthesis route of pRGO@MS(DOX)-HA nanocomposite and (b) the combined chemo-photothermal targeted therapy of tumors. Right: *In vitro* anti-tumor activity against HeLa cells. (c) Cytotoxicity assay of HeLa cells in the presence of GO, pRGO, pRGO@MS, and pRGO@MS-HA at various concentrations. (d) Cell viability of HeLa cells incubated with different concentration of DOX and pRGO@MS (DOX)-HA (equivalent concentration of DOX). Cell viability of HeLa cells treated with different forms of nanocomposites with or without NIR laser irradiation at 37 °C for (e) 24 h or (f) 48 h. The total drug content in all groups was kept at 5 micromoles. (Reproduced with permission from [211]. Copyright 2017 American Chemical Society.)

attracted great attention in the field of TDDSs, since they have several advantages, such as their highly biocompatible nature, almost zero premature drug release, efficient cellular uptake, high surface area (>700 m^2 g^{-1}), tuneable pore diameter (2–10 nm), and controllable size and morphology of the nanoparticles [209, 210]. In recent years, mesoporous silica coated graphene oxide nanostructures have been utilized effectively in synergistic targeted chemo-photothermal therapy [210, 211]. The anti-cancer drug DOX was loaded on mesoporous silica coated polydopamine functionalized reduced graphene oxide (pRGO@MS-HA) for the effective use of synergistic targeted chemo-photothermal therapy in cancer treatment. The photothermal efficiency of the prepared pRGO and pRGO@MS-HA materials was tested under 808 nm laser irradiation with 1.5 W cm^{-2} and the result revealed that the solution temperature was increased to 61.8 °C and 57.8 °C, respectively, within 10 min. The drug loading capacity of pRGO@MS-HA showed 145% ± 25% for loading of DOX molecules, whereas GO, RGO, and pRGO exhibited 40%, 22%, and 58%, respectively, as confirmed by fluorescence spectra. The pH and laser irradiation dependent DOX release behavior has been analysed and it was found that the release efficacy of DOX is highly pH responsive manner. Furthermore, the mono-therapy groups, DOX and pRGO@MS(DOX)-HA with an equivalent dose of DOX, exhibited apparent cytotoxicity to HeLa cells, illustrating the chemotherapy effect of pRGO@MS(DOX)-HA nanocomposites [211] (figure 8.52).

8.6 Concluding remarks and outlook

This chapter provided an outlook on the applications of graphene based materials, specifically on the new insights and to inspire broader interest across two different disciplines. Solar cell fabrication with the inclusion of graphene based materials plays an important role and offers cost effectiveness, high flexibility/stability, improved optical transmittance, and enhancement of the overall cell performance. Graphene based materials exhibited extraordinary characteristics towards targeted drug delivery and excellent NIR photothermal conversion efficiency. These properties can result in synergistic therapeutic effects during cancer treatment and less damage to the normal cells, which are a major concern in conventional cancer treatment. Overall, through the enhancement of photoconversion and photothermal efficiency, graphene nanostructures have the potential to become the next generation of photovoltaic technology and cancer therapy treatment in the near future.

Although graphene based materials possess multifunctional properties and have been utilized as Schottky junctions and electron/hole extraction layers, their performance is still lower than that of conventional solar cell materials, however, there is hope to improve this because of their extraordinary physical and chemical properties. Therefore, there is a plenty of room and need for extensive studies for the improvement of photovoltaic device performance. From the point of view of safety of graphene based materials in cancer therapy, many researchers have studied the biocompatibility of graphene nanostructures, and a few studies have revealed some degree of toxic effects on the immune function of biological living organisms. Therefore, it is necessary to carry out in-depth analysis of the biological safety or graphene, such as long-term toxicity, dose-dependent toxicity, and biodegradation, for the use of graphene materials in phototherapeutic cancer treatment.

References

[1] Panwar N L, Kaushik S C and Kothari S 2011 Role of renewable energy sources in environmental protection: a review *Renew. Sustain. Energy Rev.* **15** 1513–24

[2] Kralova I and Sjöblom J 2010 Biofuels—renewable energy sources: a review *J. Dispersion Sci. Technol.* **31** 409–25

[3] Kannan N and Vakeesan D 2016 Solar energy for future world: a review *Renew. Sustain. Energy Rev.* **62** 1092–105

[4] Hadadian M, Smått J H and Correa-Baena J P 2020 The role of carbon-based materials in enhancing the stability of perovskite solar cells *Energy Environ. Sci.* **13** 1377–407

[5] Zhu H, Wei J, Wang K and Wu D 2009 Applications of carbon materials in photovoltaic solar cells *Sol. Energy Mater. Sol. Cells* **93** 1461–70

[6] Wild C P, Weiderpass E and Stewart B W 2020 World Cancer Report: Cancer Research for Cancer Prevention *International Agency for Research on Cancer* https://iarc.fr/featured-news/new-world-cancer-report/

[7] Markovic Z M, Harhaji-Trajkovic L M, Todorovic-Markovic B M, Kepic D P, Arsikin K M, Jovanovic S P, Pantovic A C, Dramicanin M D and Trajkovic V S 2011 *In vitro* comparison of the photothermal anticancer activity of graphene nanoparticles and carbon nanotubes *Biomaterials* **32** 1121–9

[8] Hashemi M, Omidi M, Muralidharan B, Smyth H, Mohagheghi M A, Mohammadi J and Milner T E 2017 Reduced graphene oxide/arginine nanostructure for near-infrared absorption *ACS Appl. Mater. Interfaces* **9** 32607–20

[9] Freise E J 1962 Structure of graphite *Nature* **193** 671–2

[10] Bernal J D 1924 The structure of graphite *Proc. R. Soc.* A **106** 749
[11] Novoselov K S, Falko V I, Colombo L, Gellert P R, Schwab G and Kim K 2012 The structure of graphite *Nature* **490** 192–200
[12] Geim A K 2009 Graphene: status and prospects *Science* **324** 1530–4
[13] Chen H, Muller M B, Gilmore K J, Wallace G G and Li D 2008 Mechanically strong, electrically conductive, and biocompatible graphene paper *Adv. Mater.* **20** 3557–61
[14] Eda G, Fanchini G and Chhowalla M 2008 Large-area ultrathin films of reduced graphene oxide as a transparent and flexible electronic material *Nat. Nanotechnol.* **3** 270–4
[15] Iijima S 1991 Helical microtubes of graphitic carbon *Nature* **354** 56–8
[16] Iijima S 2002 Carbon nanotubes: past, present, and future *Physica* B **323** 1–5
[17] Mizutani W, Choi N, Uchihashi T and Tokumoto H 2001 Carbon nanotube tip for scanning tunneling microscope *Jpn J. Appl. Phys.* **40** 4328
[18] Stevens R M 2009 New carbon nanotube AFM probe technology *Mater. Today* **12** 42–5
[19] Terrones H 1994 Curved graphite and its mathematical transformations *J. Math. Chem.* **15** 143–56
[20] Ajima K, Yudasaka M, Murakami T, Maigne A, Shiba K and Ijima S 2005 Carbon nanohorns as anticancer drug carriers *Mol. Pharm.* **2** 475–80
[21] Wen D, Deng L, Zhou M, Guo S J, Shang L, Xu G B and Dong S J 2010 A biofuel cell with a single-walled carbon nanohorn-based bioanode operating at physiological condition *Biosens. Bioelectron.* **25** 1544–7
[22] Liu X Q, Shi L H, Niu W X, Li H J and Xu G B 2008 Amperometric glucose biosensor based on single-walled carbon nanohorns *Biosens. Bioelectron.* **23** 1887–90
[23] Murakami T, Sawada H, Tamura G, Yudasaka M, Iijima S and Tsucbida K 2008 Water-dispersed single-wall carbon nanohorns as drug carriers for local cancer chemotherapy *Nanomedicine* **3** 453–63
[24] Karousis N, Martinez I S, Ewels C P and Tagmatarchis N 2016 Structure, properties, functionalization, and applications of carbon nanohorns *Chem. Rev.* **116** 4850–83
[25] Kroto H W, Heath J R, O'Brien S C, Curl R F and Smalley R E 1985 C60: buckminsterfullerene *Nature* **318** 162–3
[26] Novoselov K S, Falko V I, Colombo L, Gellert P R, Schwab M G and Kim K 2012 A roadmap for graphene *Nature* **490** 192–200
[27] Wallace P R 1947 The band theory of graphite *Phys. Rev.* **71** 622–34
[28] Boehm H P, Clauss A, Fischer G O and Hofmann U 1962 Thin carbon leaves *Z. Naturforsch.* **17b** 150–3
[29] Novoselov K S, Geim A K, Morozov S V, Jiang D, Zhang Y, Dubonos S V, Grigorieva I V and Firsov A A 2004 Electric field effect in atomically thin carbon films *Science* **306** 666–9
[30] Yi M and Shen Z 2015 A review on mechanical exfoliation for the scalable production of graphene *J. Mater. Chem.* A **3** 11700–15
[31] Novoselov K S, Jiang D, Schedin F, Booth T J, Khotkevich V V, Morozov S V and Geim A K 2005 Two-dimensional atomic crystals *Proc. Natl Acad. Sci. USA* **102** 10451–3
[32] Meyer J C, Geim A K, Katsnelson M I, Novoselov K S, Booth T J and Roth S 2007 The structure of suspended graphene sheets *Nature* **446** 60–3
[33] Rajendra Kumar G, Jayasankar K, Das S K, Dash T, Dash A, Jena B K and Mishra B K 2016 Shear-force-dominated dual-drive planetary ball milling for scalable production of graphene and its electrocatalytic application with Pd nanostructures *RSC Adv.* **6** 20067–73

[34] Knieke C, Berger A, Voigt M, Taylor R N K, Röhrl J and Peukert W 2010 Scalable production of graphene sheets by mechanical delamination *Carbon* **48** 3196–320
[35] Zhao W, Fang M, Wu F, Wu H, Wang L and Chen G 2010 Preparation of graphene by exfoliation of graphite using wet ball milling *J. Mater. Chem.* **20** 5817
[36] Zhao W, Wu F, Wu H and Chen G 2010 Preparation of colloidal dispersions of graphene sheets in organic solvents by using ball milling *J. Nanomater.* **2010** 1–5
[37] Rajendra Kumar G, Jayasankar K, Das S K, Dash T, Dash A, Jena B K and Mishra B K 2016 Shear-force-dominated dual-drive planetary ball milling for scalable production of graphene and its electrocatalytic application with Pd nanostructures *RSC Adv.* **6** 20067–73
[38] Xing T, Sunarso J, Yang W, Yin Y, Glushenkov A M, Li L H, Howlett P C and Chen Y 2013 Ball milling: a green mechanochemical approach for synthesis of nitrogen doped carbon nanoparticles *Nanoscale* **5** 7970–6
[39] Chettri P, Vendamani V S, Tripathi A, Pathak A P and Tiwari A 2016 Self-assembly of functionalised graphene nanostructures by one step reduction of graphene oxide using aqueous extract of *Artemisia vulgaris Appl. Surf. Sci.* **362** 221
[40] Sathishkumar G, Pradeep K J, Vignesh V, Rajkuberan C, Jeyaraj M, Selvakumar M, Rakhi J and Sivaramakrishnan S 2016 Cannonball fruit (*Couroupita guianensis*, Aubl.) extract mediated synthesis of gold nanoparticles and evaluation of its antioxidant activity *J. Mol. Liq.* **215** 229
[41] González V J, Rodríguez A M, León V, Frontiñan-Rubio J, Fierro J L G, Durán-Prado M, Muñoz-García A B, Pavone M and Vázquez Er 2013 Sweet graphene: exfoliation of graphite and preparation of glucose-graphene cocrystals through mechanochemical treatments *Green Chem.* **20** 358–92
[42] Zhao W, Fang M, Wu F, Wu H, Wang L and Chen G 2010 Preparation of graphene by exfoliation of graphite using wet ball milling *J. Mater. Chem.* **20** 5817–9
[43] Calderón-Ayala G, Cortez-Valadez M, Mani-Gonzalez P G, Britto Hurtado R, Contreras-Rascón J I, Carrillo-Torres R C, Zayas M E, Castillo S J, Hernández-Martínez A R and Flores-Acosta M 2017 Green synthesis of reduced graphene oxide using ball milling *Carbon Letters* **21** 93–7
[44] Parsegian V A 2005 *Van der Waals Forces: A Handbook for Biologists, Chemists, Engineers, and Physicists* (New York: Cambridge University Press)
[45] McAllister M J *et al* 2007 Single sheet functionalized graphene by oxidation and thermal expansion of graphite *Chem. Mater.* **19** 4396–404
[46] Zhang C, Lv W, Xie X, Tang D, Liu C and Yang Q H 2013 Towards low temperature thermal exfoliation of graphite oxide for graphene production *Carbon* **62** 11–24
[47] Ahmed M M M, Imae T, Hill J P, Yamauchi Y, Ariga K and Shrestha L K 2018 Defect-free exfoliation of graphene at ultra-high temperature *Colloids Surf.* A **538** 127–32
[48] Wu W, Liu M, Gu Y, Guo B, Ma H, Wang P, Wang X and Zhang R 2020 Fast chemical exfoliation of graphite to few-layer graphene with high quality and large size via a two-step microwave-assisted process *Chem. Eng. J.* **381** 122592
[49] Zhu Y *et al* 2011 Carbon-based supercapacitors produced by activation of graphene *Science* **332** 1537–41
[50] Zhang C, Lv W, Xie X, Tang D, Liu C and Yang Q H 2013 Towards low temperature thermal exfoliation of graphite oxide for graphene production *Carbon* **62** 11–24
[51] Lv W *et al* 2009 Low temperature exfoliated graphenes: vacuum-promoted exfoliation and electrochemical energy storage *ACS Nano* **3** 3730–6

[52] Zhang H B, Wang J W, Yan Q, Zheng W G, Chen C and Yu Z Z 2011 Vacuum-assisted synthesis of graphene from thermal exfoliation and reduction of graphite oxide *J. Mater. Chem.* **21** 5392–7

[53] Kaniyoor A, Baby T T and Ramaprabhu S 2010 Graphene synthesis via hydrogen induced low temperature exfoliation of graphite oxide *J. Mater. Chem.* **20** 8467–9

[54] Shen B, Lu D D, Zhai W T and Zheng W G 2013 Synthesis of graphene by low-temperature exfoliation and reduction of graphite oxide under ambient atmosphere *J. Mater. Chem.* C **1** 50–3

[55] Singh C, Mishra A K and Paul A 2015 Highly conducting reduced graphene synthesis via low temperature chemically assisted exfoliation and energy storage application *J. Mater. Chem.* A **3** 18557–63

[56] Liu Z, Lin L, Ren H and Sun X 2017 CVD synthesis of graphene, thermal transport in carbon-based nanomaterials *Micro and Nano Technologies* (New York: Elsevier) ch 2 pp 19–56

[57] Yan K, Fu L, Peng H and Liu Z 2013 Designed CVD growth of graphene via process engineering *Acc. Chem. Res.* **46** 2263–74

[58] Li X *et al* 2009 Large-area synthesis of high-quality and uniform graphene films on copper foils *Science* **324** 1312–4

[59] Reina A, Jia X, Ho J, Nezich D, Son H, Bulovic V, Dresselhaus M S and Kong J 2009 Large area, few-layer graphene films on arbitrary substrates by chemical vapor deposition *Nano Lett.* **9** 30–5

[60] Bae S *et al* 2010 Roll-to-roll production of 30-inch graphene films for transparent electrodes *Nat. Nanotechnol.* **5** 574–8

[61] Dai B, Fu L, Zou Z, Wang M, Xu H, Wang S and Liu Z 2011 Rational design of a binary metal alloy for chemical vapour deposition growth of uniform single-layer graphene, *Nat. Commun.* **2** 522

[62] Zhang Y, Tang T T, Girit C, Hao Z, Martin M C, Zettl A, Crommie M F, Shen Y R and Wang F 2009 Direct observation of a widely tunable bandgap in bilayer graphene *Nature* **459** 820–3

[63] Liu L, Zhou H, Cheng R, Yu W J, Liu Y, Chen Y, Shaw J, Zhong X, Huang Y and Duan X 2012 High-yield chemical vapor deposition growth of high-quality large-area AB-stacked bilayer, graphene *ACS Nano* **6** 8241–9

[64] Liu Q, Gong Y, Wilt J S, Sakidja R and Wu J 2015 Synchronous growth of AB-stacked bilayer graphene on Cu by simply controlling hydrogen pressure in CVD process *Carbon* **93** 199–206

[65] Chen Y C, Lin W H, Tseng W S, Chen C C, Rossman G R, Chen C D, Wu Y S and Yeh N C 2020 Direct growth of mm-size twisted bilayer graphene by plasma-enhanced chemical vapor deposition *Carbon* **156** 212–24

[66] Vishwakarma R, Zhu R, Abuelwafa A A, Mabuchi Y, Adhikari S, Ichimura S, Soga T and Umeno M 2019 Direct synthesis of large-area graphene on insulating substrates at low temperature using microwave plasma CVD *ACS Omega* **4** 11263–70

[67] Chen Z, Lin Y, Rooks M and Avouris P 2007 Graphene nano-ribbon electronics *Physics* E **40** 228–32

[68] Han M Y, Özyilmaz B, Zhang Y and Kim P 2007 Energy band-gap engineering of graphene nanoribbons *Phys. Rev. Lett.* **98** 206805

[69] Jiao L, Zhang L, Wang X, Diankov G and Dai H 2009 Narrow graphene nanoribbons from carbon nanotubes *Nature* **458** 877–80
[70] Vo T H, Shekhirev M, Kunkel D A, Morton M D, Berglund E, Kong L, Wilson P M, Dowben P A, Enders A and Sinitskii A 2014 Large-scale solution synthesis of narrow graphene nanoribbons, *Nat. Commun.* **5** 3189
[71] Narita A *et al* 2014 Bottom-up synthesis of liquid-phase-processable graphene nanoribbons with near-infrared absorption *ACS Nano* **8** 11622–30
[72] https://www.iea.org/reports/sdg7-data-and-projections/access-to-electricity
[73] Ali N, Hussain A, Ahmed R, Wang M K, Zhao C, Ul Haq B and Fu Y Q 2016 Advances in nanostructured thin film materials for solar cell applications *Renew. Sustain. Energy Rev.* **59** 726–37
[74] Perez M and Perez R 2015 4A fundamental look at supply side energy reserves for the planet *IEA-SHCP-Newsletter* **62** 1–4
[75] Iqbal M Z and Rehman A U 2018 Recent progress in graphene incorporated solar cell devices *Sol. Energy* **169** 634–47
[76] Das S, Pandey D, Thomas J and Roy T 2018 The role of graphene and other 2D materials in solar photovoltaics *Adv. Mater.* **31** 1802722
[77] Fonash S 1981 *Solar Cell Device Physics* (New York: Academic)
[78] Janani M, Srikrishnarka P, Nair S V and Sreekumaran Nair A 2013 An in-depth review on the role of carbon nanostructures in dye-sensitized solar cells *J. Mater. Chem.* A
[79] The National Renewable Energy Laboratory Best Research Cell Efficiency Chart https://nrel.gov/pv/cell-efficiency.html
[80] Yin Z, Zhu J, He Q, Cao X, Tan C, Chen H, Yan Q and Zhang H 2013 Graphene-based materials for solar cell applications *Adv. Energy Mater.* **4** 1–9
[81] Wei D and Kivioja J 2013 Graphene for energy solutions and its industrialization *Nanoscale* **5** 10108–26
[82] Chen J, Li C, Eda G, Zhang Y, Lei W, Chhowalla M, Milne W I and Deng W Q 2011 Incorporation of graphene in quantum dot sensitized solar cells based on ZnO nanorods *Chem. Commun.* **47** 6084–6
[83] Kuma A 2017 Predicting efficiency of solar cells based on transparent conducting electrodes *J. Appl. Phys.* **121** 014502
[84] Betz U, Kharrazi Olsson M, Marthy J, Escolá M F and Atamny F 2006 Thin films engineering of indium tin oxide: large area flat panel displays application *Surf. Coat. Technol.* **200** 5751–9
[85] Nair R R, Blake P, Grigorenko A N, Novoselov K S, Booth T J, Stauber T, Peres N M R and Geim A K 2008 Fine structure constant defines visual transparency of graphene *Science* **320** 1308
[86] Kuzmenko A B, van Heumen E, Carbone F and van der Marel D 2008 Universal optical conductance of graphite *Phys. Rev. Lett.* **100** 117401
[87] Ma Y and Zhi L 2018 Graphene-based transparent conductive films: material systems, preparation and applications *Small Methods* **3** 1800199
[88] Li X, Zhu Y, Cai W, Borysiak M, Han B, Chen D, Piner R D, Colombo L and Ruoff R S 2009 Transfer of large-area graphene films for high-performance transparent conductive electrodes *Nano Lett.* **9** 4359–63

[89] Kim K S, Zhao Y, Jang H, Lee S Y, Kim J M, Kim K S, Ahn J H, Kim P, Choi J Y and Hong B H 2009 Large-scale pattern growth of graphene films for stretchable transparent electrodes *Nature* **457** 706–10
[90] Nair R R, Blake P, Grigorenko A N, Novoselov K S, Booth T J, Stauber T, Peres N M R and Geim A K 2008 Fine structure constant defines visual transparency of graphene *Science* **320** 1308
[91] Eda G, Fanchini G and Chhowalla M 2008 Large-area ultrathin films of reduced graphene oxide as a transparent and flexible electronic material *Nat. Nanotechnol.* **3** 270–4
[92] Hu X, Meng X, Xiong J, Huang Z, Yang X, Tan L and Chen Y 2017 Roll-to-roll fabrication of flexible orientated graphene transparent electrodes by shear force and one-step reducing post-treatment *Adv. Mater. Technol.* **2** 1700138
[93] Eda G, Lin Y Y, Miller S, Chen C W, Su W F and Chhowalla M 2008 Transparent and conducting electrodes for organic electronics from reduced graphene oxide *Appl. Phys. Lett.* **92** 233305
[94] Wu J, Becerril H A, Bao Z, Liu Z, Chen Y and Peumans P 2008 Organic solar cells with solution-processed graphene transparent electrodes *Appl. Phys. Lett.* **92** 263302
[95] Zhao Z, Fite J D, Haldar P and Lee J U 2012 Enhanced ultraviolet response using graphene electrodes in organic solar cells *Appl. Phys. Lett.* **101** 063305
[96] Cote L J, Kim F and Huang J 2009 Langmuir–Blodgett assembly of graphite oxide single layers *J. Am. Chem. Soc.* **131** 1043–9
[97] Jung Y U, Na S I, Kim H K and Kang S J 2012 Organic photovoltaic devices with low resistance multilayer graphene transparent electrodes *J. Vac. Sci. Technol.* A **30** 050604
[98] Zhu Y, Sun Z, Yan Z, Jin Z and Tour J M 2011 Rational design of hybrid graphene films for high-performance transparent electrodes *ACS Nano* **5** 6472–9
[99] Ahmada M S, Pandeya A K and Abd Rahim N 2017 Advancements in the development of TiO_2 photoanodes and its fabrication methods for dye sensitized solar cell (DSSC) applications. a review *Renew. Sustain. Energy Rev.* **77** 89–108
[100] Mathew S *et al* 2014 Dye-sensitized solar cells with 13% efficiency achieved through the molecular engineering of porphyrin sensitizers *Nat. Chem.* **6** 242–7
[101] Sadikin S N, Rahman M Y A, Umar A A and Aziz T H T 2019 Improvement of dye-sensitized solar cell performance by utilizing graphene-coated TiO_2 films photoanode *Superlattices Microstruct.* **128** 92–8
[102] Georgakilas V, Otyepka M, Bourlinos A B, Chandra V, Kim N, Kemp K C, Hobza P, Zboril R and Kim K S 2012 Functionalization of graphene: covalent and non-covalent approaches, derivatives and applications *Chem. Rev.* **112** 6156–214
[103] Taki M, Rezaei B, Fani N, Borandeh S, Abdolmaleki A and Ensafi A A 2017 Beneficial effects of amino acid-functionalized graphene nanosheets incorporated in the photoanode material of dye-sensitized solar cells: a practical and theoretical study *Appl. Surf. Sci.* **403** 218–29
[104] Yin Z, Sun S, Salim T, Wu S, Huang X, He Q, Lam Y M and Zhang H 2010 Organic photovoltaic devices using highly flexible reduced graphene oxide films as transparent electrodes *ACS Nano* **4** 5263–8
[105] Basu K, Selopal G S, Mohammadnezad M, Akilimali R, Wang Z M, Zhao H, Vetrone F and Rosei F 2020 Hybrid graphene/metal oxide anodes for efficient and stable dye sensitized solar cell *Electrochim. Acta* **349** 136409

[106] Kusumawati Y, Koussi-Daoud S and Pauporte T 2016 TiO_2/graphene nanocomposite layers for improving the performances of dye-sensitized solar cells using a cobalt redox shuttle *J. Photochem. Photobiol.* A **329** 54–60
[107] Sharifa N F M, Kadirc M Z A A, Shafiea S, Rashidb S A, Wan Hasana W Z and Shaban S 2019 Charge transport and electron recombination suppression in dye-sensitized solar cells using graphene quantum dots *Results Phys.* **13** 102171
[108] Subramaniam M R, Kumaresan D, Jothi S, McGettrickc J D and Watson T M 2018 Reduced graphene oxide wrapped hierarchical TiO_2 nanorod composites for improved charge collection efficiency and carrier lifetime in dye sensitized solar cells *Appl. Surf. Sci.* **428** 439–47
[109] Kundu S *et al* 2017 Enhancing the efficiency of DSSCs by the modification of TiO_2 photoanodes using N, F and S, co-doped graphene quantum dots *Electrochim. Acta* **242** 337–43
[110] Shu W, Liu Y, Peng Z, Chen K, Zhang C and Chen W 2013 Synthesis and photovoltaic performance of reduced graphene oxide–TiO_2 nanoparticles composites by solvothermal method *J. Alloys Compd.* **563** 229–33
[111] Cai H, Li J, Xu X, Tang H, Luo J, Binnemans K, Fransaer J and De Vos D E 2017 Nanostructured composites of one-dimensional TiO_2 and reduced graphene oxide for efficient dye sensitized solar cells *J. Alloys Compd.* **697** 132–7
[112] Zhang H, Lv Y, Yang C, Chen H and Zhou X 2018 One-step hydrothermal fabrication of TiO_2/reduced graphene oxide for high-efficiency dye-sensitized solar cells *J. Electron. Mater.* **47** 1630–7
[113] Akilimali R, Selopal G S, Benetti D, Esparza I S, Algarabel P A, Teresa J M D, Wang Z M, Stansfiel B, Zhao H and Rose F 2018 Hybrid TiO_2-graphene nanoribbon photoanodes to improve the T photoconversion efficiency of dye sensitized solar cells *J. Power Sources* **396** 566–73
[114] Patil J V, Mali S S, Shaikh J S, Patil A P, Patil P S and Hong C K 2019 Influence of reduced graphene oxide-TiO_2 composite nanofibers in organic indoline DN350 based dye sensitized solar cells *Synth. Met.* **256** 116146
[115] Wei L, Wang P, Yang Y, Dong Y, Fan R, Song W, Qiu Y, Yang Y and Luan T 2017 Enhanced performance of dye sensitized solar cells by using a reduced graphene oxide/TiO_2 blocking layer in the photoanode *Thin Solid Films* **639** 12–21
[116] Yu J-S 2012 High efficient Pt counter electrode prepared by homogeneous deposition method for dye-sensitized solar cell *Appl. Energy* **100** 132–7
[117] Seema H, Zafar Z and Samreen A 2020 Evaluation of solution processable polymer reduced graphene oxide transparent films as counter electrodes for dye-sensitized solar cells *Arabian J. Chem.* **13** 4978–86
[118] Ma J, Yuan S, Yang S, Lu H and Li Y 2018 Poly(3,4-ethylenedioxythiophene)/reduced graphene oxide composites as counter electrodes for high efficiency dye-sensitized solar cells *Appl. Surf. Sci.* **440** 8–15
[119] Lukaszkowicz K, Szindler M, Drygała A, Dobrzański L A, Prokopiuk vel Prokopowicz M, Pasternak I, Przewlokam A, Szindler M M and Domański M 2017 Graphene-based layers deposited onto flexible substrates: used in dye-sensitized solar cells as counter electrodes *Appl. Surf. Sci.* **424** 157–63
[120] Shahid M U, Mohamed N M, Muhsan A S, Bashiri R, Shamsudind A E and Azella Zaine S N 2019 Few-layer graphene supported polyaniline (PANI) film as a transparent counter electrode for dye-sensitized solar cells *Diamond Relat. Mater.* **94** 242–51

Liu Y-C, Zhai P, Lu M-N, Reddy K S K, Lee C-C, Tingare Y, Yeh C-Y and Wei T-C 2017 Platinum-free counter electrode using polymer-capped graphene nanoplatelets for cobalt (II)/(III)-mediated porphyrin-sensitized solar cells *Energy Technol.* **5** 756–64

[121] Roy-Mayhew J D, Bozym D J, Punckt C and Aksay I A 2010 Functionalized graphene as a catalytic counter electrode in dye-sensitized solar cells *ACS Nano* **4** 6203–11

[122] Choi H, Kim H, Hwang S, Han Y and Jeon M 2011 Graphene counter electrodes for dye-sensitized solar cells prepared by electrophoretic deposition *J. Mater. Chem.* **21** 7548–51

[123] Liu C Y, Huang K C, Chung P H, Wang C C, Chen C Y, Vittal R, Wu C G, Chiu W Y and Ho K C 2012 Graphene-modified polyaniline as the catalyst material for the counter electrode of a dye-sensitized solar cell *J. Power Sources* **217** 152–7

[124] He B, Tang Q, Wang M, Ma C and Yuan S 2014 Complexation of polyaniline and graphene for efficient counter electrodes in dye-sensitized solar cells: enhanced charge transfer ability *J. Power Sources* **256** 8–13

[125] Cai H, Tang Q, He B, Wang M, Yuan S and Chen H 2014 Self-assembly of graphene oxide/polyaniline multilayer counter electrodes for efficient dye-sensitized solar cells *Electrochim. Acta* **121** 136–42

[126] Wang M, Tang Q, Xu P, He B, Lin L and Chen H 2014 Counter electrodes from polyaniline–graphene complex/graphene oxide multilayers for dye-sensitized solar cells *Electrochim. Acta* **137** 1–20

[127] Ramasamy M S, Nikolakapoulou A, Raptis D, Dracopoulos V, Paterakis G and Lianos P 2015 Reduced graphene oxide/polypyrrole/PEDOT composite films as efficient Pt-free counter electrode for dye-sensitized solar cells *Electrochim. Acta* **173** 276–81

[128] Lee C P, Lai K Y, Lin C A, Li C T, Ho K C, Wu C I, Lau S P and He J H 2017 A paper-based electrode using a graphene dot/PEDOT:PSS composite for flexible solar cells *Nano Energy* **36** 260–7

[129] Nagavolu C, Susmitha K, Raghavender M, Giribabu L, Sankara Rao K B, Smith C T G, Mills C A, Silva S R P and Srikanth V V S S 2016 Pt-free spray coated reduced graphene oxide counter electrodes for dye sensitized solar cells *Sol. Energy* **137** 143–7

[130] Sahito I A, Sun K C, Arbaba A A and Jeong S H 2019 Synergistic effect of thermal and chemical reduction of graphene oxide at the counter electrode on the performance of dye-sensitized solar cells *Sol. Energy* **190** 112–8

[131] Kishore Kumar D, Kumar Swami S, Dutta V, Chen B, Bennett N and Upadhyaya H M 2019 Scalable screen-printing manufacturing process for graphene oxide platinum free alternative counter electrodes in efficient dye sensitized solar cells *Flat Chem.* **15** 100105

[132] Seema H, Zafar Z and Samreen A 2020 Evaluation of solution processable polymer reduced graphene oxide transparent films as counter electrodes for dye-sensitized solar cells *Arabian J. Chem.* **13** 4978–86

[133] Ma J, Yuan S, Yang S, Lu H and Li Y 2018 Poly(3,4-ethylenedioxythiophene)/reduced graphene oxide composites as counter electrodes for high efficiency dye-sensitized solar cells *Appl. Surf. Sci.* **440** 8–15

[134] Salleh S A, Rahman M Y A, Yumni Z and Aziz T H T 2020 Nickel sulphide-reduced graphene oxide composites as counter electrode for dye-sensitized solar cells: influence of nickel chloride concentration *Arabian J. Chem.* **13** 5191–7

[135] Shin S, Dao V D and Choi H S 2018 Incorporating MoFe alloys into reduced graphene oxide as counter electrode catalysts for dye-sensitized solar cells *Arab. J. Chem.* **13** 2414–24

[136] Givalou L, Tsichlis D, Zhang F, Karagianni C S, Terrones M, Kordatos K and Falara P 2019 Transition metal–graphene oxide nanohybrid materials as counter electrodes for high efficiency quantum dot solar cells *Catal. Today* **355** 860–9

[137] Pang B, Lin S, Shi Y, Wang Y, Chen Y, Ma S, Feng J, Zhang C, Yu L and Dong L 2019 Synthesis of $CoFe_2O_4$/graphene composite as a novel counter electrode for high performance dye-sensitized solar cells *Electrochim. Acta* **297** 70–6

[138] Murugadoss V, Panneerselvam P, Yan C, Guo Z and Angaiah S 2019 A simple one-step hydrothermal synthesis of cobalt-nickel selenide/graphene nanohybrid as an advanced platinum free counter electrode for dye sensitized solar cell *Electrochim. Acta* **312** 157–67

[139] Koo D *et al* 2020 Flexible organic solar cells over 15% efficiency with polyimide-integrated graphene electrodes *Joule* **4** 1–4

[140] Ma J, Shen W and Yu F 2017 Graphene-enhanced three-dimensional structures of MoS_2 nanosheets as a counter electrode for Pt-free efficient dye-sensitized solar cells *J. Power Sources* **351** 58–66

[141] Ramar A, Saraswathia R, Vilian A T E, Chen S M and Wang F M 2017 Polyisothianaphthene/graphene nanocomposite as a new counter electrode material for high performance dye sensitized solar cell *Synth. Met.* **230** 58–64

[142] Jeon Y J, Yun J M, Kim D Y, Na S I and Kim S S 2012 High-performance polymer solar cells with moderately reduced graphene oxide as an efficient hole transporting layer *Sol. Energy Mater. Sol. Cells* **105** 96

[143] Amolloa T A, Mola G T and Nyamori V O 2018 High-performance organic solar cells utilizing graphene oxide in the active and hole transport layers *Sol. Energy* **171** 83–91

[144] Lee D Y, Na S I and Kim S S 2016 Graphene oxide/PEDOT:PSS composite hole transport layer for efficient and stable planar heterojunction perovskite solar cells *Nanoscale* **8** 1513–22

[145] Liu J, Xue Y, Gao Y, Yu D, Durstock M and Dai L 2012 Hole and electron extraction layers based on graphene oxide derivatives for high-performance bulk heterojunction solar cells *Adv. Mater.* **24** 2228–33

[146] Wang D H, Kim J K, Seo J H, Park I, Hong B H, Park J H and Heeger A J 2013 Transferable graphene oxide by stamping nanotechnology: electron-transport layer for efficient bulk-heterojunction solar cells *Angew. Chem. Int. Ed.* **52** 2874–80

[147] Smith C T G, Rhodes R W, Beliatis M J, Imalka Jayawardena K D G, Rozanski L J, Mills C A, Ravi S and Silva P 2014 Graphene oxide hole transport layers for large area, high efficiency organic solar cells *Appl. Phys. Lett.* **105** 073304

[148] Cheng X, Long J, Wu R, Huang L, Tan L, Chen L and Chen Y 2017 Fluorinated reduced graphene oxide as an efficient hole-transport layer for efficient and stable polymer solar cells *ACS Omega* **2** 2010–6

[149] Ozcan S, Erer M C, Vempati S, Uyar T, Toppare L and Çırpan A 2020 Graphene oxide-doped PEDOT:PSS as hole transport layer in inverted bulk heterojunction solar cell *J. Mater. Sci., Mater. Electron.* **31** 3576–84

[150] Wang S, Huang X, Sun H and Wu C 2017 Hybrid UV-ozone-treated rGO-PEDOT:PSS as an efficient hole transport material in inverted planar perovskite solar cells *Nanoscale Res. Lett.* **12** 619

[151] Lee D Y, Na S I and Kim S S 2016 Graphene oxide/PEDOT:PSS composite hole transport layer for efficient and stable planar heterojunction perovskite solar cells *Nanoscale* **8** 1513–22

[152] Collazo J N, Maldonado J L, Cruz J S, Argüelles D B, Quintana I C, Espinosa C V and Borja D R 2019 Functionalized and reduced graphene oxide as hole transport layer and for use in ternary organic solar cell *Opt. Mater.* **98** 109434
[153] Kwon S N, Jung C H and Na S I 2016 Electron-beam-induced reduced graphene oxide as an alternative hole-transporting interfacial layer for high-performance and reliable polymer solar cells *Org. Electron.* **34** 67–74
[154] Rafique S, Abdullah S M, Shahid M M, Ansari M O and Sulaiman K 2017 Significantly improved photovoltaic performance in polymer bulk heterojunction solar cells with graphene oxide/PEDOT:PSS double decked hole transport layer *Sci. Rep.* **7** 39555
[155] Li D, Cui J, Li H, Huang D, Wang M and Shen Y 2016 Graphene oxide modified hole transport layer for $CH_3NH_3PbI_3$ planar heterojunction solar cells *Sol. Energy* **131** 176–82
[156] Agresti A, Pescetelli S, Taheri B, EsaDel Rio Castillo A, Cinà L, Bonaccorso F and di Carlo A 2016 Graphene–perovskite solar cells exceed 18% efficiency: a stability study *ChemSusChem* **9** 1–2
[157] Li X and Zhu H 2016 The graphene–semiconductor Schottky junction *Phys. Today* **69** 46
[158] Ye Y and Dai L 2012 Graphene-based Schottky junction solar cells *J. Mater. Chem.* **22** 24224–9
[159] Kong X, Zhang L, Liu B, Gao H, Zhang Y, Yan H and Song X 2019 Graphene/Si Schottky solar cells: a review of recent advances and prospects *RSC Adv.* **9** 863–77
[160] Li X, Zhu H, Wang K, Cao A, Wei J, Li C, Jia Y, Li Z, Li X and Wu D 2010 Graphene-on-silicon Schottky junction solar cells *Adv. Mater.* **22** 2743–8
[161] Xie C *et al* 2011 Monolayer graphene film/silicon nanowire array Schottky junction solar cells *Appl. Phys. Lett.* **99** 133113
[162] Feng T *et al* 2011 Graphene based Schottky junction solar cells on patterned silicon-pillar-array substrate *Appl. Phys. Lett.* **99** 233505
[163] Lin Y, Li X, Xie D, Feng T, Chen Y, Song R, Tian H, Ren T, Wang K and Zhu H 2013 Graphene/semiconductor heterojunction solar cells with modulated antireflection and graphene work function *Energy Environ. Sci.* **6** 108–15
[164] Li Y F, Yang W, Tu Z Q, Liu Z C, Yang F, Zhang L Q and Hatakeyama R 2014 Schottky junction solar cells based on graphene with different numbers of layers *Appl. Phys. Lett.* **104** 043903
[165] Zhang X, Xie C, Jie J, Zhang X, Wu Y and Zhang W 2013 High-efficiency graphene/Si nanoarray Schottky junction solar cells via surface modification and graphene doping *J. Mater. Chem.* A **1** 6593–601
[166] Feng T *et al* 2012 Efficiency enhancement of graphene/silicon-pillar-array solar cells by HNO_3 and PEDOT-PSS *Nanoscale* **4** 2130–3
[167] Rehman M A, Akhtar I, Choi W, Akbar K, Farooq A, Hussain S, Shehzad M A, Chun S H, Jung J and Seo Y 2018 Influence of an Al_2O_3 interlayer in a directly grown graphene-silicon Schottky junction solar cell *Carbon* **132** 157–64
[168] Xie C *et al* 2012 Schottky solar cells based on graphene nanoribbon/multiple silicon nanowires junctions *Appl. Phys. Lett.* **100** 193103
[169] Jie W, Zheng F and Hao J 2013 Graphene/gallium arsenide-based Schottky junction solar cells *Appl. Phys. Lett.* **103** 233111
[170] Wen L, Gao F, Yu Y, Xu Z, Liu Z, Gao P, Zhang S and Li G 2018 Enhancing the photovoltaic performance of GaAs/graphene Schottky junction solar cells by interfacial modification with self assembled alkyl thiol monolayer *J. Mater. Chem.* A **6** 17361–70

[171] Luo Y, Yan X, Zhang J, Li B, Wu Y, Lu Q, Jin C, Zhang X and Ren X 2018 A graphene/single GaAs nanowire Schottky junction photovoltaic device *Nanoscale* **10** 9212–7
[172] Liu X, Zhang X W, Meng J H, Yin Z G, Zhang L Q, Wang H L and Wu J L 2015 High efficiency Schottky junction solar cells by co-doping of graphene with gold nanoparticles and nitric acid *Appl. Phys. Lett.* **106** 233901
[173] Ho P H *et al* 2015 Sunlight-activated graphene-heterostructure transparent cathodes: enabling high-performance n-graphene/p-Si Schottky junction photovoltaics *Energy Environ. Sci.* **8** 2085–92
[174] Xie C, Zhang X, Ruan K, Shao Z, Dhaliwal S S, Wang L, Zhang Q, Zhang X and Jie J 2013 High-efficiency, air stable graphene/Si micro-hole array Schottky junction solar cells *J. Mater. Chem.* A **1** 15348–54
[175] Gao Z, Jin W, Li Y, Song Q, Wang Y, Zhang K, Wang S and Dai L 2015 Flexible solar cells based on CdSe nanobelt/graphene Schottky junctions *J. Mater. Chem.* C **3** 4511–4
[176] Ye Y, Gan L, Dai L, Dai Y, Guo X, Meng H, Yu B, Shi Z, Shanga K and Qin G 2011 A simple and scalable graphene patterning method and its application in CdSe nanobelt/graphene Schottky junction solar cells *Nanoscale* **3** 1477–81
[177] Alnuaimi A, Almansouri I, Saadat I and Nayfeh A 2018 High performance graphene-silicon Schottky junction solar cells with HfO_2 interfacial layer grown by atomic layer deposition *Sol. Energy* **164** 174–9
[178] Jaque D, Martınez Maestro L, del Rosal B, Haro-Gonzalez P, Benayas A, Plaza J L, Martın Rodrıguez E and Garcıa Sole J 2014 Nanoparticles for photothermal therapies *Nanoscale* **6** 9494–530
[179] Sahadev N and Anappara A A 2017 Enhanced photothermal effect in reduced graphene oxide in solid-state *J. Appl. Phys.* **122** 185103
[180] Jaque D, Martınez Maestro L, del Rosal B, Haro-Gonzalez P, Benayas A, Plaza J L, Martın Rodrıguez E and Garcıa Sole J 2014 Nanoparticles for photothermal therapies *Nanoscale* **6** 9494–530
[181] Hessel C M, Pattani V P, Rasch M, Panthani M G, Koo B, Tunnell J W and Korgel B A 2011 Copper selenide nanocrystals for photothermal therapy *Nano Lett.* **11** 2560–6
[182] Zhao Y, Dunn A, Lin J and Shi D 2019 Photothermal effect of nanomaterials for efficient energy applications, novel nanomaterials for biomedical, environmental and energy applications *Micro and Nano Technologies: Novel Nanomaterials for Biomedical, Environmental and Energy Applications* (Amsterdam: Elsevier) pp 415–34
[183] Liu J, Liang H, Li M, Luo Z, Zhang J, Guo X and Cai K 2018 Tumor acidity activating multifunctional nanoplatform for NIR-mediated multiple enhanced photodynamic and photothermal tumor therapy *Biomaterials* **157** 107–24
[184] Wang S *et al* 2015 Plasmonic copper sulfide nanocrystals exhibiting near-infrared photothermal and photodynamic therapeutic effects *ACS Nano* **9** 1788–800
[185] Chen Y, Khan A R, Yu D, Zhai Y, Ji J, Shid Y and Zhai G 2019 Pluronic F127-functionalized molybdenum oxide nanosheets with pH-dependent degradability for chemo-photothermal cancer therapy *J. Colloid Interface Sci.* **553** 567–80
[186] Peng H, Tang S, Tian Y, Zheng R, Zhou L and Yang W 2016 Highly ligand-directed and size-dependent photothermal properties of magnetite particles *Part. Part. Syst. Charact.* **33** 332–40
[187] Fernandes N, Rodrigues C F, Moreira A F and Correia I J 2020 Overview of the application of inorganic nanomaterials in cancer photothermal therapy *Biomater. Sci.* **8** 2990–3020

[188] Augustine S, Singh J, Srivastava M, Sharma M, Das A and Malhotra B D 2017 Recent advances in carbon based nanosystems for cancer theranostics *Biomater. Sci.* **5** 901–52
[189] Hui Y, Chang H-C, Dong H and Zhang X 2019 *Carbon Nanomaterials for Bioimaging, Bioanalysis, and Therapy* 1st edn ed Yuen (New York: Wiley)
[190] Chen Y W, Su Y L, Hu S H and Chen S Y 2016 Functionalized graphene nanocomposites for enhancing photothermal therapy in tumor treatment *Adv. Drug Delivery Rev.* **105** 190–204
[191] Nan H Y, Ni Z H, Wang J, Zafar Z, Shi Z X and Wang Y Y 2013 The thermal stability of graphene in air investigated by Raman spectroscopy *J. Raman Spectrosc.* **44** 1018–21
[192] Balandin A A, Ghosh S, Bao W, Calizo I, Teweldebrhan D, Miao F and Lau C N 2008 Superior thermal conductivity of single-layer graphene, *Nano Lett.* **8** 902–7
[193] Melamed J R, Edelstein R S and Day E S 2015 Elucidating the fundamental mechanisms of cell death triggered by photothermal therapy *ACS Nano* **9** 6–11
[194] Wu M C, Deokar A R, Liao J H, Shih P Y and Ling Y C 2013 Graphene-based photothermal agent for rapid and effective killing of bacteria *ACS Nano* **7** 1281–90
[195] Yang K, Hu L, Ma X, Ye S, Cheng L, Shi X, Li C, Li Y and Liu Z 2012 Multimodal imaging guided photothermal therapy using functionalized graphene nanosheets anchored with magnetic nanoparticles *Adv. Mater.* **24** 1868–72
[196] Prise K M and O'Sullivan J M 2009 Radiation-induced bystander signalling in cancer therapy *Nat. Rev. Cancer* **9** 351–60
[197] Yu J, Lin Y H, Yang L, Huang C C, Chen L, Wang W C, Chen G W, Yan J, Sawettanun S and Lin C H 2016 Improved anticancer photothermal therapy using the bystander effect enhanced by antiarrhythmic peptide conjugated dopamine-modified reduced graphene oxide nanocomposite *Adv. Healthcare Mater.* **6** 1–9
[198] Brookmeyer R, Johnson E, Ziegler-Graham K and Arrighi H M 2007 Forecasting the global burden of Alzheimer's disease *Alzheimers Dement.* **3** 186–91
[199] Li M, Zhao A, Dong K, Li W, Ren J and Qu X 2015 Chemically exfoliated WS_2 nanosheets efficiently inhibit amyloid β-peptide aggregation and can be used for photothermal treatment of Alzheimer's disease *Nano Res.* **8** 3216–27
[200] Yin T, Xie W, Sun J, Yang L and Liu J 2016 Penetratin peptide-functionalized gold nanostars: enhanced BBB permeability and NIR photothermal treatment of Alzheimer's disease using ultralow irradiance *ACS Appl. Mater. Interfaces* **8** 19291–302
[201] Ma M, Gao N, Sun Y, Du X, Ren J and Qu X 2018 Redox-activated near-infrared-responsive polyoxometalates used for photothermal treatment of Alzheimer's disease *Adv. Healthcare Mater.* **7** 1800320
[202] Liu D *et al* 2019 Using near-infrared enhanced thermozyme and *scFv* dual-conjugated Au nanorods for detection and targeted photothermal treatment of Alzheimer's disease *Theranostics* **9** 2268–81
[203] Li M, Yang X, Ren J, Qu K and Qu X 2012 Using graphene oxide high near-infrared absorbance for photothermal treatment of Alzheimer's disease *Adv. Mater.* **24** 1722–8
[204] Yang Z, Ge C, Liu J, Chong Y, Gu Z, Jimenez-Cruz C A, Chai Z and Zhou R 2015 Destruction of amyloid fibrils by graphene through penetration and extraction of peptides *Nanoscale* **7** 18725–37
[205] Chen J, Liu H, Zhao C, Qin G, Xi G, Li T, Wang X and Chen T 2014 One-step reduction and PEGylation of graphene oxide for photothermally controlled drug delivery *Biomaterials* **35** 4986–95

[206] Hai L, He D, He X, Wang K, Yang X, Liu J, Cheng H, Huang X and Shangguan J 2017 Facile fabrication of a resveratrol loaded phospholipid@reduced graphene oxide nano-assembly for targeted and near-infrared laser-triggered chemo/photothermal synergistic therapy of cancer *in vivo J. Mater. Chem.* B **5** 5783–92

[207] Vivero-Escoto J L, Slowing I I, Trewyn B G and Lin V S Y 2010 Mesoporous silica nanoparticles for intracellular controlled drug delivery *Small* **6** 1952–67

[208] Chen Y, Chen H and Shi J 2013 *In vivo* bio-safety evaluations and diagnostic/therapeutic applications of chemically designed mesoporous silica nanoparticles *Adv. Mater.* **25** 3144–76

[209] Wang T T, Lan J, Zhang Y, Wu Z L N, Li C M, Wang J and Huang C Z 2015 Reduced graphene oxide gated mesoporous silica nanoparticles as a versatile chemo-photothermal therapy system through pH controllable release *J. Mater. Chem.* B **3** 6377–84

[210] Yao X, Tian Z, Liu J, Zhu Y and Hanagata N 2017 Mesoporous silica nanoparticles capped with graphene quantum dots for potential chemo-photothermal synergistic cancer therapy *Langmuir* **33** 591–9

[211] Shao L, Zhang R, Lu J, Zhao C, Deng X and Wu Y 2017 Mesoporous silica coated polydopamine functionalized reduced graphene oxide for synergistic targeted chemo-photothermal therapy *ACS Appl. Mater. Interfaces* **9** 1226–36

[212] Rahmati M and Mozafari M 2019 Biological response to carbon-family nanomaterials: interactions at the nano-bio interface *Front. Bioeng. Biotechnol.* **7** 4

[213] Ameri T, Lia N and Brabec C J 2013 Highly efficient organic tandem solar cells: a follow up review *Energy Environ. Sci.* **6** 2390

[214] Hu C, Liu D, Xiao Y and Dai L 2018 Functionalization of graphene materials by heteroatom-doping for energy conversion and storage *Prog. Natural Sci.: Mater. Int.* **28** 121–32

Part II

Environmental Remediation

Chapter 9

Metal doped iron (III) oxide nanomaterials for wastewater treatment

N Jayaprakash, E Sundaravadivel, S Rajalakshmi and R Suresh

Metal ion doped iron(III) oxide (Fe_2O_3) nanoparticles are promising materials in the field of environmental remediation. Because dopants significantly influence the phase control, active surface area, band gap, band potentials, magnetic behavior, electrical properties, morphology, and facets of Fe_2O_3 it is thus possible to tune the performances of doped Fe_2O_3 in pollution removal studies. Moreover, Fe_2O_3 based materials are non-toxic, highly stable, and cheap. In this chapter, the different kinds of synthesis methods reported for metal doped Fe_2O_3 are outlined. The effects of dopants on the crystallite phase, size, band gap energy, morphology, magnetic behavior, and electrical conductivity of Fe_2O_3 particles are described. Then, applications of metal doped Fe_2O_3 as an adsorbent, catalyst, Fenton catalyst, photocatalyst, and electrocatalyst in the removal of inorganic (such as arsenic, mercury, lead, and copper) and organic pollutants (dyes and emerging contaminants) in water and as a disinfectant for pathogens are discussed with reference to the recent literature. The important factors responsible for the observed better performance of doped Fe_2O_3 are also mentioned. Finally, a summary on the research efforts on metal ion doped Fe_2O_3 in pollution removal is provided.

9.1 Introduction

In recent decades, due to the human population explosion, deforestation, and rapid industrial development, pollution has occurred in the air, water, and soil ecosystems and has become a serious problem in the world. It has been reported that nine million deaths occur per year due to diseases related to pollution [1–3]. Hence, advanced treatments are needed for minimization of water pollution. The available pollution removal techniques, including ion-exchange, membrane process, activated sludge process, reverse osmosis, and precipitation methods, are not sufficient to cure

doi:10.1088/978-0-7503-5143-0ch9

environmental pollution. Therefore effective methods must be developed to remove pollutants. In this context, techniques such as adsorption processes, catalysis, photocatalysis, electrocatalysis, and advanced oxidation processes are being developed for the treatment of wastewater. Due to the rapid growth of nanotechnology, various kinds of nanomaterials such as metal oxides, plasmonic metals, carbon based materials, metal organic frameworks, black phosphorous based materials, alloys, inorganic–organic hybrid materials, MXenes, and zero-valent iron nanoparticles are used in the detoxification of toxic pollutants in water, air, and soil [4–8]. Among them, iron(III) oxide materials have received widespread attention from researchers because of their large surface area, abundant material, ease of synthesis, visible light absorption properties, high reactivity, magnetic properties, non-toxicity, and redox properties [9]. There are four forms of iron(III) oxides: α-Fe_2O_3, β-Fe_2O_3, γ-Fe_2O_3, and ε-Fe_2O_3. Among them, α-Fe_2O_3 and γ-Fe_2O_3 are well-known materials that are used in the removal of pollutants in water and air. A few examples of the application of pure Fe_2O_3 nanoparticles in the detoxification of pollutants in water/air are listed in table 9.1 [10–25].

Table 9.1. Applications of pure Fe_2O_3 in the decontamination of pollutants.

Material	Role	Condition	Pollutant/efficiency	Reference
β-Fe_2O_3	Photo-Fenton reagent	Visible light (<650 nm)/H_2O_2	Rhodamine B (94.8%)	[10]
α-Fe_2O_3	Catalyst		H_2S	[11]
α-Fe_2O_3 mesoporous network	Fenton reagent	1 ml 30% H_2O_2	Methylene blue (~100%)	[12]
Coral α-Fe_2O_3	Photo-Fenton reagent	32 W ultraviolet/ H_2O_2	Methylene blue (95%), methyl orange (94%), bromo green (94%), methyl red (76%)	[13]
Flower α-Fe_2O_3	Adsorbent		As(V) (41.46 mg g^{-1}) Cr(VI) (33.82 mg g^{-1})	[14]
γ-Fe_2O_3 nanoparticles	Adsorbent		Pb(II) (10.55 mg g^{-1}) Zn(II) (4.79 mg g^{-1}) Cd(II) (1.75 mg g^{-1})	[15]
α-Fe_2O_3 microsphere	Catalyst	Persulfate	Rhodamine B (100%)	[16]
α-Fe_2O_3 nanotube arrays	Photoelectrocatalyst	Visible light	Azo dye	[17]
α-Fe_2O_3 nanotubes	Catalyst	Heat energy	Ammonium perchlorate	[18]
Fe_2O_3	Thermocatalyst	Thermal energy	CO_2 decomposition	[19]

Fe_2O_3 hollow box	Photo-Fenton reagent		Malachite green (99.9%)	[20]
γ-Fe_2O_3	Adsorbent		Vanadium (97.6%)	[21]
Fe_2O_3	Adsorbent		Anionic dye (91%)	[22]
Fe_2O_3	Photocatalyst	Light/H_2O_2	Crystal violet (~100%) Methylene blue (~100%)	[23]
Fe_2O_3	Fenton catalyst	$NaHSO_3$	Orange II (90%)	[24]
Porous Fe_2O_3	Adsorbent and photocatalyst		Pb(II), (97.8 mg g^{-1}) Cr(VI) 175.5 (mg g) Humic acid (159.4 mg g^{-1}) Methyl blue (425.9 mg g^{-1})	[25]

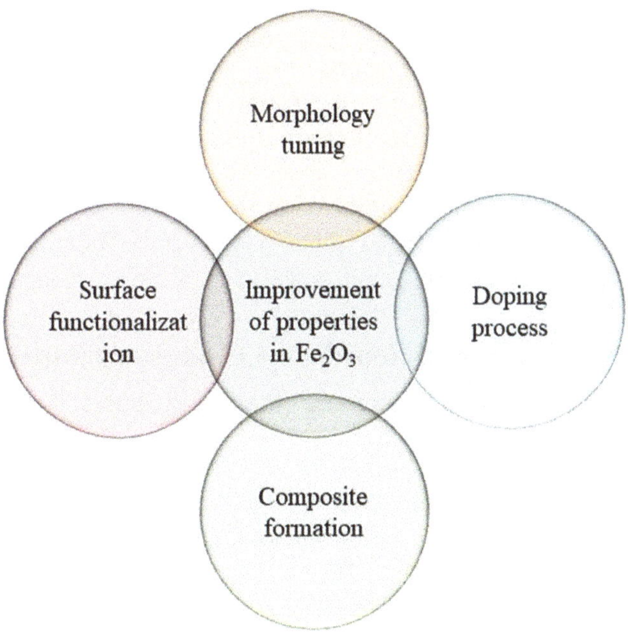

Figure 9.1. Improvement strategies for enhanced performance of Fe_2O_3 in environmental remediation.

However, pure Fe_2O_3 nanomaterials suffer in practical applications due to their instability in experimental conditions and low removal efficiency. To eliminate these issues, the following four strategies have been developed (figure 9.1): (i) morphology tuning; (ii) surface functionalization; (iii) metal/non-metal doping; and (iv) composite formation. For example, Wang *et al* prepared α-Fe_2O_3 polyhedrons, rods, ellipses, and cylinders and used these for photocatalytic decomposition of methylene blue in the presence of H_2O_2. The photocatalytic efficiency followed the order polyhedrons (98%) > rods (75%) > ellipses (50%) > cylinders (44%). The reason for the enhanced performance of α-Fe_2O_3 polyhedrons is due to the higher surface energy of the {101} facets while α-Fe_2O_3 rods, ellipses, and cylinders show {001} facets with low surface energy [26]. Sun *et al* used phosphate modified α-Fe_2O_3 nanoparticles (–Fe–O–P–OH) as visible light driven photocatalysts for degradation of phenol in liquid form and acetaldehyde in gaseous form with greater degradation efficiency. They found that the phosphate modified α-Fe_2O_3 surface significantly promotes the adsorbed O_2 to capture photogenerated electrons [27]. Bin *et al* synthesized carbon-doped Fe_2O_3 octahedrons and used them as heterogeneous Fenton catalysts to oxidize methylene blue in the presence of H_2O_2. The removal efficiency was found to be 96% under visible light illumination. The greater observed Fenton-like performance is attributed to the large surface area and the synergistic effects between the Fe_2O_3 and the carbon dopant in Fe_2O_3 [28]. Co^{2+} doped Fe_2O_3 magnetic nanoparticles were also reported as photocatalysts assisted by visible light for the degradation of pararosaniline in a solution. The photocatalytic efficiency was attributed mainly to the Co concentration. Among other materials, 5% Co doped Fe_2O_3 exhibits the highest photocatalytic degradation efficiency toward pararosaniline [29]. In 2020 Shen *et al* developed an Fe_2O_3@microalgae composite for adsorption of Cr(VI), Cu(II), Pb(II), and Cd(II) ions from water. The adsorption capacity of the Fe_2O_3@microalgae composite for metal ions was in the following order: Cr(VI) > Pb(II) > Cd(II) > Cu(II). The hydroxyl (–OH) and carboxyl functional groups (–COOH) were involved in the adsorption process [30].

In this chapter the vital role of metal ion doped Fe_2O_3 in environmental remediation is discussed.

9.2 The synthesis of metal doped Fe_2O_3 nanostructures

Metal ion doped Fe_2O_3 nanostructures are obtained by different synthesis methods, which are listed below:

(i) *Precipitation method.*

In the precipitation method, metal doped Fe_2O_3 nanoparticles are precipitated in an aqueous solution using precipitating agents. For example, Mansour *et al* prepared Sn doped α-Fe_2O_3 nanospheres by using $FeCl_3{\cdot}6H_2O$ and $SnCl_4{\cdot}5H_2O$ and ammonia solution as the precipitating agent. The diameter of the Sn doped α-Fe_2O_3 nanospheres was determined to be 20–35 nm, which is smaller in size compared to pure α-Fe_2O_3 [31]. Similarly, Ce doped α-Fe_2O_3 nanoparticles have also been synthesized using a one-pot co-precipitation method [32]. The Ga doped α-Fe_2O_3 with different molar ratios of Ga/Fe were prepared by Xiao *et al* using a facile parallel flow co-precipitation method [33]. Guo *et al* recently synthesized calcium-doped α-Fe_2O_3 nanoparticles by a precipitation–calcination method, in which non-toxic precursors were used. The mean nanoparticle size of the calcium-doped α-Fe_2O_3 was in the range of 10–20 nm [34].

(ii) *Surfactant assisted precipitation.*

The use of surfactants in the precipitation method determines the morphology of the Fe_2O_3 particles. For example, Mohapatra *et al* prepared Mg doped Fe_2O_3 nanoparticles using a CTAB as surfactant agent [35]. The shape of the particles was found to be in the range of 40–200 nm.

(iii) *Hydrolysis method.*

Fe_2O_3 nanostructures can be synthesized easily by a hydrolysis method in which aqueous ferric chloride solution is heated (~100 °C) followed by calcination at about 400 °C. This method is also applied to prepare Co doped α-Fe_2O_3 nanoparticles using an appropriate amount of cobaltous chloride [36].

(iv) *Hydrothermal method.*

Generally, hydrothermal synthesis methods have a number of advantages, such as fewer impurities in the final products, shorter time, and control of the shape and size of particles. This method is also used to prepare metal doped α-Fe_2O_3 nanostructures. For example, Baig *et al* have prepared V doped Fe_2O_3 particles using $FeCl_3$ and VCl_3 as iron and vanadium precursors. The precursor aqueous solution was kept in an autoclave at 100 °C for 24 h and the resulting powder was annealed at 550 °C for 2 h. The V doped α-Fe_2O_3 cubic particles were obtained with size range of 350 nm to 400 nm [37]

(v) *Solid state method.*

Sr doped Fe_2O_3 was prepared using a solid state method by using Fe_2O_3 and strontium carbonate powders. The homogeneous mixture was annealed at 900 °C to obtain the doped samples [38].

(vi) *Solution combustion method.*

Jawad *et al* prepared Cr^{3+} doped γ-Fe_2O_3 nanostructures using a solution combustion method. Along with iron and chromium nitrates, citric acid was used as an oxidizing agent and fuel. The crystallite size was found to be 16.9 nm to 25.6 nm for samples with different Fe/Cr ratios [39].

(vii) *Microwave method.*

The microwave irradiation method has many advantages. It is simple and cost effective, has high yields and high purity, takes less time, and reduces energy waste. Sangaiya *et al* adopted a microwave irradiation process for the preparation of Sn doped Fe_2O_3 nanoparticles, in which $FeCl_3$ and $SnCl_2$ aqueous solutions were irradiated with microwaves for 15 min. The morphology of the final product was spherical in shape with a 24 nm to 5 nm size in diameter [40].

(viii) *Electrochemical method.*

Qi *et al* reported an electrochemical method for the preparation of Zn doped α-Fe_2O_3 nanotube arrays [41]. Similarly, Wang *et al* developed Ti and Sn co-doped anodic α-Fe_2O_3 films using a simple anodic deposition technique by combining a $TiCl_4$ treatment and Sn doping on FeOOH layers [42].

(ix) *Chemical vapor deposition.*

Lim *et al* synthesized Pt doped γ-Fe_2O_3 thin film using a plasma-enhanced chemical vapor deposition technique [43]. Pulsed laser deposition has also been carried out recently for the synthesis of Ti doped α-Fe_2O_3 films from Fe_3O_4 and TiO_2 by Franceschi *et al* [44].

(x) *Electro-spinning method.*

Santangelo *et al* fabricated Ti or Si doped Fe_2O_3 fibers using an electro-spinning method over conductive fluorine doped tin oxide/glass substrates [45].

(xi) *Ball milling method.*

Lemine *et al* prepared Ni doped α-Fe_2O_3 nanostructures using a high-energy ball milling method from a mixture of nickel metal (Ni) and α-Fe_2O_3 [46].

9.3 Properties of metal doped Fe_2O_3 nanostructures

(i) *Crystallite size.*

The average crystallite size of Fe_2O_3 is strongly affected by the dopant elements. In Cr doped γ-Fe_2O_3 ($Cr_xFe_{2-x}O_3$), a decrease in crystallite size (16.91 nm to 25.61 nm) with the increase in Cr^{3+} concentration ($x = 0.00$ to 0.10) was observed. It is due to the smaller ionic radius of Cr^{3+} (0.061 nm) compared to Fe^{3+} (0.064 nm) [39]. Similarly, doping of Ce into α-Fe_2O_3 crystal lattice causes a significant decrease in average crystalline size [32]. However, in case of Mg doped Fe_2O_3, the average crystallite size increases from 29 nm to 87 nm with the increase of Mg concentration [47].

(ii) *Crystalline phase.*

The phase of Fe_2O_3 is controlled by the dopants. For example, in the absence of dopant, α-Fe_2O_3 fibers were obtained. Even Ti doping does not cause any changes in the phase of Fe_2O_3, while Si doping induces the formation of γ-Fe_2O_3 [45]. Similarly, Ni dopant also causes the growth of γ-Fe_2O_3 [48].

(iii) *Band gap.*

The optical band gap of Fe_2O_3 is altered by the dopants. For example, an increase in the band gap with an increase of Mg concentration was observed in Mg doped Fe_2O_3 thin films [47]. On the other hand, the band gap decreases with increasing concentration of cobalt ions in Fe_2O_3 [29].

(iv) *Morphology.*

The dopant element has major role on the morphology of Fe_2O_3. For example, pure α-Fe_2O_3 powder displays a pseudo-cubic morphology, while 2.4% Eu^{3+} doped α-Fe_2O_3 powder exhibits rice-grain morphology [49].

(v) *Magnetic property.*

The magnetic property of pure Fe_2O_3 can be tuned by using appropriate dopant concentrations. For example, Bhat *et al* investigated the effect of Dy^{3+} ion doping on the magnetic property of α-Fe_2O_3 (α-$Dy_xFe_{2-x}O_3$). They found that the saturation magnetization increases through doping of

($x = 0.03$) Dy^{3+} in α-Fe_2O_3. Therefore, α-$Dy_{0.03}Fe_{1.97}O_3$ particles saturate at 10 kOe while pure α-Fe_2O_3 does not saturate up to 20 kOe [50].

(vi) *Conductivity.*

α-Fe_2O_3 has poor electrical conductivity. Hence, metal ion doping in α-Fe_2O_3 makes it either a p-type or n-type conductor. For example, according to Liao *et al*, Zr, Si, or Ge doping can induce n-type nature in α-Fe_2O_3. These dopants do not act as electron trapping sites within α-Fe_2O_3. Also, the covalent interactions between dopants and oxygen are greater in number and provide more charge carriers [51].

(vii) *Oxygen vacancy defects.*

The doping process can induce surface oxygen defects on Fe_2O_3 particles. For example, Wang *et al* prepared Ce doped α-Fe_2O_3 particles with high oxygen defects. Oxygen defects can improve the carrier density and provide more oxygen adsorption sites which will be beneficial for its catalytic activity [52].

9.4 Applications of doped Fe_2O_3 in environmental remediation

Environmental remediation processes involve the capturing, degradation, and mineralization of pollutants that are harmful to humans and the ecosystem. As mentioned earlier, various kinds of nanomaterials are used in this process. Among others, metal ion doped Fe_2O_3 nanomaterials have been widely used in pollution remediation processes. The major methods adopted in the pollutant removal processes are adsorption and catalysis (figure 9.2).

9.4.1 Adsorption

Adsorption is an effective method for the removal of heavy metals and metalloid ions and organic dyes from water. Fe_2O_3 based magnetic materials are widely used as adsorbents. In this process, polluted water is mixed with adsorbent material and shaken continuously for an appropriate time by using a mechanical shaker. During this process,

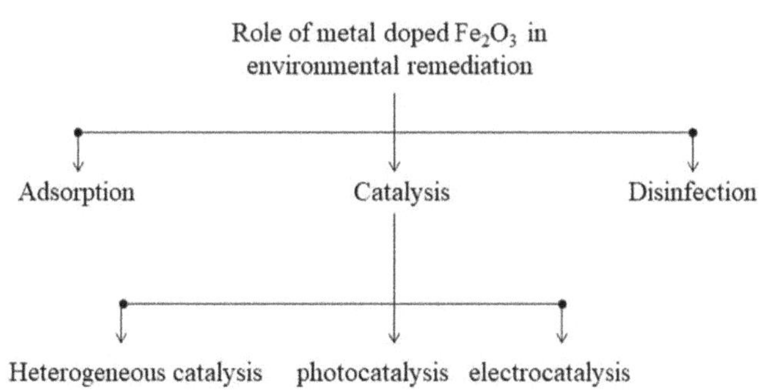

Figure 9.2. Application of metal doped Fe_2O_3 in environmental remediation.

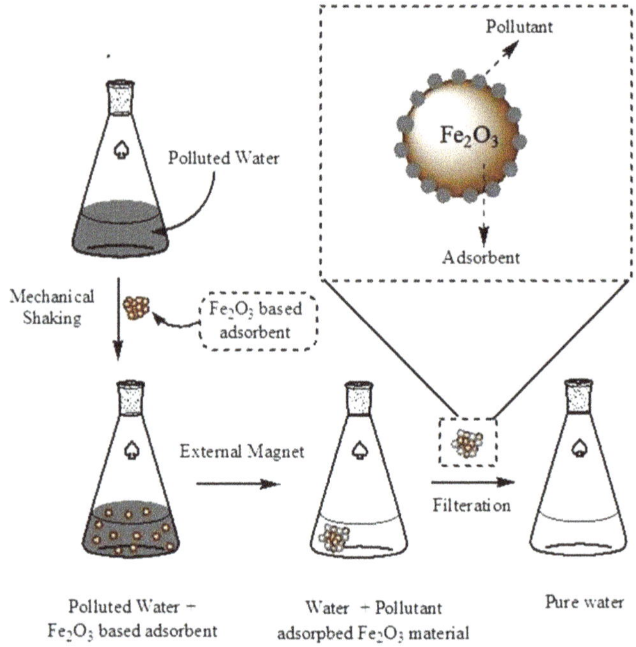

Figure 9.3. Treatment of polluted water using an adsorption process.

the pollutant molecules are specifically adsorbed at the surface of the adsorbent particles. By using an external magnetic field, the adsorbent can be separated easily and thus pure water can be obtained (figure 9.3). The adsorbed contaminants can be eliminated from the adsorbent by using an acid or base solution. The following factors are responsible for the adsorption properties of metal ions in Fe_2O_3:

(i) *Dopant concentration.*

The concentration of dopant in Fe_2O_3 affects the adsorption property of Fe_2O_3 significantly. For example, Lemine *et al* synthesized α-Fe_2O_3 nanoparticles doped with different concentrations of nickel (Ni^{2+}) and utilized them as an adsorbent for the removal of Cd(II), Co(II), and Ni(II) ions from their polluted water. When compared to pure and other doped α-Fe_2O_3, the 15wt% Ni doped sample showed greater adsorption efficiency towards Cd^{2+} (65.46 mg g^{-1}), Co^{2+} (43.29 mg g^{-1}), and Cd^{2+} (42 mg g^{-1}). However, Ni doped α-Fe_2O_3 did not act as an efficient adsorbent for Cr^{4+} ion removal [46].

(ii) *Magnetic behavior.*

In general, α-Fe_2O_3 is considered an efficient adsorbent material for the removal of toxic inorganic ions and organic dyes. However, α-Fe_2O_3 is an antiferromagnetic material. Hence, it is difficult to recover it from a solution. In order to eliminate this practical difficulty, a doping process was adopted in α-Fe_2O_3. For example, Cui *et al* synthesized highly porous Mn doped α-Fe_2O_3 adsorbent material and achieved effective removal of Pb(II), Cr(VI), and As (III) species from wastewater with easy magnetic separation [53].

(iii) *Facet orientation.*

The facet dependent adsorption behavior of Mn doped α-Fe_2O_3 sorbent was also reported. It was found that the increase of Mn concentration in α-Fe_2O_3 crystal lattices predominantly leads to the growth of {116}-faceted saucer-shaped nanosheets to {001}-faceted hexagonal nanosheets. For adsorption of Pb^{2+} ions, {001}-faceted hexagonal nanosheets exhibit high adsorption efficiency and selectivity. In the case of Cd(II) and Hg(II) ions, {116}-faceted saucer-shaped nanosheets show strong adsorption with high selectivity [54].

(iv) *Surface functional groups.*

Surface functional groups play an important role in the selective binding of pollutants from its solution. For example, Li *et al* have synthesized Al doped Fe_2O_3 nanoparticles to remove As(III) from aqueous solution without pre-oxidation. It was found that it had a greater adsorption capacity toward As(III). This is mainly due to the presence of a surface hydroxyl group (–OH) and specific surface area of the adsorbent material [55].

(v) *Nature of dopant.*

Cobalt (Co) and nickel (Ni) doped Fe_2O_3 powders were prepared for the removal of Pb(II) ions from solution. The adsorption batch experimental results showed that the Pb(II) removal capacity of Co doped Fe_2O_3 and Ni doped Fe_2O_3 are 136.0 mg g^{-1} and 97.5 mg g^{-1}, respectively. Compared to the Ni doped sample, the Co doped Fe_2O_3 clearly showed good adsorption performance toward Pb(II) ions under identical experimental conditions [56].

(vi) *pH of the medium.*

The pH of the solution is an important parameter to achieve high adsorption efficiency with selectivity. For example, Wong *et al* performed the removal of Cu(II) ions from solution using polyvinyl alcohol–alginate beads containing Co doped Fe_2O_3 particles. The Cu(II) ion adsorption capacity of the adsorbent was mainly dependent on pH the solution. The optimum pH was found to be 6 [57].

9.4.2 Catalysis

9.4.2.1 Heterogeneous catalysis

Metal ion doped Fe_2O_3 nanoparticles have been used as a catalyst for the removal of organic pollutants in solution. For example, Yan *et al* used Si doped α-Fe_2O_3 nanocatalyst for the removal of nitrobenzene in an aqueous solution, assisted by ozone. This means that the nanocatalyst facilitates the decomposition of ozone into hydroxyl radicals which effectively decompose the nitrobenzene in an aqueous solution. The best catalytic performance was observed for samples with an Si/Fe ratio of 0.5. The catalyst dosage, solution pH, and temperature also play a major role in the efficiency of nitrobenzene removal [58]. Lv *et al* studied the catalytic properties of Co and Mn doped γ-Fe_2O_3 in the degradation of 2,4-dichlorophenoxy-acetic acid, 2,4-diclorophenol, and 2,4,6-trichlorophenol in water along with ozone.

The ozone interacts with chemisorbed water and hydroxyl groups on the catalyst surface and generates reactive oxygen species which decompose the targeted organic molecules effectively. The Mn dopant increased the surface Lewis acid sites of the doped γ-Fe_2O_3, creating surface hydroxyl groups and chemisorbed water, and also enhancing the interfacial electron transfer, resulting in higher catalytic property [59].

Guo *et al* synthesized Ca doped α-Fe_2O_3 with abundant oxygen vacancies and used it for the degradation of rhodamine B assisted by peroxymonosulfate. The decomposition of rhodamine B in the presence of 5% Ca doped α-Fe_2O_3 is eight times greater than that of pure α-Fe_2O_3. The enhanced activity of 5% Ca doped α-Fe_2O_3 is due to the efficient generation of $^1O_2$ and $O_2^{\bullet -}$, and electrical conductivity. Further, 5% Ca doped α-Fe_2O_3/peroxymonosulfate also showed good catalytic efficiency in natural water samples. The Ca doping process in Fe_2O_3 generates active oxygen species in the solution due to the replacement of strongly bound oxygen by relatively weakly bound oxygen atoms. In the presence of peroxymonosulfate, active oxygen is converted into $^1O_2$, which absorbs electrons and forms a superoxide radical. On the other hand, hydroxyl and sulfate radicals are also produced by a Ca doped α-Fe_2O_3–peroxymonosulfate system. The produced reactive radicals degrade rhodamine B effectively (equations (9.1)–(9.9)) [60]:

$$\text{-}Fe^{3+}\text{-}O^{2-}\text{-}Fe^{3+}\text{-} \longrightarrow \text{-}Fe^{2+}\text{-}O_{vac}\text{-}Fe^{2+}\text{-}1/2O_2 \tag{9.1}$$

$$Ovac \longrightarrow O^* \tag{9.2}$$

$$O^* + HSO_5^- \longrightarrow HSO_4^- + {}^1O_2 \tag{9.3}$$

$$O_2 + e^- \longrightarrow O_2^{\bullet -} \tag{9.4}$$

$$Fe(II) + HSO_5^- \longrightarrow Fe(III) + SO_4^{\bullet -} + OH^- \tag{9.5}$$

$$Fe(III) + HSO_5^- \longrightarrow Fe(II) + SO_5^{\bullet -} + H^+ \tag{9.6}$$

$$Fe(III) + e^- \longrightarrow Fe(II) \tag{9.7}$$

$$SO_4^{\bullet -} + OH^- \longrightarrow {}^{\bullet}OH + SO_4^{2-} \tag{9.8}$$

$$HSO_5^- + e^- \longrightarrow SO_4^{\bullet -} + OH^- \tag{9.9}$$

Ma *et al* studied the catalytic properties of Ti doped Fe_2O_3 nanoparticles in a 1,2-dichlorobenzene oxidation reaction. Among the other doped samples, 10 mol% Ti doped Fe_2O_3 nanoparticles show the best catalytic performance along with higher stability. This is due to the smaller crystallite size, excellent low-temperature reducibility, and high surface active oxygen concentration. The 1,2-dichlorobenzene undergoes a two-step redox process on the surface of the Ti doped Fe_2O_3 catalyst [61]. Su *et al* have performed degradation of decachlorobiphenyl on Ca doped Fe_2O_3

catalyst at 300 °C. The by-product of decachlorobiphenyl degradation is meta- and para-isomers of hydrodechlorination products. Only 1% Ca doped Fe_2O_3 powder showed the maximum efficiency among the four prepared catalysts [62].

Oliveira *et al* reported V doped Fe_2O_3 powder as catalysts to oxidize methylene blue in water containing H_2O_2. The doped Fe_2O_3 powder showed remarkable catalytic activity toward the degradation of methylene blue dye. They observed that the H_2O_2 pretreatment of the V doped Fe_2O_3 powder significantly improves the surface and structural property of the Fe_2O_3, which enhance the catalytic properties of the material [63]. The same research group also found that 3% niobium (Nb) doped α-Fe_2O_3 is an effective catalyst for the treatment of methylene blue contaminated water. They found that H_2O_2 pre-treated Nb doped α-Fe_2O_3 could degrade 75% of methylene blue dye in aqueous solution [64].

From the above studies, the following requirements are derived for doped Fe_2O_3 as an ideal catalyst in the degradation reactions:

(i) An optimum dopant concentration in Fe_2O_3 is essential.
(ii) Active surface functional groups such as hydroxyl groups on the catalyst are required.
(iii) H_2O_2 pretreatment of catalyst is required.
(iv) Larger specific surface area and mesoporous structure is needed.
(v) Oxygen vacancy is desirable.

9.4.2.2 Photocatalysis

Photocatalysis is a benign technology for the removal of various kinds of environmental pollutants in water and air. In photocatalysis, solar light, a natural renewable resource, is used together with a catalyst. To achieve efficient photocatalysis, the catalyst should possess a band gap of 1.3–1.4 eV for absorption of solar light. Fe_2O_3 has a band gap of ~2.5 eV and, hence, it is capable of absorbing visible in the solar spectrum, nevertheless, it suffers in acting as a good photocatalyst due to its fast recombination rate and high photo-corrosion behavior. Also, the band gap of an ideal photocatalyst should be 1.3–1.4 eV. These drawbacks were eliminated via a metal doping process. Therefore, in recent decades, Fe_2O_3 doped with various metal ions was synthesized and their photocatalytic performance toward a number of pollutants was studied. Table 9.2 shows a list of metal ion doped Fe_2O_3 along with their photocatalytic performance. The following aspects affect the properties of photocatalysis:

(i) *Nature of dopant.*

The nature of elements has significant effect on the photocatalytic property of Fe_2O_3. For instance, the Cu or Ni or Co doped Fe_2O_3 nanoparticles were synthesized and used as photocatalyst for degradation of Acid Red-27 dye in aqueous solution under visible light irradiation. The obtained experimental results suggest that the Cu doped Fe_2O_3 exhibited higher photocatalytic efficiency than that of pure Fe_2O_3, Ni and Co doped Fe_2O_3 [65].

Table 9.2. Photocatalytic performance of metal ion doped Fe_2O_3 nanomaterials.

Photocatalyst	Pollutant (concentration)	Light source	Efficiency (%)	Reason	Reference
Cu^{2+}–Fe_2O_3	Acid red-27 (2.5×10^{-5} M, 200 ml)	150 W tungsten lamp	98%/90 min	(i) Redox potential of Cu^{2+}/Cu (+0.34 versus SHE). (ii) Morphology and surface area. (iii) Surface negative charge.	[65]
6 mol% Sn doped α-Fe_2O_3/0.05 g	Methylene blue (50 ml)	Medium pressure xenon lamp (150 W)	~100%/150 min	(i) Reduction in band gap energy. (ii) Transformation of $Sn^{3+/5+}$ into Sn^{4+}.	[31]
$Mn^{2+/3+}$ doped α-Fe_2O_3/30 mg + 0.15 ml of 30% H_2O_2	Methyl orange (2.5×10^{-5} mol L^{-1})	1 sun lamp 2.5 W,	80%/360 min	(i) Defect states of Fe^{2+}.	[66]
1.78 v% Ti^{4+}-doped γ-Fe_2O_3/10 mg	Methylene blue (5 mg L^{-1}/50 ml)	UV (T5–8 W) or visible (xenon lamp, 55 W)	~100%/120 min	(i) Surface negative charge. (ii) Narrow band gap.	[67]
3% Ga^{3+}-α-Fe_2O_3/1 g L^{-1}	Methyl orange (50 ml/10 mg L^{-1})	500 W high-pressure mercury lamp and 350 W Xe lamp	40%/60 min ~80%/12 h	(i) Specific surface area. (ii) Separation efficiency.	[33]
5% Sn-α-Fe_2O_3/0.05 g + 1 ml 30% H_2O_2	Methylene blue, rhodamine B, methyl orange (100 ml/20 mg L^{-1})	250 W halogen spotlight	~100%/70 min	(i) Surface area. (ii) Ability to absorb light in visible light region. (iii) Ability to reduce recombination of e-/h+ pairs.	[68]

$Sn\text{-}Fe_2O_3$/75 mg	Methylene blue (100 ml)	Sunlight (639.3 W m^{-2})	92.5%/4 h	(i) Change in band gap energy.	[40]
(Al^{3+}, Zn^{2+}, Ti^{4+})-α-Fe_2O_3/0.05 g + 2 ml 30 wt% H_2O_2	Methylene blue (100 ml, 10 mg l^{-1}).	Xenon lamp (300 W, 300–1000 nm, 80 mW cm^{-2})	72.2%/150 min	(i) Improved charge transfer efficiency.	[69]
20% $Zn\text{-}Fe_2O_3$/50 mg	Methylene blue (100 ml, 1×10^{-5} mol l^{-1})	Xe arc lamp (300 W)	84.4%/75 min	(i) Decrease in the band gap. (ii) Increase in electron effective mass.	[70]
5% Co^{2+}-Fe_2O_3/0.05 g	Pararosaniline (100 ml, 5×10^{-5} M)	Visible light	89.8%/60 min	(i) Co^{2+} acts as trapping center for electrons.	[29]

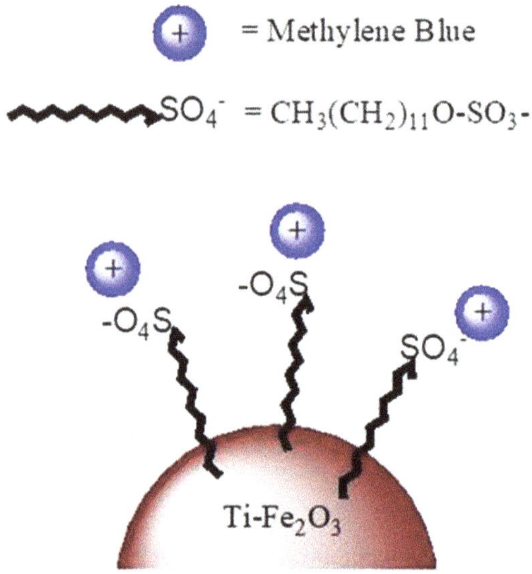

Figure 9.4. Surface adsorption of methylene blue on Ti doped Fe_2O_3/sodium dodecyl sulfate. (Reproduced with permission from [67]. Copyright 2018 Elsevier. Redraw with modification.)

(ii) *Concentration of dopant.*
The optimum concentration is necessary to achieve better photocatalysis. For example, (1 mol% to 6 mol%) Sn doped α-Fe_2O_3 nanoparticles were prepared and examined their photocatalytic activities using degradation of methylene blue under visible light illumination. Compared to other samples, 6 mol% Sn doped α-Fe_2O_3 has improved capacity to produce hydroxyl radicals, which were responsible for the obvious increase in the photocatalytic activity [31]. Similarly, Cha *et al* have reported Mn doped α-Fe_2O_3 with photocatalytic properties towards photodegradation of methyl orange dye under visible light irradiation [66].

(iii) *Active surface area.* The active surface is essential for a good photocatalyst. For example, Mirbagheri *et al* synthesized Ti doped γ-Fe_2O_3 photocatalyst for the degradation of methylene blue in an aqueous solution (figure 9.4). Compared to pure γ-Fe_2O_3, the 1.78 v% Ti doped γ-Fe_2O_3 had a higher surface area and thus showed greater photocatalytic activity towards the degradation of methylenbe blue. Probably, a shorter contact time is sufficient between the methylene blue and the photocatalyst due to the generation of a higher concentration of charge carriers. However, the pH has a remarkable effect on the removal efficiency, due to the negative charges of sodium dodecyl sulfate coated Ti doped γ-Fe_2O_3 particles [67].

(iv) *Improved separation efficiency of photocharges.* Metal ion doping significantly improves the separation of photogenerated charges, since dopant elements act as trap centers within the band structure of Fe_2O_3 (figure 9.5). For example, doping of Ga^{3+} ions into α-Fe_2O_3 increases the separation efficiency of the photo-induced charges. Therefore, 3% Ga^{3+} doped Fe_2O_3

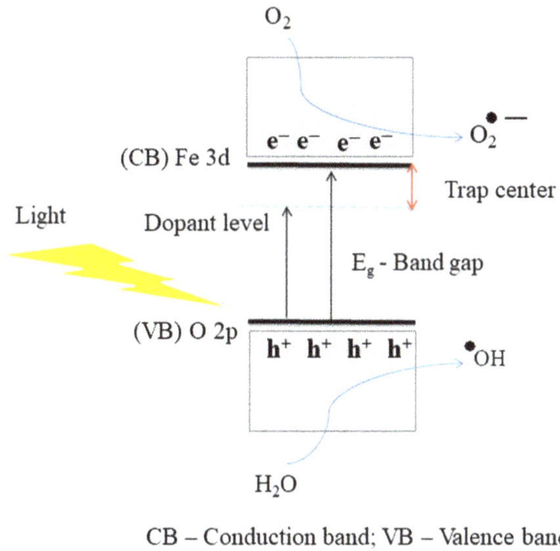

Figure 9.5. Radical generation by M^{2+} doped Fe_2O_3 under visible light illumination.

showed greater photocatalytic efficiency for decolorization of methyl orange aqueous solution [33].

(v) *Improved visible light absorption ability.*

Cao *et al* prepared Sn doped α-Fe_2O_3 powder with mesoporous structures and applied it as a photocatalyst for the degradation of methylene blue dye in solution under visible light illumination. The 5% Sn doped α-Fe_2O_3 showed the highest photocatalytic activity and long-term stability. This is because Sn doping causes smaller particle size, an increased ability to absorb in visible light, efficient charge separation, and an improved electron transfer tendency [68].

(vi) *Improved hydroxyl radical generation.*

Under visible light illumination, a Sn doped Fe_2O_3 photogenerated electron–hole pair will develop. Sn^{4+} ions within the lattices of Fe_2O_3 absorb electrons and forms Sn^{3+} ions, whereas Sn^{4+} ions react with holes to form Sn^{5+} ions. Both Sn^{3+} and Sn^{5+} ions are unstable and thus they react with O_2 and hydroxide to form superoxide and hydroxyl radicals, respectively. Finally, the generated radicals decompose the methylene blue dye (equations (9.10)–(9.15)) [31].

Similarly, the band gap of Fe_2O_3 photocatalysts was tuned from 2.21 eV to 2.44 eV through doping of Sn in an Fe_2O_3 lattice. For this reason, Sn doped Fe_2O_3 shows better photocatalytic activity in methylene blue dye degradation under sunlight [40]:

$$\alpha - Fe_2O_3 + h\nu \longrightarrow h^+_{VB} + e^-_{CB} \quad (9.10)$$

$$Sn^{4+} + e^-_{CB} \longrightarrow Sn^{3+} \quad (9.11)$$

$$Sn^{3+} + O_2 \longrightarrow Sn^{4+} + O_2^{\bullet -} \quad (9.12)$$

$$Sn^{4+} + h^+_{VB} \longrightarrow Sn^{5+} \quad (9.13)$$

$$Sn^{5+} + OH^- \longrightarrow Sn^{4+} + {}^{\bullet}OH \quad (9.14)$$

$$\text{MB dye} + OH \longrightarrow \text{Degradation products} \quad (9.15)$$

The influence of multi-elemental doping on the photocatalytic activity of Fe_2O_3 was also investigated. For example, Wu *et al* reported Al, Zn, and Ti doped α-Fe_2O_3 using blast furnace dust and evaluated its catalytic performance using photodegradation of methylene blue dye. The photodegradation efficiency was found to be 72.2% within 150 min under simulated sunlight illumination [69].

9.4.2.3 Electrocatalysis

Electrochemical processes are also used to achieve degradation or complete mineralization of organic pollutants in water (figure 9.6). For example, Luo *et al* reported the degradation of tetracycline in an aqueous solution using a heterogeneous electro-Fenton process using a copper doped Fe@Fe_2O_3 nanoparticle loaded nickel foam cathode. The catalyst exhibited 78.1% tetracycline degradation within 2 h. The mineralization efficiency was found to be 89.8% after 6 h. The mechanism of tetracycline degradation was explained in the following way [71]. By three pathways, hydroxyl radicals are produced during this treatment process. (i) The Fe^{2+} reacts with *in situ* generated H_2O_2 to form hydroxyl radicals and Fe^{3+} (equations (9.16)–(9.20)). The Fe^{3+} receive an electron on the cathode to form Fe^{2+} (equation (9.21)). (ii) Cu^{2+} is reduced by H_2O_2 to form hydroxyl radicals and Cu^{1+}. Meanwhile, a reaction between Cu^+ and Fe^{3+} occurs and forms Fe^{2+} and Cu^{2+}. In addition, a

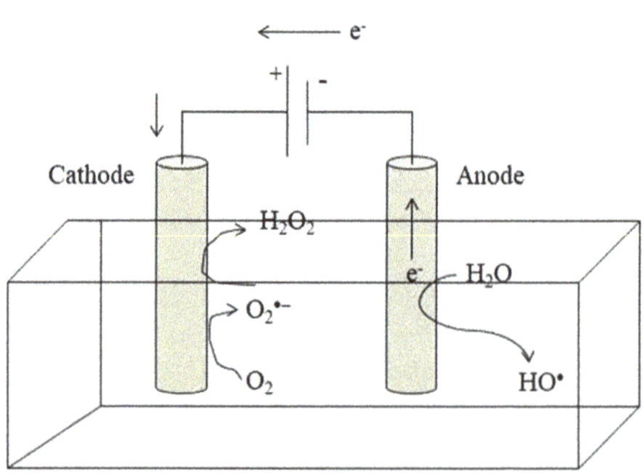

Figure 9.6. Generation of free radicals using a doped Fe_2O_3 electrode.

replacement reaction occurs between Cu^{2+} and Fe^0, promoting the generation of Fe^{2+} (equation (9.22)). Consequently, additional generation of hydroxyl radicals is promoted. (iii) $M(^{\bullet}OH)$ is generated *in situ* at the surface of the anode, which can only attack the tetracycline around the surface area. In contrast, the Fe^{2+} and Cu^{2+} leach out into the bulk solution to participate in homogeneous reactions. In addition, the dissolved oxygen absorbed on Ni foam could be reduced to $^{\bullet}O_2^-$ species which subsequently generate H_2O_2 and hydroxyl radicals to attack tetracycline. Overall, hydroxyl radicals attack tetracycline molecules, leading to complete mineralization:

$$M + H_2O \longrightarrow M(OH) + H^+ + e^- \tag{9.16}$$

$$O_2 + 2H^+ + 2e^- \longrightarrow H_2O_2 \tag{9.17}$$

$$Ni + 2O_2 \longrightarrow Ni^{2+} + 2\ O_2^{\bullet -} \tag{9.18}$$

$$^{\bullet}O_2^- + 2H^+ + e^- \longrightarrow H_2O_2 \tag{9.19}$$

$$Fe^{2+} + H_2O_2 + H^+ \longrightarrow Fe^{3+} + H_2O + {}^{\bullet}OH \tag{9.20}$$

$$Fe^{3+} + e- \longrightarrow Fe^{2+} \tag{9.21}$$

$$Cu^{2+} + Fe \longrightarrow Cu + Fe^{2+} \tag{9.22}$$

9.4.3 Disinfection

Disinfection is one of the important steps in water treatment processes, in which micro-organisms such as bacteria, viruses, and protozoa were inactivated using chemical or physical methods. In chemical disinfection methods, a chemical agent, including ozone, chlorine, calcium oxychloride, and chloramine compounds are used, while in physical disinfection UV light illumination, size exclusion, or adsorption are used. Chemical disinfection is the most commonly used method. In recent decades, Fe_2O_3 nanostructures have been studied as disinfectant agents for varieties of bacteria, viruses, and fungi [72] because Fe_2O_3 particles produce reactive oxygen species that decompose microbial cells effectively (figure 9.7). For example, Mn doped α-Fe_2O_3 thin films were fabricated used to destroy *Staphylococcus aureus* bacteria. Pure α-Fe_2O_3 shows 16.66% antibacterial efficiency while 8 wt% Mn doped α-Fe_2O_3 shows 58.33% efficiency. When bacteria come into contact with doped α-Fe_2O_3 particles, a reactive oxygen species is formed which can react with the peptide bonds in the cell wall of bacteria and disrupt them [73]. Bhushan *et al* have prepared Ag^+ ion doped α-Fe_2O_3 nanospheres and applied them as antibacterial agents against *Bacillus subtilis, S. aureus, Pseudomonas fluorescens*, and *Escherichia coli*. The Ag^+ ion doped α-Fe_2O_3 particles showed remarkable activity against *B. subtilis* and *P. fluorescens*. The antibacterial activity of Ag^+ doped α-Fe_2O_3 samples was also comparable to the standard antibiotic streptomycin [74].

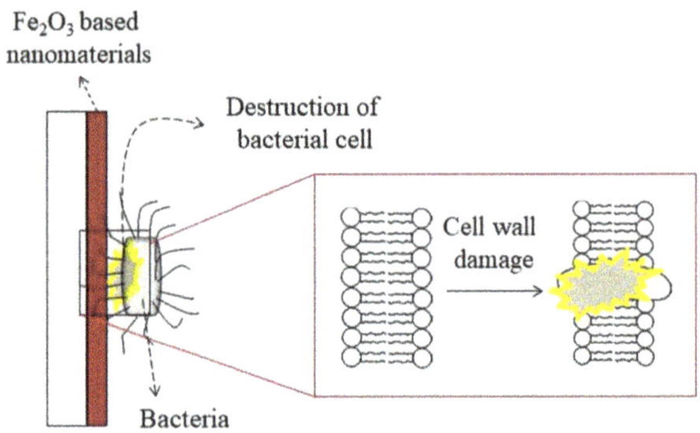

Figure 9.7. Possible disinfectant activity of doped Fe_2O_3 nanoparticles.

9.5 Summary

From the literature review it is clear that metal ion doping will tune the active surface area, morphology, and band potential values of Fe_2O_3. Therefore several research groups have attempted to used metal ions, including transition and inner transition metal ions, as dopants in Fe_2O_3 (mainly α- and γ-Fe_2O_3) nanoparticles and studied their activities in the abatement of organic and inorganic pollutants in water. Pollutant abatement methods include adsorption, catalysis, photocatalysis, electrocatalysis, and disinfection. In the adsorption method, dopant concentration, magnetic properties, facet orientation, and surface functional groups are the key factors to achieve better adsorbent materials. In catalysis, the optimum dopant concentration, active functional group, specific surface area, and oxygen vacancies are the important parameters to obtain the best catalyst. Moreover, metal ion doped Fe_2O_3 nanoparticles are also used as disinfecting agents. Therefore, the application of metal ion doped Fe_2O_3 nanoparticles in existing wastewater treatment techniques will be an excellent option in the future.

References

[1] Moreno-Brush M, McLagan D S and Biester H 2019 Fate of mercury from artisanal and small-scale gold mining in tropical rivers: hydrological and biogeochemical controls. a critical review *Crit. Rev. Environ. Sci. Technol.* **50** 437–75

[2] Dong F, Wang Y, Su B, Hua Y and Zhang Y 2019 The process of peak CO_2 emissions in developed economies: a perspective of industrialization and urbanization *Resour. Conserv. Recycl.* **141** 61–75

[3] Ibrahim M, Labaki M, Giraudon J M and Lamonier J F 2020 Hydroxyapatite, a multifunctional material for air, water and soil pollution control: a review *J. Hazard. Mater.* **383** 121139

[4] Feng M, Dionysiou D D and Sharma V K 2020 Black phosphorous-based nanostructures in environmental remediation: current status and future perspectives *Chem. Eng. J.* **389** 123460

[5] Feng L and Didier A 2020 Nanocatalysts and other nanomaterials for water remediation from organic pollutants *Coord. Chem. Rev.* **408** 213180

[6] Suresh R, Mangalaraja R V, Mansilla H D, Santander P and Yáñez J 2020 Reduced graphene oxide based photocatalysis *Green Photocatalysts, Environmental Chemistry for a Sustainable World* ed M Naushad *et al* vol 34 (Cham: Springer) ch 6 pp 145–66

[7] Wu Y, Pang H, Liu Y, Wang X, Yu S, Fu D, Chen J and Wang X 2019 Environmental remediation of heavy metal ions by novel-nanomaterials: a review *Environ. Pollut.* **246** 608

[8] Dawei W, Suresh C P, Shih-Hsin H, Jingbin Z, Yi L and Dionysios D D 2018 Plasmonic-based nanomaterials for environmental remediation *Appl. Catal.* B **237** 721

[9] Maneesha M and Doo-Man C 2015 α-Fe_2O_3 as a photocatalytic material: a review *Appl. Catal.* A **498** 126

[10] Yuanming Z, Ningsi Z, Tiantian W, Huiting H, Yong C, Zhaosheng L and Zhigang Z 2019 Heterogeneous degradation of organic contaminants in the photo-Fenton reaction employing pure cubic β-Fe_2O_3 *Appl. Catal.* B **245** 410

[11] Changfeng L, Wu Q and Changqing D 2016 H_2S adsorption and decomposition on the gradually reduced α-Fe_2O_3 (001) surface: a DFT study *Appl. Surf. Sci.* **387** 720

[12] Shilpa S, Sujith K and Anish M A 2020 Green synthesized α-Fe_2O_3 mesoporous network for heterogeneous Fenton oxidation of thiazine dye *Mater. Lett.* X **5** 100037

[13] Nishesh Kumar G, Yasaman G, Jiyeol B and Kwang S K 2020 Synthesis of coral-like α-Fe_2O_3 nanoparticles for dye degradationat neutral pH *J. Mol. Liq.* **301** 112473

[14] Hanfeng L, Binbin X and Zhoucheng W 2013 Self-assembled 3D flower-like α-Fe_2O_3 microstructures and their superior capability for heavy metal ion removal *Mater. Chem. Phys.* **141** 727–34

[15] Ali A, Shahriar H, Ahmad M R, Esmaeel D and Houshang A H 2014 Optimization of heavy metal removal from aqueous solutions by maghemite (γ-Fe_2O_3) nanoparticles using response surface methodology *J. Geochem. Explor.* **147** 151

[16] Fanyue M, Min S, Bing S, Yuexing W, Qi C and Yue C 2020 Enhanced degradation of rhodamine B via α-Fe_2O_3 microspheres induced persulfate to generate reactive oxidizing species *Chemosphere* **243** 125322

[17] Zhonghai Z, Faruk M H and Takakazu T 2010 Self-assembled hematite (α-Fe_2O_3) nanotube arrays for photoelectrocatalytic degradation of azo dye under simulated solar light irradiation *Appl. Catal.* B **95** 423

[18] Limin S, Shujuan Z, Bin C, Jingjie G and Xicheng J 2010 A hydrothermal method for preparation of α-Fe_2O_3 nanotubes and their catalytic performance for thermal decomposition of ammonium perchlorate *Colloids Surf.* A **360** 1

[19] Khedr M H, Bahgat M, Nasr M I and Sedeek E K 2007 CO_2 decomposition over freshly reduced nanocrystalline Fe_2O_3 *Colloids Surf.* A **302** 517

[20] De Bin J, Xiaoying L, Xuan X and Yu Xin Z 2018 Double-shell Fe_2O_3 hollow box-like structure for enhanced photo-Fenton degradation of malachite green dye *J. Phys. Chem. Solids* **112** 209

[21] Cheng Y, Jialiang Z, Xinrui Z, Yubo L, Yongqiang C and Chengyan W 2020 Deep and efficient removal of vanadium from molybdate solution using magnetic γ-Fe_2O_3 nanoparticles *Appl. Surf. Sci.* **529** 147060

[22] Saima N, Ghulam M, Ibrahim S M, Saima N, Munawar I, Muhammad Y, Tariq J and Jan N 2020 Iron oxide (Fe_2O_3) prepared via green route and adsorption efficiency evaluation for an anionic dye: kinetics, isotherms and thermodynamics studies *J. Mater. Res. Technol.* **9** 4206

[23] Abdelrahman E A, Hegazey R M, Yousra H K and Ahmed A 2019 Facile synthesis of Fe_2O_3 nanoparticles from Egyptian insecticide cans for efficient photocatalytic degradation of methylene blue and crystal violet dyes *Spectrochim. Acta* A **222** 117195

[24] Yu M, Jinchuan Z, Mengying S, Jianfeng M and Sridhar K 2020 A novel Fenton-like system of Fe_2O_3 and $NaHSO_3$ for orange II degradation *Sep. Purif. Technol.* **230** 115866
[25] Xing L *et al* 2018 Porous Fe_2O_3 microcubes derived from metal organic frameworks for efficient elimination of organic pollutants and heavy metal ions *Chem. Eng. J.* **336** 241
[26] Xia W, Jianqiang W, Zhentao C, Shuguang W and Minhua C 2014 Facet effect of α-Fe_2O_3 crystals on photocatalytic performance in the photo-Fenton reaction *RSC Adv.* **4** 34387
[27] Wanting S, Qingqiang M, Liqiang J, Dening L and Yue C 2013 Facile synthesis of surface-modified nanosized α-Fe_2O_3 as efficient visible photocatalysts and mechanism insight *J. Phys. Chem.* C **117** 1358
[28] Bin R, Yuelong X, Chenghuai Z, Lihui Z, Junping Z and Zhenfa L 2019 Degradation of methylene blue by a heterogeneous Fenton reaction using an octahedron-like, high-graphitization, carbon doped Fe_2O_3 catalyst *J. Taiwan Inst. Chem.* E **97** 170
[29] Suresh R, Giribabu K, Manigandan R, Mangalaraja R V, Solorza J Y, Stephen A and Narayanan V 2017 Synthesis of Co^{2+}-doped Fe_2O_3 photocatalyst for degradation of pararosaniline dye *Solid State Sci.* **68** 39
[30] Li S, Junjun W, Zhanfei L, Ling F, Ran C, Xueling W, Jiaokun L and Weimin Z 2020 A high-efficiency Fe_2O_3@microalgae composite for heavy metal removal from aqueous solution *J. Water Process Eng.* **33** 101026
[31] Mansour H, Bargougui R, Autret-Lambert C, Gadri A and Ammar S 2018 Co-precipitation synthesis and characterization of tin-doped α-Fe_2O_3 nanoparticles with enhanced photocatalytic activities *J. Phys. Chem. Solids* **114** 1
[32] Cardillo D, Konstantinov K and Devers T 2013 The effects of cerium doping on the size, morphology, and optical properties of α-hematite nanoparticles for ultraviolet filtration *Mater. Res. Bull.* **48** 4521
[33] Zhenghua X, Jianzhang L, Junbo Z, Wei H, Jun Z, Shengtian H, Xiaolin L, Jinjin H and Minjiao L 2014 Enhanced photocatalytic decolorization of methyl orange by gallium-doped α-Fe_2O_3 *Mater. Sci. Semicond. Process.* **24** 104
[34] Guo S, Wang H, Yang W, Fida H, You L and Zhou K 2020 Scalable synthesis of Ca-doped-Fe_2O_3 with abundant oxygen vacancies for enhanced degradation of organic pollutants through peroxymonosulfate activation *Appl. Catal.* B **262** 118250
[35] Mohapatra M, Padhi T, Anand S and Mishra B K 2012 CTAB mediated Mg-doped nano Fe_2O_3: synthesis, characterization, and fluoride adsorption behavior *Desalin. Water Treat.* **50** 376–86
[36] Suresh R, Prabu R, Vijayaraj A, Giribabu K, Stephen A and Narayanan V 2012 Facile synthesis of cobalt doped hematite nanospheres: magnetic and their electrochemical sensing properties *Mater. Chem. Phys.* **134** 590–6
[37] Faisal B, Yousaf H K, Safa J, Bernabé M, S and Saira B 2019 Hydrothermal syntheses of vanadium doped α-Fe_2O_3 cubic particles with enhanced photoelectrochemical activity *Sol. Energy* **182** 332
[38] Tulliani J M, Baroni C, Zavattaro L and Grignani C 2013 Strontium-doped hematite as a possible humidity sensing material for soil water content determination *Sensors* **13** 12070
[39] Jawad A and Ashraf S S Z 2011 Structural and magnetic properties of Cr^{+3} doped nano-structured γ-Fe_2O_3 synthesized by a modified solution combustion technique *Eur. Phys. J. Appl. Phys.* **54** 10402
[40] Sangaiya P and Jayaprakash R 2018 Tuning effect of Sn doping on structural, morphological, optical, electrical and photocatalytic properties of iron oxide nanoparticles *Mater. Sci. Semicond. Process.* **85** 40

[41] Xiaopeng Q, Guangwei S, Meng W, Lixuan M and Wensheng S 2013 Electrochemical synthesis of p-type Zn-doped α-Fe_2O_3 nanotube arrays for photoelectrochemical water splitting *Chem. Commun.* **49** 5742

[42] Lei W, Chong-Yong L and Patrik S 2013 Ti and Sn co-doped anodic α-Fe_2O_3 films for efficient water splitting *Electrochem. Commun.* **30** 21

[43] Il-Sung L, Gun-Eik J, Chang Kyo K and Dae-Ho Y 2001 Fabrication and gas sensing characteristics of pure and Pt-doped γ-Fe_2O_3 thin film *Sens. Actuators* B **77** 215

[44] Giada F, Florian K, Matthias M, Gareth S P, Michael S, Ulrike D and Michele R 2020 A model system for photocatalysis: Ti-doped α-Fe_2O_3 $(1\bar{1}02)$ single-crystalline films *Chem. Mater.* **32** 3753

[45] Saveria S, Patrizia F, Fabiola P, Sara S, Marcello M, Salvatore P, Francesco M, Vladimiro D S and Pierluigi A 2017 Effect of Ti- or Si-doping on nanostructure and photo-electro chemical activity of electro-spun iron oxide fibres *Int. J. Hydrog. Energy* **42** 28070

[46] Lemine O M, Ghiloufi I, Bououdina M, Khezami L, Ould M'hamed M and Hassan A T 2014 Nanocrystalline Ni doped α-Fe_2O_3 for adsorption of metals from aqueous solution *J. Alloys Compd.* **588** 592

[47] Sharmin M and Podder J 2020 Band gap tuning, n-type to p-type transition and ferromagnetic properties of Mg doped α-Fe_2O_3 nanostructured thin films *J. Alloys Compd.* **818** 152850

[48] Suresh R, Giribabu K, Manigandan R, Stephen A and Narayanan V 2014 Fabrication of Ni–Fe_2O_3 magnetic nanorods and application to the detection of uric acid *RSC Adv.* **4** 17146–55

[49] Allia P, Barrera G, Bonelli B, Freyria F S and Tiberto P 2013 Magnetic properties of pure and Eu-doped hematite nanoparticles *J. Nanopart. Res.* **15** 2118

[50] Bhat R, Qayoom M, Dar G N and Want B 2020 Structural, dielectric, optical and magnetic studies of dysprosium doped iron oxide nanostructures *Mater. Chem. Phys.* **245** 122764

[51] Liao P, Toroker M C and Carter E A 2011 Electron transport in pure and doped hematite *Nano Lett.* **11** 1775–81

[52] Xiao W, Tongkai W, Guangkuo S, Yang L, Shouwei Z, Xiaolong D and Xijin X 2020 Oxygen vacancy defects engineering on Ce-doped α-Fe_2O_3 gas sensor for reducing gases *Sens. Actuators* B **302** 127165

[53] Cui H J, Cai J K, Shi J W, Yuan B, Ai C L and Fu M L 2014 Fabrication of 3D porous Mn doped α-Fe_2O_3 nanostructures for the removal of heavy metals from wastewater *RSC Adv.* **4** 10176–9

[54] Yuan Q, Li P, Liu J, Lin Y, Cai Y, Ye Y and Liang C 2017 Facet-dependent selective adsorption of Mn-doped α-Fe_2O_3 nanocrystals toward heavy-metal ions *Chem. Mater.* **29** 10198–205

[55] Li R, Li Q, Gao S and Shang J K 2011 Enhanced arsenite adsorption onto litchi-like Al-doped iron oxides *J. Am. Ceram. Soc.* **94** 584–91

[56] Chen W, Lu Z, Xiao B, Gu P, Yao W, Xing J, Asiri A M, Alamry K A, Wang X and Wang S 2019 Enhanced removal of lead ions from aqueous solution by iron oxide nanomaterials with cobalt and nickel doping *J. Clean. Prod.* **211** 1250–8

[57] Wong E T, Chan K H, Irfan M and Idris A 2016 Investigation of Cu(II) removal by cobalt doped iron oxide captured in PVA-alginate beads *Desalin. Water Treat.* **57** 2052–63

[58] Yan P, Shen J, Yuan L, Kang J, Wang B, Zhao S and Chen Z 2019 Catalytic ozonation by Si-doped α-Fe_2O_3 for the removal of nitrobenzene in aqueous solution *Sep. Purif. Technol.* **228** 115766

[59] Lv A, Hu C, Nie Y and Qu J 2010 Catalytic ozonation of toxic pollutants over magnetic cobalt and manganese co-doped γ-Fe_2O_3 *Appl. Catalysis* B **100** 62–7
[60] Guo S, Wang H, Yang W, Fida H, You L and Zhou K 2020 Scalable synthesis of Ca-doped α-Fe_2O_3 with abundant oxygen vacancies for enhanced degradation of organic pollutants through peroxymonosulfate activation *Appl. Catalysis* B **262** 118250
[61] Ma X, Suo X, Cao H, Guo J, Lv L, Sun H and Zheng M 2014 Deep oxidation of 1,2-dichlorobenzene over Ti-doped iron oxide *Phys. Chem. Chem. Phys.* **16** 12731–40
[62] Su G, Huang L, Shi R, Liu Y, Lu H, Zhao Y, Yang F, Gao L and Zheng M 2016 Thermal dechlorination of PCB-209 over Ca species-doped Fe_2O_3 *Chemosphere* **144** 81–90
[63] Oliveira H S, Oliveira L C A, Pereira M C, Ardisson J D, Souza P P, Patrício P O and Moura F C C 2015 Nanostructured vanadium-doped iron oxide: catalytic oxidation of methylene blue dye *New J. Chem.* **39** 3051–8
[64] Oliveira H S, Almeida L D, Freitas V A A, Moura F C C, Souza P P and Oliveira L C A 2015 Nb-doped hematite: highly active catalyst for the oxidation of organic dyes in water *Catal. Today* **240** 176–81
[65] Satheesh R, Vignesh K, Suganthi A and Rajarajan M 2014 Visible light responsive photocatalytic applications of transition metal (M = Cu, Ni and Co) doped α-Fe_2O_3 nanoparticles *J. Environ. Chem. Eng.* **2** 1956–68
[66] Cha H G, Noh H S, Kang M J and Kang Y S 2013 Photocatalysis: progress using manganese-doped hematite nanocrystals *New J. Chem.* **37** 4004–9
[67] Mirbagheri N S and Sabbaghi S 2018 A Ti-doped γ-Fe_2O_3/SDS nano-photocatalyst as an efficient adsorbent for removal of methylene blue from aqueous solutions *J. Environ. Manage.* **213** 56–65
[68] Cao Z, Qin M, Gu Y, Jia B, Chen P and Qu X 2016 Synthesis and characterization of Sn-doped hematite as visible light photocatalyst *Mater. Res. Bull.* **77** 41–7
[69] Wu Z J, Wang L C, Gao Z F, Liu W M and Wu X R 2016 Recycling blast furnace dust into metals (Al, Zn and Ti)-doped hematite with enhanced photocatalytic activity *J. Environ. Chem. Eng.* **4** 341–5
[70] Sivaranjani R, Thayumanavan A and Sriram S 2019 Photocatalytic activity of Zn-doped Fe_2O_3 nanoparticles: a combined experimental and theoretical study *Bull. Mater. Sci.* **42** 185
[71] Luo T, Feng H, Tang L, Lu Y, Tang W, Chen S, Yu J, Xie Q, Ouyang X and Chen Z 2020 Efficient degradation of tetracycline by heterogeneous electro-Fenton process using Cu-doped Fe@Fe_2O_3: mechanism and degradation pathway *Chem. Eng. J.* **382** 122970
[72] Rufus A, Sreeju N and Philip D 2016 Synthesis of biogenic hematite (α-Fe_2O_3) nanoparticles for antibacterial and nanofluid applications *RSC Adv.* **6** 94206–17
[73] Belkhedkar M R, Ubale A U, Sakhare Y S, Zubair N and Musaddique M 2016 Characterization and antibacterial activity of nanocrystalline Mn doped Fe_2O_3 thin films grown by successive ionic layer adsorption and reaction method *J. Assoc. Arab Univ. Basic Appl. Sci.* **21** 38–44
[74] Bhushan M, Muthukamalam S, Sudharani S and Viswanath A K 2015 Synthesis of α-$Fe_{2-x}Ag_xO_3$ nanocrystals and study of their optical, magnetic and antibacterial properties *RSC Adv.* **5** 32006–14

IOP Publishing

Nanostructured Materials for Sustainable Energy and Environmental Remediation

R V Mangalaraja, R Udayabhaskar, P Sathishkumar and Joydeep Dutta

Chapter 10

Nanoferrites for green magnetic refrigeration

T Prabhakaran, F Béron, A M G Carvalho, R V Mangalaraja and J C Denardin

Finding an alternative solution to the existing conventional toxic refrigerants has a lot of scope in the refrigeration industries, where magnetic materials are gaining interest for cooling applications. The heating and cooling behaviour of magnetic materials have been explored under magnetization and demagnetization cycles. In the search for environmentally friendly and low cost materials, spinel ferrites have attracted researchers' attention for refrigeration. Various single phases, mixed phases, and nanocomposites have been synthesized through physical and chemical routes and their magnetocaloric effect has been analysed using indirect methods towards the possibility of using them as an alternative refrigerant to fossil fuels. In this chapter we provide a brief introduction to the magnetocaloric effect, the measurement approaches, and the investigation of the magnetocaloric parameters of different spinel ferrites and ferrite nanocomposites for magnetic refrigeration applications.

10.1 Introduction

Worldwide, the terms 'energy and energy conversion', 'green and eco-friendly', 'abundant and inexpensive materials', 'high efficiency', and 'low power consumption' have become of common interest and these concepts are enforced by the limitations of conventional systems. The current research trends are moving toward finding an alternative solution to improve the efficiency of existing conventional devices, for example, home appliances, air conditioners, coolers, refrigerators, etc. As per the website *Our World in Data*, the demand for energy and the power consumption of home appliances increases every year [1]. Among the various energy consumption devices, refrigerators and air conditioners are being given particular attention as they are widely required not only in homes but also in research laboratories, clinical laboratories, food industries, supermarkets, transport vehicles,

doi:10.1088/978-0-7503-5143-0ch10

etc. These devices contain gas or liquid forms of coolant known as refrigerants, which can be any of the following materials: hydrofluorocarbons (HFCs), hydrochlorofluorocarbons (HCFCs), chlorofluorocarbons (CFCs), methane (CH_3), etc. The global warming potential of these refrigerants is high as they emit toxic greenhouse gases into the environment, and thus the elimination of fossil fuels is an important issue [2, 3]. For example, we need to reduce CO_2 emissions by 15% each year to limit global warming to 1.5 °C by 2040 without any negative emission technologies [1]. This demand is critical and the usage of fossil fuels needs to be reduced significantly by a suitable solid-state refrigerant for the welfare of society.

Magnetic materials play a crucial role in this field where their heating behaviour can be controlled by magnetization and demagnetization cycles under adiabatic conditions. The adiabatic temperature change under magnetization and demagnetization cycles is called the magnetocaloric effect (MCE) and the materials that exhibit this effect are called magnetic refrigerants (MRs) [4]. Weiss and Piccard discovered the magnetocaloric effect experimentally in 1917 by observing a sizeable and reversible temperature change in nickel near its Curie temperature [5], while later, in 1976, Brown demonstrated the first room temperature magnetic refrigeration in Gd plates [6]. Magnetic refrigeration is attractive for cooling devices due to its eco-friendly nature and the fact that the refrigerants are made up of solids that do not emit any greenhouse gases. Moreover, it is found that the energy efficiency of magnetic refrigerators is 30% to 60% of the Carnot cycle, which is better than the conventional gas compression and expansion cooling devices. After the discovery of a giant MCE in $Gd_5Si_2Ge_2$ compounds near room temperature, the research towards the development of various MRs increased tremendously [6], with the number of research articles with the keyword 'magnetic refrigeration' increasing exponentially year to year (see figure 10.1).

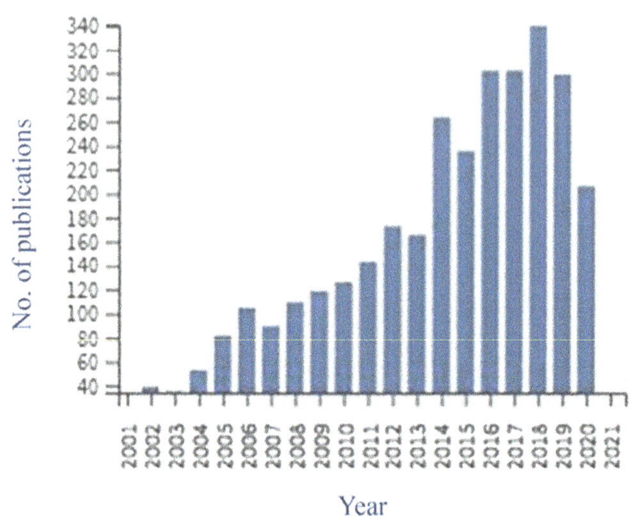

Figure 10.1. Publication numbers (the total number of publications is 3434) in the last twenty years for a keyword search for 'magnetic refrigeration' on the Web of Science.

For the aim of achieving the best MCE near room temperature, various materials have been proposed and reported in the literature. For example, rare earth based crystalline materials, Laves phases, La[Fe(SiAl)]$_{13}$, $Gd_5(Si,Ge)_4$ phases, RE–X–Mn–X–RE (X = Si, Ge or B) intermetallic compounds, ferromagnetic lanthanum manganites, Y–Fe alloys, molecular magnets, and 0% to 75 at% of rare earth based alloys [7]. In rare earth free crystalline compounds, Heusler alloys, Mn(Fe, Co, Ni) (Si, Ge), (Mn, Fe, Co)$_{59}$ (Ge, Si, Sb)$_3$, MnAs alloys, MnF(P, As) alloys, Mn antiperovskites, FeRh alloys, Fe_2P alloys, and Fe-transition metal compounds, abundant $AlFe_2B_2$ compounds and spinel compounds were found [7]. Rare earth and transition metal-based amorphous materials have also been developed. Furthermore, different strategies have been attempted to enhance the refrigerant MCE, such as the formation of multiphase and composite materials to achieve a table-like MCE, rapid solidification of intermetallic alloys to achieve better mechanical properties, solid-state metallurgical processes to make fine powders of alloys, etc. Apart from the intermetallic bulk materials, the nanostructures such as nanoparticles, thin films, and self-assembled nanostructures have also been developed and their MCE was examined in a range of field and temperature. A detailed classification of materials and their properties for magnetic refrigeration can be found in [7].

This chapter presents an overview of the MCE in spinel ferrite nanoparticles (SF NPs) and nanocomposites. High performance ferrites are promising candidates for numerous technological applications. To mention a only few, they are being used as permanent magnets for toys, electrical motors, electrical transformers, and heavy weight lifters in ports. They can also be used in sensors, catalysts, magnetic recording media, magnetic resonance imaging (MRI), targeted drug delivery, magnetic hyperthermia, spintronics, heat transfer fluids, and magnetic refrigerants [8–10]. Ferrites are attractive in various fields owing to their abundant nature, ease of processing, eco-friendliness, inexpensive raw materials, etc. The above factors are the key parameters for developing ferrite-based refrigerants for cooling devices. Basically, ferrites do not have a higher magnetic moment compared to metallic alloys. However, the moderate MCE of ferrites would be feasible for designing a cooling device for microelectronics. This chapter is divided into the following sections. We start with the basic working principles of a conventional refrigerator and of magnetic refrigeration, followed by the direct and indirect methods for estimating the MCE. Then a brief introduction to spinel ferrites, their advantages, and MCE performance is provided with focus on their cooling applications. Finally, we provide a brief summary and some future perspectives.

10.1.1 Conventional refrigerators (CRs) and the need for solid-state refrigerants

In earlier days of human civilization, people did not have the option to store food such as meat and vegetables as fresh. Later, the need for cooling devices became essential in various sectors, which motivated researchers to design the CR. It works based on the compression and expansion cycles of gas (or coolant) molecules. The main parts of a CR are the compressor, expansion valve, condenser, and the coolant.

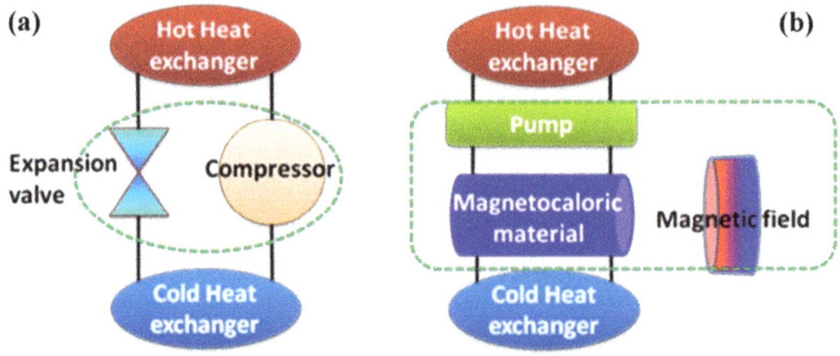

Figure 10.2. Schematic diagrams of (a) CR and (b) MR devices with their main parts.

During the cycle, the coolant transforms from liquid to gas and gas to liquid states. The low pressure cold gas molecules passing through the storage cabin are sucked in by the compressor and are transformed into high pressure hot gas. Then the gas flows through the condenser coils behind the device where the heat is dissipated into the surroundings. Finally, the condensed liquid cools down through the expansion valve and low pressure cold gas circulates inside the cabin where it absorbs heat. This process continues until a defined temperature is reached [3]. The operating principles of a CR and air conditioners (ACs) is similar. However, compared to an AC, a CR runs 24 h a day, 7 days a week. Coolers, chillers, and ACs are still using the global warming potential gases HFCs, HCFCs, and CFCs as refrigerants. Although they are slowly being phased out in home appliances, they are in high demand in large scale industries. Therefore, finding a suitable alternative refrigerant to CRs is an urgent need, where the MRs play a role.

The basic differences between CRs and MRs are depicted in figure 10.2. The magnetic refrigerator replaces the high power consumption parts, such as the compressor and expansion valve in a CR [3, 11], and, most importantly, the fossil fuels are replaced by a solid-state magnetic material as the coolant. Magnetic materials have a special property that undergoes a change in temperature upon magnetization and demagnetization. To achieve cooling, the solid magnetic refrigerants are exposed under a moving permanent magnet (as a source of magnetic field) with the help of a motor.

10.1.2 Magnetic refrigeration

The magnetic refrigerator works based on the following thermodynamic cycle (figure 10.3). The main parts of the MR are the active refrigerant (the magnetocaloric material (MCM)), hot and cold sinks, heat transfer fluid/air, and magnetic field source (a permanent magnet). The cycle consists of two adiabatic and two isothermal processes [4, 7, 12]. The MCM temperature at zero magnetic field in stage I is represented as $T = T_i$. Then it is adiabatically magnetized to stage II, where the temperature increases from T_i to $T_i + \Delta T$ due to the change in MCM magnetic entropy. In the presence of an external magnetic field, the excess heat (Q) is

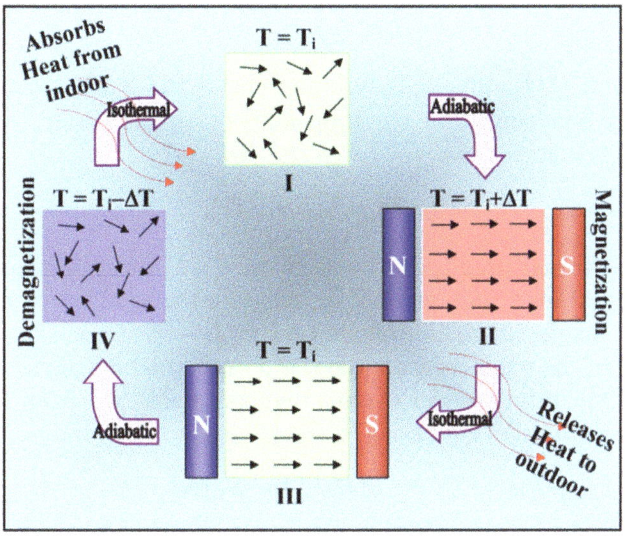

Figure 10.3. Magnetic refrigeration cycle [12].

isothermally dissipated into heat sinks via a heat transfer medium and reaches the temperature $T = T_i$ at stage III. Further, in stage IV the temperature of the MCM reduces to $T_i - \Delta T$ (below the initial temperature) in an adiabatic demagnetization process where the magnetic material returns to its disordered magnetic state. During this process the MCM is capable of absorbing heat from the cabin (where the food is stored) via a heat transfer medium to the cold sink isothermally and complete the cycle at $T = T_i$. The above described change in temperature is a manifestation of the magnetocaloric effect. This cycle is analogous to the gas compression and expansion in a CR. However, the power consumption of an MR is 20% to 30% lower than for a CR.

10.1.2.1 Magnetocaloric effect parameters

The magnetocaloric effect is characterized by two parameters: the isothermal entropy change (ΔS_T) and adiabatic temperature change (ΔT_S). All magnetic materials exhibit the MCE but its strength varies according to the material type and properties. The entropy of magnetic materials can be changed by the magnetic field, temperature, and other thermodynamic parameters [4, 7]. The total entropy of the system (the magnetic material) at a constant pressure can be represented as [4]

$$S(H, T) = S_M(H, T) + S_l(H, T) + S_e(H, T), \tag{10.1}$$

where S_M, S_l, and S_e are, respectively, the magnetic, lattice, and electronic entropies. This equation is valid for rare earth materials, while it is possible for 3D transition materials only if the electron–phonon interaction is not considered. If the S_l and S_M of magnetic materials are independent of the magnetic field, then the total entropy (S) of the system can be written as

$$S(H_i, T_i) = S_M(H_f, T_f) + S_l(T_f) + S_e(T_f), \tag{10.2}$$

where H_i (H_f) and T_i (T_f) are the initial (final) fields and temperatures, respectively. Under adiabatic magnetization the total entropy of the system is constant and equation (10.2) becomes

$$S(H_i, T_i) = S_M(H_f, T_f), \tag{10.3}$$

where the magnetic entropy S_M decreases. However, S_l and S_e increase the magnetic material's temperature simultaneously to maintain the total entropy of the system. As a result, the magnetic material attains $\Delta T = T_f - T_i$.

On the basis of the magnetization data, the isothermal magnetic entropy change (at a constant temperature) under varying magnetic field (ΔH) from $H_i \rightarrow H_f$ can be estimated using Maxwell's equations [4, 7]:

$$\Delta S_T(H, T) = S_M(H_i, T) - S_M(H_f, T), \tag{10.4}$$

$$\Delta S_T(H, T) = \int_{H_i}^{H_f} \left(\frac{\partial S_M(H, T)}{\partial H}\right)_T dH, \tag{10.5}$$

$$\Delta S_T(H, T) = \int_{H_i}^{H_f} \left(\frac{\partial M(H, T)}{\partial T}\right)_H dH, \tag{10.6}$$

$$\Delta S_T(H, T) = S_M(H_f, T) - S_M(H_i, T). \tag{10.7}$$

The sign of the isothermal magnetic entropy change (ΔS_M) can be either positive or negative and it defines the inverse or normal MCE, respectively. From the thermodynamic relation the heat capacity (C) at a constant magnetic field is

$$C = \left(\frac{dQ}{dT}\right)_H = T\left(\frac{dS}{dT}\right)_H. \tag{10.8}$$

If $\left(\frac{\partial S}{\partial H}\right)_T = \left(\frac{\partial M}{\partial T}\right)_H$ then the equation (10.8) will be

$$dT = \frac{-T}{C}\left(\frac{\partial M}{\partial T}\right)_H dH. \tag{10.9}$$

Therefore, the total temperature change under an external magnetic field change from H_i to H_f is

$$\Delta T_S(H, T) = -\int_{H_i}^{H_f} \frac{T}{C}\left(\frac{\partial M}{\partial T}\right)_H dH. \tag{10.10}$$

In addition to these, the refrigerant efficiency, known as refrigerant capacity (RC), should be estimated according to the amount of heat transferred between the cold and hot reservoirs in a cycle by the equation

$$\mathrm{RC}(H) = \int_{T_{\mathrm{cold}}}^{T_{\mathrm{hot}}} \Delta S_{\mathrm{M}} dT, \tag{10.11}$$

where T_{cold} and T_{Hot} are the temperatures at the cold and hot reservoirs, respectively.

The suitability of the magnetic refrigerant for cooling technology is defined by the parameters given in equations (10.6), (10.10), and (10.11).

10.1.3 Estimation of the MCE by direct and indirect methods

In practice, the MCE of the materials is estimated in two ways, the direct and indirect methods. For example, in the direct method, ΔT_{S} of the MCM is recorded directly by a temperature sensor under an external magnetic fields. In contrast, ΔS_{T} and heat capacity are used to calculate ΔT_{S} via an indirect method.

10.1.3.1 Direct measurement

The magnetocaloric material temperature change ($T_f - T_i$) in the magnetic field variation was first proposed by Weiss and Forer in 1926, where they used a thermocouple to measure the temperature. Since then, different methods have been developed to measure the MCE. A superconducting solenoid with a maximum field of 70 kOe has been used, where the temperature change was monitored by five thermocouples after the maximum field was reached [13]. The pulsed magnetic field produced by a solenoid has been utilized where a thermocouple or magneto-optical technique was used to measure the temperature change. It has been found that Copper-constantan thermocouple in the form of wires was better than for strips. A superconducting solenoid is a static magnetic field method that requires more than 10 s to reach the maximum magnetic field above 30 K, whereas the time must be shorter at 10–20 K to measure the MCE [14]. This drawback is overcome by suddenly inserting the sample once the maximum field is achieved. The sample insert, temperature sensor, magnetic field of the solenoid, apparatus sensitivity, heat loss, and possible errors are considered in the design and measurements. Thermoacoustics is a non-contact method where the temperature change is measured by the generation of thermoacoustic waves during the expansion and contraction of a gas layer in the measuring cell due to the MCE [15]. The waves are received by a microphone and the corresponding voltage variation is directly proportional to the temperature change in the sample. In this method, the signals from eddy currents, hysteresis losses, and magnetostriction effects were found to be negligible. In addition to these methods, a differential scanning calorimeter and other calorimetric Peltier methods can be used to measure ΔS_{T} and ΔT_{S} [4, 16, 17]. Calorimetric methods are more suitable for first-order phase transition (FOPT) materials, since the heat flux under the magnetic field is evaluated directly.

10.1.3.2 Indirect methods

In indirect methods, the MCE parameter ΔS_{M} is derived from the isothermal magnetization data and this method is more appropriate for a second-order phase transition (SOPT) material. The magnetic entropy change and heat capacity are used to estimate the MCE quantity ΔT_S. Equation (10.6) is used to find ΔS_M and the simplified form of it can be written as [12]

$$-\Delta S_M = \sum_i \left(\frac{M_{i+1} - M_i}{T_{i+1} - T_i} \right)_H H_i, \tag{10.12}$$

where M_i and M_{i+1} are the initial and final magnetizations in the initial (H_i) and final (H_f) magnetic fields, respectively. Generally, magnetometers are used in this method and follow two different measurement protocols: (i) isothermal magnetization curves $M(H)$ at different temperatures are used to estimate ΔS_M from the integrated area between two magnetization isotherms; and (ii) ΔS_M is computed from the integration (equation (10.6)) of the iso-field magnetization curves $M(T)$ for different magnetic fields. Among these, $M(H)$ would be the right choice if we consider only the measurement speed and it will give appropriate results for SOPT materials [4]. The use of inappropriate protocols might lead to artifacts in the MCE evaluation, either over estimation or under estimation [7]. Therefore, necessary precautions must be taken depending on the material type.

10.1.4 First-order and second-order phase transition materials

FOPT materials undergo a magneto-structural transition at a particular temperature and field that gives rise to giant MCE at the vicinity of this transition. These types of materials are very attractive for magnetic refrigeration. However, it is found that the magneto-structural transitions are accompanied by magnetic and structural hysteresis, which is considered as a drawback for reaching maximum efficiency in refrigeration cycles. On the other hand, SOPT materials exhibit an MCE around a ferromagnetic and ferrimagnetic transition temperature called the Curie temperature [7, 18–20]. An SOPT material does not show the hysteresis effect as observed in FOPT materials. However, the MCE of these system is comparatively lower than for FOPT materials.

10.1.4.1 Spinel ferrites

Among the various SOPT materials, we are interested in spinel ferrites due to their ease of preparation, chemical and thermal stability, abundant nature, low cost compared to rare earth based refrigerants, and particle-size-dependent magnetic properties. Spinel ferrites have the general molecular formula of AB_2O_4 where A^{2+} and B^{3+} are divalent and trivalent cations, respectively, occupying tetrahedral (A) and octahedral (B) interstitial positions of a face-centred cubic (fcc) lattice formed by O^{2-} ions. According to the occupancy of divalent and trivalent cations among the A and B sites, it is defined as a normal, inverse, or mixed spinel structure [21, 22]. As an example, the cubic structure of the magnetite (Fe_3O_4) phase is presented in figure 10.4. The ferrite magnetic properties depend on the distribution of cations between the A and B interstitial positions and the cation type. Moreover, the cation distribution and the corresponding magnetic properties are different in nanocrystalline spinel ferrites compared to bulk ones.

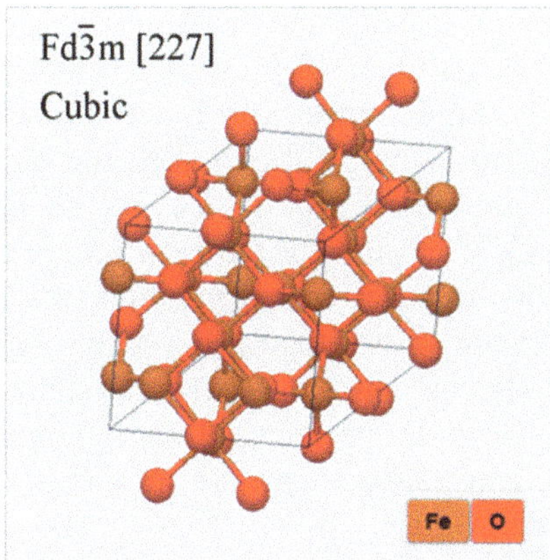

Figure 10.4. The structure of Fe_3O_4. (Image obtained from the Materials Project: mp-19306, https://materialsproject.org/materials/mp-193066/# [23–25].)

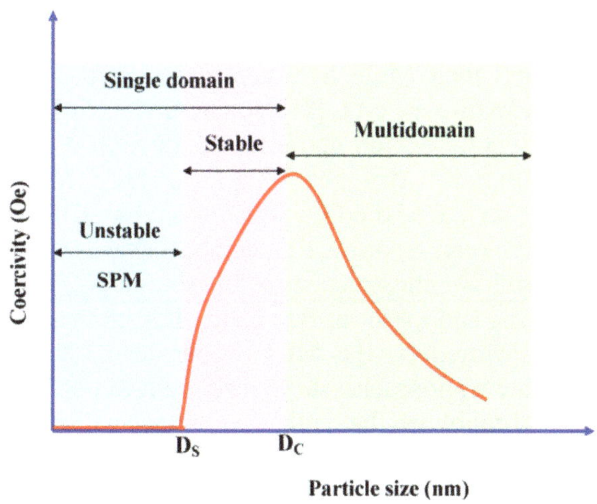

Figure 10.5. Particle-size-dependent magnetic properties and their magnetic states [28, 29].

Spinel ferrites show particle-size-dependent magnetic properties, as depicted in figure 10.5. They transform from multidomain to single domain states at a critical size (D_C). Further size reduction leads to a change from a stable multidomain to unstable single domain superparamagnetic (SPM) state below the particle size (D_S). In addition, the critical sizes D_C and D_S depend on the materials. Below D_S, the nanoparticle magnetization can randomly flip direction under the influence of temperature. In a zero magnetic field, when the time of magnetization measurement is longer than the Néel relaxation time (the time between two flips), the average magnetization becomes zero and the condition is said to be a superparamagnetic state [26]. In this case, the nanoparticle magnetization can be enhanced, similarly to paramagnets. However, the nanoparticles attain a giant magnetic moment compared to paramagnets [27]. Nanoferrites show a transition from superparamagnetic to ferro/ferrimagnetism below a temperature called the blocking temperature (T_B). Modulating the particle size could modify T_B, which may shift the maximum magnetic entropy change close to ambient temperature. Superparamagnetic materials are attracting attention because of the possibility of using them as magnetic refrigerants due to the following advantages: (i) a high magnetic moment compared to paramagnets; (ii) no hysteresis; (iii) the superparamagnetic blocking temperature (T_B) is below the usual Curie temperature of the ferrites; and (iv) the magnetic moment and transition temperature can be modulated by particle size and substitution of foreign metal ions.

10.2 The performance of ferrites as magnetic refrigerants

Small-sized magnetic nanoferrites usually exhibit low magnetic entropy change (ΔS_M) but with a large span of working temperature, which can enhance the overall refrigerant capacity. So far various compositions of spinel ferrites have been obtained with and without substitution of foreign elements to fine-tune the magnetic properties and blocking temperature and their MCE has been investigated. Plain ferrites, such as magnetite Fe_3O_4 [30], hematite α-Fe_2O_3 [31], cobalt ferrite ($CoFe_2O_4$) [32–34], nickel ferrite ($NiFe_2O_4$) [30], $BiFeO_3$ [35], and zinc ferrite ($ZnFe_2O_4$) [36, 37], have been found. In the case of substitution, $Mn_{0.68}Zn_{0.25}Fe_{2.07}O_4$ [32], $Mn_{0.3}Zn_{0.7}Fe_2O_4$ [38], $Ni_{0.3}Zn_{0.7}Fe_2O_4$ [39], $Zn_{0.6}Cu_{0.4}Fe_2O_4$ [40], $Zn_{0.4}Ni_{0.2}Cu_{0.4}Fe_2O_4$ [40], and $Cu_{0.4}Zn_{0.6}Fe_2O_4$ [41] have been reported. Researchers have also examined Fe/γ-Fe_2O_3 core–shell and γ-Fe_2O_3 hollow nanostructures as refrigerants [42, 43]. It is observed from the literature that the ferrite nanoparticles exhibit an MCE close to 300 K, or well below the room temperature. Moreover, the MCE of ferrites of the same composition synthesized through different approaches showed different values. Therefore, in addition to the particle size and morphology, the synthesis approach is also an influencing factor. A large number of synthesis and physical and chemical strategies can be found in the literature, such as the solid-state route, mechanical milling, co-precipitation, microwave assisted co-precipitation, hydrothermal route, microwave hydrothermal route, sonochemical method, combustion, sol–gel method, Pechini method, citric sol–gel method, glycine nitrate process, glycine-assisted combustion method, thermal decomposition, microemulsion method, etc [7, 22, 44, 45]. Though various approaches have been used to increase MCE of ferrites, we have synthesized ferrites and ferrite nanocomposites by microwave assisted co-precipitation method for refrigeration applications.

In this chapter, we focus on the MCE of single phase $ZnFe_2O_4$ and $Ni_{0.3}Zn_{0.7}Fe_2O_4$ nanoparticles synthesized through a microwave assisted co-precipitation method and composites of $CoFe_2O_4$–$CoFe_2$ and $Ni_{0.3}Zn_{0.7}Fe_2O_4$–graphene nanocomposites synthesized by using the thermal reduction and microwave assisted co-precipitation methods, respectively.

10.2.1 The MCE of $ZnFe_2O_4$ nanoparticles

Zinc ferrite (ZFO) nanoparticles were synthesized through a microwave assisted co-precipitation method as follows [45]. Aqueous zinc and iron nitrate solutions were mixed in a 1:2 ratio and then 2.52 M of aqueous sodium hydroxide solution was slowly added during vigorous stirring. The resulting suspension was then subjected to microwave irradiation for 7 min continuously and allowed to cool down. The final precipitates were washed thoroughly and dried at 100 °C for 4 h to yield ultrasmall ZFO nanoparticles.

The spherical ZFO nanoparticles synthesized through this approach exhibit a single phase fcc cubic spinel structure with a mean particle size of 5.07 nm. Non-hysteresis superparamagnetic ZFO nanoparticles showing an unsaturated curve with magnetization of 13.3 emu g^{-1} were used for the magnetocaloric test [45]. In order to estimate the MCE indirectly, a set of isothermal magnetization curves were recorded from 5 K to 305 K with an interval of 10 K, as shown in figure 10.6. The antiferromagnetic ordering of ZFO is evidenced by the magnetization jump at 15 K and then the drop with temperature (the inset of figure 10.6). The magnetic entropy change is calculated using equation (10.12) and its variation with respect to the magnetic field at different temperatures is illustrated in figure 10.7. The ZFO nanoparticles exhibit an inverse magnetocaloric effect in the low temperature region, which is attributed to the antiferromagnetic ordering of the ZFO. The magnetic entropy change varies from −0.15 to −0.65 J kg^{-1} K^{-1} with an increase of the field from 10 to 40 kOe, respectively [45]. It is also observed that at a constant magnetic field, the magnetic entropy change increases with temperature and attains maximum

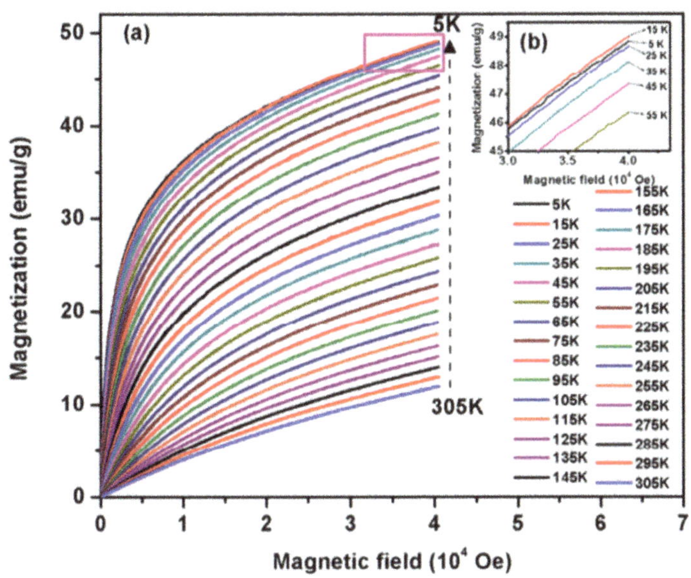

Figure 10.6. Isothermal magnetization curves of ZFO nanoparticles. (Reproduced with permission from [45]. Copyright 2018 Wiley.)

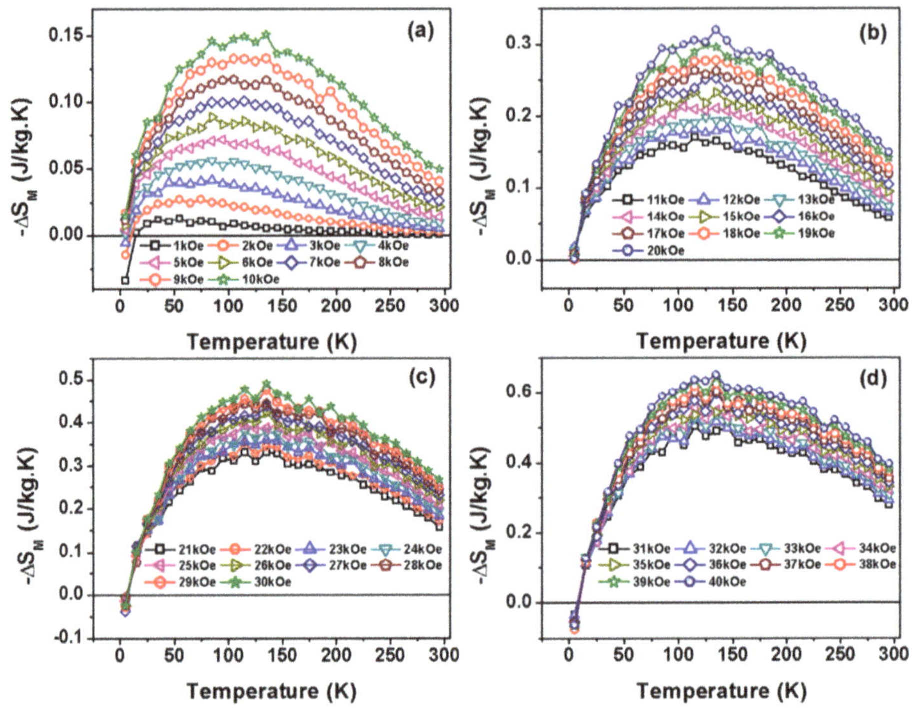

Figure 10.7. Magnetic entropy change with respect to temperature and field. (Reproduced with permission from [45]. Copyright 2018 Wiley.)

at a temperature four to five times higher than the T_B of ZFO [45]. Most of the superparamagnetic ferrites show a similar trend with an increase of temperature at a constant magnetic field. It is demonstrated that nanoferrites follow the magnetic field up to a certain temperature and later the thermal energy overcomes the anisotropy energy that causes the magnetic moments to flip randomly that leads to a decrease in (magnetization) magnetic entropy at higher temperatures [46].

The trend shown in figure 10.7 is impressive for a refrigeration application as it exhibits a large span of working temperatures. The refrigerant capacity calculated from equation (10.11) conveys that the working temperature span (ΔT_{FWHM}) increases from 104 K to 286 K with a field from 10 to 40 kOe. Moreover, the maximum RC of 187 J kg^{-1} is achieved for 40 kOe and it is higher than the value reported for Gd and $(Fe_{70}Ni_{30})_{89}Zr_7B_4$ [45, 47, 48]. To achieve optimal MCE and RC in nanoferrites, strong interparticle interactions, magnetocrystalline anisotropy energy, and a narrow particle size distribution might be useful. The MCE presented for ZFO is comparatively higher than the MCE reported for ZFO synthesized using wet ball milling and sonochemical methods [36, 37]. The field dependence of the maximum magnetic entropy change and RC is represented by the expressions $\Delta S_M^{max} \propto H^n$ and $RC \propto H^N$ respectively where n and N are the local exponents. The values obtained from the slope are 1.066 and 1.205 for the exponents n and N,

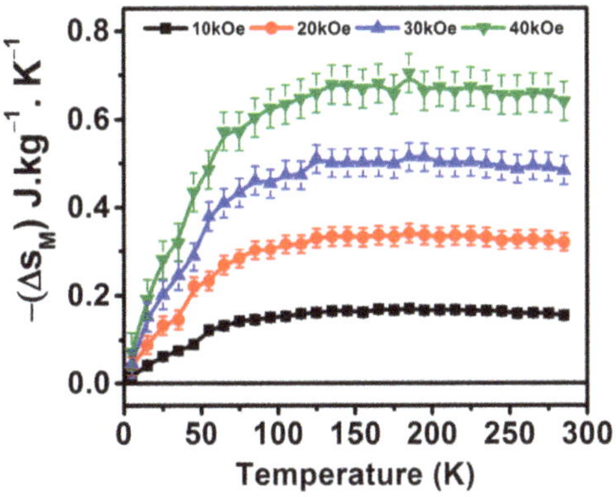

Figure 10.8. Magnetic entropy change of NZF nanoparticles with respect to temperature and fields. (Reproduced with permission from [49]. Copyright 2019 American Chemical Society.)

respectively, and corroborates the second-order phase transition and falls within the ferromagnetic regime.

10.2.1.1 The MCE of nickel zinc ferrite nanoparticles

In zinc ferrite, the substitution of Zn by other divalent transition elements influences the magnetic properties. As a result, appreciable modification has been evidenced [49]. Also, the MCE of bulk nickel zinc ferrite ($Ni_{0.3}Zn_{0.7}Fe_2O_4$, NZF) has been reported and it shows a maximum magnetic entropy change under ambient conditions [50]. Therefore, the NZF nanoparticles were synthesized through the microwave assisted co-precipitation method, as we mentioned for ZFO.

The single phase NZF nanoparticles show fcc cubic spinel structure and nearly spherical nanoparticles with an average particle size of 9.5 nm. We obtain superparamagnetic NZF nanoparticles with a saturation magnetization of 20.1 emu g^{-1} [49]. A protocol to estimate the MCE of the NZF was performed and the magnetic entropy change and refrigerant capacity were calculated. The magnetic entropy change as a function of temperature and field is shown in figure 10.8. As we observed in ZFO nanoparticles, the MCE of the NZF also increases with field. It shows a maximum ΔS_M of -0.68 J kg^{-1} K^{-1} at 135 K which is two to three times higher than the T_B value. The ΔS_M trend of NZF is more interesting for magnetic refrigeration, as it shows a constant ΔS_M from 135 K to temperatures above room temperature [49]. Therefore, the NZF exhibits an RC of 217 J kg^{-1} with a working temperature span of 320 K, which is higher than the 286 K reported for ZFO [45, 49]. The results reveal that the substitution of Zn by Ni cations improves both the magnetic entropy change as well as RC. The choice of suitable cation substitution would be beneficial for the enhancement of the MCE parameters.

10.2.2 Ferrite nanocomposites

In addition to the substitution of foreign elements, other materials, such as the formation of nanostructures, nanocomposites, two phase systems, multilayer structures, and thin films, can also be implemented to study their suitability as magnetic refrigerant. In the following section we discuss two systems of nano-composites: two-phase ferrite nanocomposites and ferrite/graphene nanocomposites.

10.2.2.1 The MCE of $CoFe_2O_4$–$CoFe_2$ nanocomposites

An exchange coupled nanocomposite consisting of soft and hard magnetic phases shows the combined effects of high saturation magnetization and coercivity that is not observed independently in those phases [51]. Various synthesis approaches have been found in the literature where the preparation of hard–soft $CoFe_2O_4$–$CoFe_2$ composite was performed through a thermal reduction method under a hydrogen atmosphere. First, single phase cobalt ferrite nanoparticles were synthesized through a co-precipitation method using 0.1 M and 0.2 M of cobalt and iron nitrate solutions. Then, 0.84 M of sodium hydroxide solution was slowly added to the solution mixture preheated at 40 °C. The obtained dark precipitates were washed, filtered, and dried at 100 °C for 18 h in an oven. In the second stage of the process, the fine cobalt ferrite nanoparticles were thermally reduced at 800 °C for 2 h under a hydrogen atmosphere. During this process, the hard magnetic cobalt ferrite is reduced into the soft $CoFe_2$ magnetic phase. For example, $CoFe_2O_4 + 4H_2 \rightarrow CoFe_2 + 4H_2O$.

The $CoFe_2O_4$ reduced at 800 °C exhibits a cubic structure of highly crystalline $CoFe_2$ phase with an average crystallite size of 30 nm. However, the reduced $CoFe_2$ phase does not show any distinct morphology due to severe aggregation. The sample shows high saturation magnetization, low remanence magnetization, and coercivity of 190 emu g^{-1}, 50 Oe, and 4.2 emu g^{-1}, respectively [51]. This result confirms the formation of a hard–soft composite where $CoFe_2O_4$ is not completely reduced into the $CoFe_2$ phase. The magnetic composite with the above properties may show an appreciable MCE for refrigeration applications. In order to confirm this, a set of isothermal magnetization curves were measured from 200 K to 310 K, and the temperature versus magnetic entropy change for various magnetic fields is shown in figure 10.9. It is found that the $CoFe_2O_4$–$CoFe_2$ composite shows no considerable change in ΔS_M from 200 K to 260 K, but it shows normal and inverse MCE near the

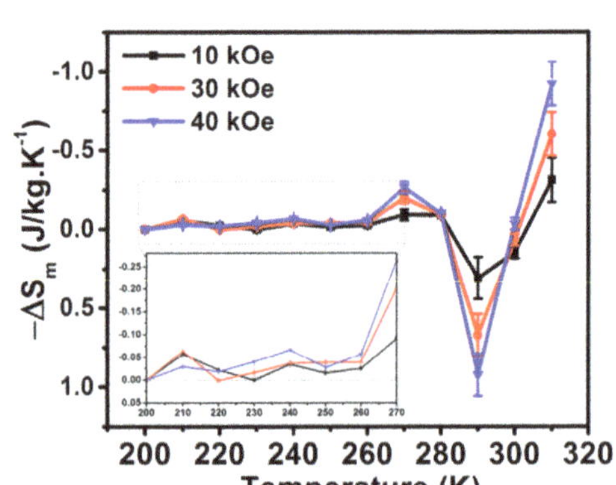

Figure 10.9. Magnetic entropy change of the $CoFe_2O_4$–$CoFe_2$ composite. (Reproduced with permission from [51]. Copyright 2017 Elsevier.)

irreversibility temperature. This confirms that larger magnetic particles display better MCE performance than the smaller ones. The maximum ΔS_M of $CoFe_2O_4$–$CoFe_2$ is −0.92 and 0.92 J kg^{-1} K^{-1} at 290 K and 310 K, respectively [51], which is higher than for the ZFO and NZF nanoparticles [45, 49]. The hard–soft composite has a ΔS_M near room temperature whereas it has a lower RC of 3.7 J kg^{-1} for $\Delta H = 40$ kOe.

10.2.2.2 The MCE of nickel zinc ferrite/graphene nanocomposite

In the surge of interest in exploring novel composites, the light-weight two-dimensional graphene attracted researchers due to its good electrical, mechanical, and chemical properties [52]. In addition, the magnetic ordering of graphene nanostructures is found to be interesting and it would unfold the limitation of magnetic based devices. Ideal graphite and carbon is non-magnetic, however, the presence of defects (intrinsic and extrinsic), impurities, and boundaries can induce magnetic orderings in graphene nanostructures [53, 54]. For example, single atom vacancies produced by the irradiation of high-energy particles and hydrogen chemisorptions defects from hydrogen plasma treatment are observed in nanographites [53]. The graphene lattice can be viewed as two interpenetrating hexagonal sublattices (α and β) of carbon atoms. If a defect is present in the α lattice, the p_z orbitals of carbon atoms in the β lattice only contributes to the quasilocalised state [53, 55] and it forms superstructures around the defects (see figure 10.10 for hydrogen chemisorptions and single-atom defects). Yazyev *et al* have shown that both hydrogen chemisorptions and single-point defects induce quasilocalised states within the carbon

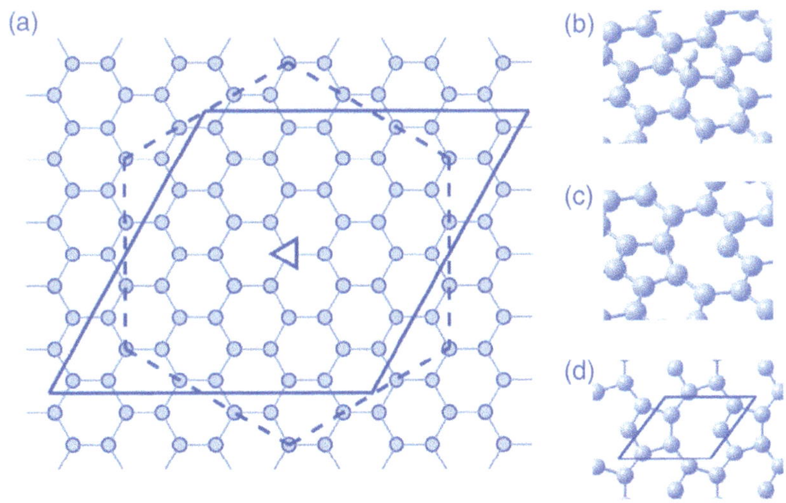

Figure 10.10. (a) The extendable two-dimensional hexagonal lattice of defects in the graphene sheet. Full and dashed lines represent the unit cell and Voronoi cell, respectively. The defective atom is indicated as a triangle as per the defect state orientation. The size of the super-cell is attributed to 9acc separation between neighbouring defects n = 3. (b) Structure of the hydrogen hemisorption defect. (c) Structure of the vacancy defect. (d) Hexagonal closest packing n = 1 of vacancy type defects with the corresponding unit cell. (Reproduced with permission from [53]. Copyright 2007 American Physical Society.)

nanostructures, whereas the presence of hydrogen chemisorption defects yields strong Stoner ferromagnetism [53]. The closest packing of vacancy type defects was found to exhibit a magnetic moment of 1.15 μ_B, whereas smaller defect concentrations show the magnetic moment in the range of 1.45 μ_B to 1.53 μ_B per defect [53]. Total magnetic moment of vacancy type defect is calculated from the contribution of the localized sp^2 dangling bond state. The defect-induced magnetism also depends on the covalent bonding between the second nearest neighbour atoms due to reconstruction. Moreover, at low defect concentrations, the magnetism is sensitive to Fermi energy change from self-doping, to other defects or applied bias, and to disorder-induced broadening [53]. The impurities, single-atom defects, and vacancies created by the foreign elements can induce magnetism in graphene nanostructures. Different magnetic properties of graphene have been reported. The defects present in the same hexagonal lattice lead to ferromagnetism whereas antiferromagnetism is found for the defects in different hexagonal lattices [53]. Paramagnetism has been reported for N_2 irradiated graphene flakes [56]. Other research report shows that the sublattice imbalance determines whether the defects interact ferromagnetically or antiferromagnetically [57]. Antiferromagnetic and ferrimagnetic orderings were observed in RGO-Ni–$NiFe_2O_4$ nanocomposites [57]. Therefore, we decided to analyse the defect-induced MCE on nickel zinc ferrite/graphene nanocomposites for refrigeration.

A nickel zinc ferrite–graphene nanocomposite was synthesized using a microwave assisted co-precipitation method, as discussed previously [49]. To analyse the different structural defect densities, three different NZF–G nanocomposites were synthesized at 20 mg, 60 mg, and 100 mg of graphene loading during the synthesis and are called NZF–G2, NZF–G6, and NZF–G10, respectively. The NZF and NZF–G nanocomposites exhibit cubic structure and the NZF nanoparticles are decorated over the few-layer ($\leqslant$ 4) graphene. The average size of the NZF–G nanocomposites is 7.6 nm, 6.3 nm, and 6.0 nm for NZF–G2, NZF–G6, and NZF–G10, respectively [49]. The effect of the NZF and graphene interaction on the defect density is evidenced through Raman analysis. It shows the formation of a higher degree of structural defects in NZF–G nanocomposites than in the pristine graphene. The I_D/I_G ratio estimated from two prominent vibrations located at 1343 cm^{-1} (D band) and 1586 cm^{-1} (G band) reveals that the I_D/I_G variation is from 1.01 to 1.68, 1.68 to 1.83, and 1.83 to 1.57 for NZF–G2, NZF–G6, and NZF–G10, respectively. Among the three, NZF–G6 exhibited a higher structural defect density than the others [49]. Therefore, the results show that the incorporation of NZF successfully induced defects in the graphene nanostructures.

The NZF and the NZF with different graphene loading exhibited superparamagnetism at room temperature and a magnetization that decreases with the increase of graphene. To perform the defect-induced MCE measurements, a set of isothermal magnetization curves were obtained in a range of magnetic fields and temperatures. The NZF–G nanocomposites show an anomalous MCE, which is not observed in plain NZF nanoparticles (see figure 10.11) and the effect is due to the structural defect-induced magnetic ordering of graphene. The ΔS_M of NZF–G2 decreases from −0.51 to −0.34 J kg^{-1} K^{-1} (for a magnetic field change of 40 kOe) with an

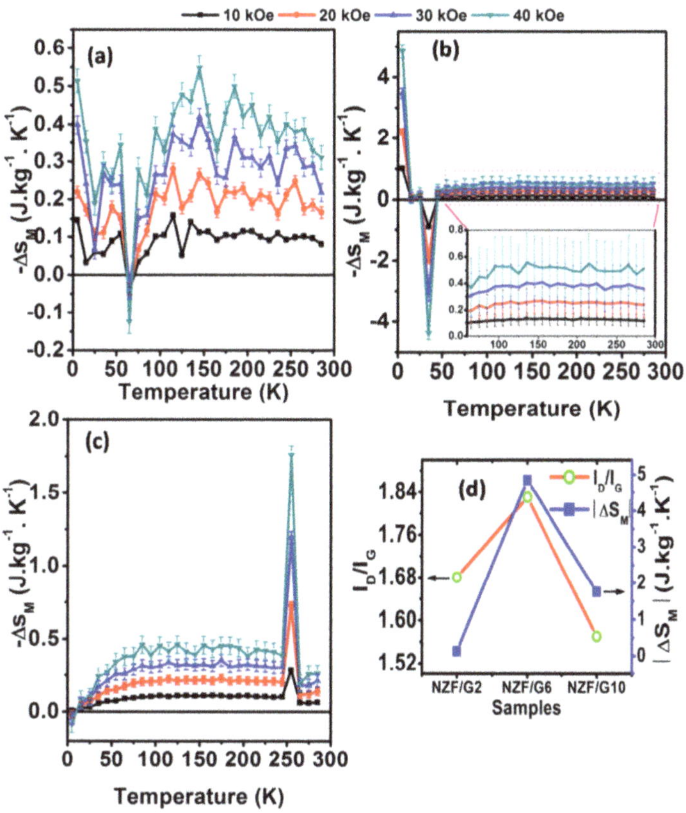

Figure 10.11. Magnetic entropy change of (a) NZF–G2, (b) NZF–G6, and (c) NZF–G10, and (d) defect related magnetic entropy change.(Reproduced with permission from [49]. Copyright 2019 American Chemical Society.)

increase of temperature from 5 K to 55 K. With a further increase of temperature, it shows an inverse MCE of 0.12 J kg^{-1} K^{-1} at 65 K and then follows the normal MCE. The MCE in the NZF–G6 sample changes from the −4.81 J kg^{-1} K^{-1} normal MCE to 4.38 J kg^{-1} K^{-1} with an increase of temperature from 5 K to 35 K. The NZF–G10 composite exhibits inverse magnetocaloric effect (IMCE) at 5 K and later it follows a normal MCE and it shows an abrupt change in entropy at 255 K with −1.76 J kg^{-1} K^{-1} for $\Delta H = 40$ kOe. Among the three composites, NZF–G6 shows a superior MCE, possibly due to the higher structural defect density ($I_D/I_G = 1.83$). Moreover, the results show that the magnetic ordering of NZF–G nanocomposites at 5–295 K varies from FiM < FiM–AFM < AFM for I_D/I_G of 1.57 < 1.68 < 1.83 [49].

10.2.3 Comparison of the MCE with previous reports

Our results for MCE and RC were compared with previous reports on ferrites, as shown in figure 10.12. The magnetic entropy change and refrigerant capacity of

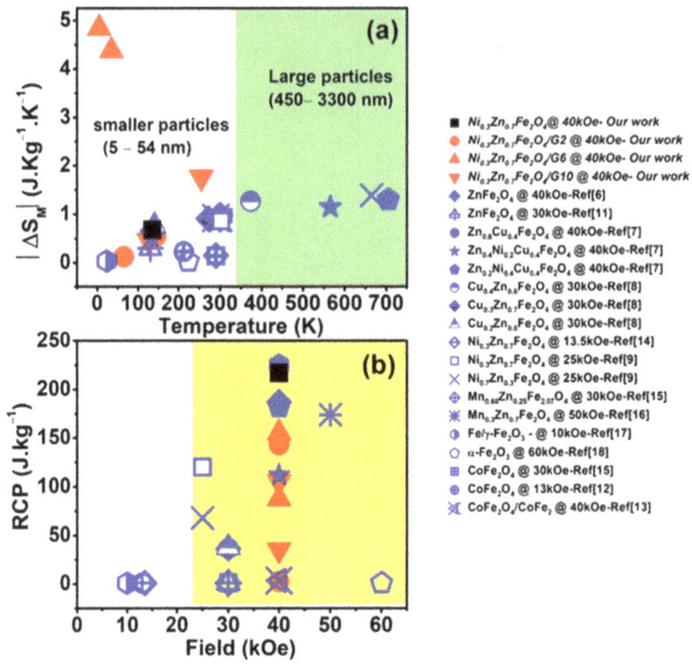

Figure 10.12. Comparison of (a) the magnetic entropy change and (b) RCP with earlier reports on spinel ferrites. (Reproduced with permission from [49]. Copyright 2019 American Chemical Society.)

ferrites are investigated for a vast range of particle sizes. The ferrites with particle sizes from 5 nm to 54 nm present significant ΔS_M values from 5 K to 305 K, whereas larger particles from 450 nm to 3300 nm present some effect above 375 K [49]. The NZF synthesized through our method exhibits a lower ΔS_M (0.68 J kg^{-1} K^{-1}) than the value reported for bulk NZF [39, 50, 52]. However, this value is higher than the values reported for $ZnFe_2O_4$, Fe–y-Fe_2O_3, α-Fe_2O_3, $Mn_{0.68}Zn_{0.25}Fe_{2.07}O_4$, and $CoFe_2O_4$ nanoparticles [31, 32, 34, 36, 42, 45]. It is important to emphasize that NZF–G nanocomposites exhibit superior MCE compared to other ferrites reported so far. For example, the present NZF–G6 displays ΔS_M comparable to the value 3.8 J kg^{-1} K^{-1} reported for $La_{1.6}Ca_{1.4}Mn_2O_7$ polycrystals and the 4.1 J kg^{-1} K^{-1} reported for Gd [58]. Moreover, the ΔS_M of NZF–G10 is higher than the 1.58 J kg^{-1} K^{-1} found in $(Fe_{70}Ni_{30})_{99}Cr_1$ nanoparticles [59].

Although the advisable range of magnetic field for the magnetic refrigeration is $\leqslant$ 20 kOe, a better refrigerant capacity was observed only above 35 kOe (see figure 10.12(b)). The comparison shows that pure NZF nanoparticles yield an RC close to the value 225 J kg^{-1} reported for $Zn_{0.6}Cu_{0.4}Fe_2O_4$ [40] and higher than the RC of bulk $Ni_{0.3}Zn_{0.7}Fe_2O_4$ and polycrystalline $La_{0.48}Ca_{0.52}MnO_3$ nanoparticles [50, 60]. The defect-induced RCs of NZF–G6 and NZF–G10 show comparable cooling power to the values for $Zn_{0.4}Ni_{0.2}Cu_{0.4}Fe_2O_4$ and $Cu_{0.4}Zn_{0.6}Fe_2O_4$ [40, 41]. The results show that the NZF–graphene nanocomposites are promising candidates both at the cryogenic and close to ambient temperatures. Therefore, by controlling the structural defect density in graphene nanocomposites, one can enhance the MCE parameters for cooling applications.

10.3 Perspectives and summary

Spinel ferrites show appreciable development in the area of magnetic refrigeration owing to their magnetic properties and abundant nature. They are one of the important second-order phase transition materials that show particle size and compositional dependent magnetic properties. Different synthesis strategies have been explored to fine-tune their properties via simple chemical and physical routes, where the chemical route is more appropriate than the physical one. Compared to normal spinel ferrites, mixed spinel structures show a good effect on the magnetocaloric effect parameters, for example, the NZF presents higher values than $CoFe_2O_4$. The substitution of a third transition metal in the lattice site of the divalent metal cation in the spinel structure enhances the magnetic entropy change and refrigerant capacity. Detailed analyses show that nanoferrites exhibit a magnetic entropy change below room temperature but with higher refrigerant capacity and working span of temperature compared to bulk ferrites and rare earth based materials. In addition to the single phase and mixed spinel ferrites, the composites, such as hard–soft and ferrite–graphene nanocomposites are found to be important as they show superior MCE compared to the ferrites reported so far. In particular, the defect-induced MCE in ferrite/graphene nanostructures is important as they are derived from low cost and highly stable materials.

For real-time cooling applications, the MCE values obtained for ferrites are comparatively lower than for the rare earth based materials. However, they can be used for microelectronic cooling devices. Different synthesis strategies of composites could be an alternative solution to enhance the required properties for cooling applications. For example, the fabrication of multiferroic nanocomposites from the combination of ferroelectric and ferromagnetic ceramics, and electroactive polymer based ferrite nanocomposites may open up new refrigerants for society. Here, the caloric effect of the multiferroic nanocomposite can be controlled by various stimulating fields such as magnetic fields, electric fields, and stress (mechanical) fields. The instrumentation to investigate these different caloric effects through a single device needs to be explored to speed up the research towards finding alternative refrigerants.

Acknowledgements

The authors gratefully acknowledge FAPESP Postdoctoral Fellowship Process Number (2018) 19096-1, FAPESP Award number: 2017/10581-1 and National Council for Scientific and Technological Development—CNPq, Government of Brazil, and FONDECYT Postdoctoral Research Project No.: 3160170, Government of Chile for financial assistance.

References

[1] Ritchie H, Roser M and Rosado P 2017 CO_2 and greenhouse gas emissions *Our World Data* https://ourworldindata.org/co2-and-other-greenhouse-gas-emissions
[2] Allen S K *et al* 2012 Summary for policymakers *Managing the Risks of Extreme Events and Disasters to Advance Climate Change Adaptation: Special Report of the Intergovernmental Panel on Climate Change* ed C Field *et al* (Cambridge: Cambridge University Press) pp 3–22
[3] Mezaal N A, Osintsev K V and Zhirgalova T B 2017 Review of magnetic refrigeration system as alternative to conventional refrigeration system *IOP Conf. Ser.: Earth Environ. Sci.* **87** 032024
[4] Tishin A M and Spichkin Y I 2016 *The Magnetocaloric Effect and Its Applications* (Boca Raton, FL: CRC Press)
[5] Smith A 2013 Who discovered the magnetocaloric effect? *Eur. Phys. J.* H **38** 507–17
[6] Gschneidner K A and Pecharsky V K 2008 Thirty years of near room temperature magnetic cooling: where we are today and future prospects *Int. J. Refrig.* **31** 945–61
[7] Franco V, Blázquez J S, Ipus J J, Law J Y, Moreno-Ramírez L M and Conde A 2018 Magnetocaloric effect: from materials research to refrigeration devices *Prog. Mater. Sci.* **93** 112–232
[8] Prabhakaran T, Mangalaraja R V, Denardin J C and Jiménez J A 2017 The effect of reaction temperature on the structural and magnetic properties of nano $CoFe_2O_4$ *Ceram. Int.* **43** 5999–606
[9] Chaudhary V and Ramanujan R V 2016 Magnetocaloric properties of Fe–Ni–Cr nanoparticles for active cooling *Sci. Rep.* **6** 35156
[10] Franco V and Conde A 2010 Scaling laws for the magnetocaloric effect in second order phase transitions: from physics to applications for the characterization of materials *Int. J. Refrig.* **33** 465–73
[11] Tušek J, Zupan S, Prebil I and Poredoš A 2009 Magnetic cooling—development of magnetic refrigerator *Stroj. Vestnik./J. Mech. Eng.* **55** 293–302
[12] Franco V, Blázquez J S, Ingale B and Conde A 2012 The magnetocaloric effect and magnetic refrigeration near room temperature: materials and models *Annu. Rev. Mater. Res.* **42** 305–42
[13] Green G, Patton W and Stevens J 1988 *The Magnetocaloric Effect of Some Rare Earth Metals BT—Advances in Cryogenic Engineering.* ed R W Fast (Boston, MA: Springer) pp 777–83
[14] Tishin A M 1988 Magnetocaloric effect in heavy rare earth metals and their alloys *PhD Thesis* Moscow State University
[15] Gopal B R, Chahine R, Foldeaki M and Bose T K 1995 Noncontact thermoacoustic method to measure the magnetocaloric effect *Rev. Sci. Instrum.* **66** 232
[16] Franco V, Blázquez J S, Ingale B and Conde A 2012 The magnetocaloric effect and magnetic refrigeration near room temperature: materials and models *Annu. Rev. Mater. Res.* **42** 305–42
[17] Barclay J A 1988 *Magnetic Refrigeration: A Review of a Developing Technology BT—Advances in Cryogenic Engineering* ed R W Fast (Boston, MA: Springer) pp 719–31
[18] Balli M, Jandl S, Fournier P and Kedous-Lebouc A 2017 Advanced materials for magnetic cooling: fundamentals and practical aspects *Appl. Phys. Rev.* **4** 021305
[19] Mao Q *et al* 2016 Large low field magnetocaloric effect in first-order phase transition compound $TlFe_3Te_3$ with low-level hysteresis *Sci. Rep.* **6** 34235

[20] Law J Y *et al* 2018 A quantitative criterion for determining the order of magnetic phase transitions using the magnetocaloric effect *Nat. Commun.* **9** 2680
[21] Goldman A 1999 *Handbook of Modern Ferromagnetic Materials.* (New York: Springer)
[22] Prabhakaran T, Mangalaraja R V, Denardin J C and Jiménez J A 2017 The effect of calcination temperature on the structural and magnetic properties of co-precipitated $CoFe_2O_4$ nanoparticles *J. Alloys Compd.* **716** 171–83
[23] Fleet M E 1986 The structure of magnetite: symmetry of cubic spinels *J. Solid State Chem.* **62** 75–82
[24] Finger L W, Hazen R M and Hofmeister A M 1986 High-pressure crystal chemistry of spinel ($MgAl_2O_4$) and magnetite (Fe_3O_4): comparison with silicate spinels *Phys. Chem. Miner.* **13** 215–20
[25] Glazyrin K 2011 Iron in oxides, silicates and alloys under extreme pressure-temperature conditions *Dissertation* University of Bayreuth 20111–123
[26] Bedanta S, Petracic O and Kleemann W 2015 Supermagnetism *Handbook of Magnetic Materials* ed K H J Buschow vol 23 (Amsterdam: Elsevier) ch 1 pp 1–83
[27] Marghussian V 2015 Magnetic properties of nano-glass ceramics *Nano-Glass Ceramics Processing, Properties and Applications* (Oxford: William Andrew) ch 4 pp 181–223
[28] Cullity B D and Graham C D 2008 *Introduction to Magnetic Materials* (Hoboken, NJ: Wiley)
[29] Li Q, Kartikowati C W, Horie S, Ogi T, Iwaki T and Okuyama K 2017 Correlation between particle size/domain structure and magnetic properties of highly crystalline Fe_3O_4 nanoparticles *Sci. Rep.* **7** 9894
[30] Gass J 2012 Functional magnetic nanoparticles *Thesis* USF Tampa https://digitalcommons.usf.edu/etd/4047
[31] Pastor J M *et al* 2012 Entropy change linked to the magnetic field induced Morin transition in hematite nanoparticles *Appl. Phys. Lett.* **100** 63102
[32] Poddar P *et al* 2006 Magnetocaloric effect in ferrite nanoparticles *J. Magn. Magn. Mater.* **307** 227–31
[33] Vázquez-Vázquez C *et al* 2008 Magnetocaloric effect and size-dependent study of the magnetic properties of cobalt ferrite nanoparticles prepared by solvothermal synthesis *Phys. Status Solidi* **205** 1358–62
[34] Gopalan E V, Al-Omari I A, Kumar D S, Yoshida Y, Joy P A and Anantharaman M R 2010 Inverse magnetocaloric effect in sol–gel derived nanosized cobalt ferrite *Appl. Phys.* A **99** 497–503
[35] Ramachandran B and Rao M S R 2009 Low temperature magnetocaloric effect in polycrystalline $BiFeO_3$ ceramics *Appl. Phys. Lett.* **95** 142505
[36] Gass J, Srikanth H, Kislov N, Srinivasan S S and Emirov Y 2008 Magnetization and magnetocaloric effect in ball-milled zinc ferrite powder *J. Appl. Phys.* **103** 07B309
[37] Cho J H, Ko S G, Ahn Y and Choi E J 2008 Magnetic entropy change in zinc-ferrite nanoparticles synthesized by using a sonochemical method *J. Korean Phys. Soc.* **53** 746–9
[38] Chaudhary V and Ramanujan R V 2014 Iron oxide-based magnetic nanoparticles for high temperature span magnetocaloric applications *MRS Proc.* **1708** 51–6
[39] Chau N, Thuan N K, Minh D and Luong N 2008 Effects of Zn content on the magnetic and magnetocaloric properties of Ni–Zn ferrites *VNU J. Sci. Math. –Phys.* **24** 155–62
[40] Oumezzine E, Hcini S, Baazaoui M, Hlil E K and Oumezzine M 2015 Structural, magnetic and magnetocaloric properties of echini sol–gel method aqueous solution *Powder Technol.* **278** 189–95

[41] Akhter S *et al* 2014 Magnetic and magnetocaloric properties of $Cu_{1-x}Zn_xFe_2O_4$ ($x = 0.6, 0.7, 0.8$) ferrites *J. Magn. Magn. Mater.* **367** 75–80
[42] Chandra S, Biswas A, Khurshid H, Li W, Hadjipanayis G C and Srikanth H 2013 Magnetic entropy change in core/shell and hollow nanoparticles *J. Phys. Condens. Matter* **25** 426003
[43] Poddar P, Srinath S, Gass J, Prasad B L V and Srikanth H 2007 Magnetic transition and large magnetocaloric effect associated with surface spin disorder in Co and $Co_{core}Ag_{shell}$ nanoparticles *J. Phys. Chem.* C **111** 14060–6
[44] Prabhakaran T, Mangalaraja R V, Denardin J C and Jiménez J A 2017 The effect of reaction temperature on the structural and magnetic properties of nano $CoFe_2O_4$ *Ceram. Int.* **43** 5599–606
[45] Thandapani P, Ramalinga Viswanathan M and Denardin J C 2018 Magnetocaloric effect and universal curve behavior in superparamagnetic zinc ferrite nanoparticles synthesized via microwave assisted co-precipitation method *Phys. Status Solidi* a **215** 1700842
[46] Baldomir D *et al* 2007 Magnetic field dependence of the magnetocaloric effect in magnetic nanoparticle systems: a Monte Carlo simulation *J. Non Cryst. Solids* **353** 793–5
[47] Mathew S P and Kaul S N 2011 Tuning magnetocaloric effect with nanocrystallite size *Appl. Phys. Lett.* **98** 172505
[48] Ipus J J, Ucar H and McHenry M E 2011 Near room temperature magnetocaloric response of an (FeNi)ZrB Alloy *IEEE Trans. Magn.* **47** 2494–7
[49] Prabhakaran T *et al* 2019 Probing the defect-induced magnetocaloric effect on ferrite/graphene functional nanocomposites and their magnetic hyperthermia *J. Phys. Chem.* C **123** 25844–55
[50] Anwar M S, Ahmed F and Koo B H 2014 Enhanced relative cooling power of $Ni_{1-x}Zn_xFe_2O_4$ ($0.0 \leqslant x \leqslant 0.7$) ferrites *Acta Mater.* **71** 100–7
[51] Prabhakaran T, Mangalaraja R V and Denardin J C 2017 The structural, magnetic and magnetic entropy changes on $CoFe_2O_4/CoFe_2$ composites for magnetic refrigeration application *J. Magn. Magn. Mater.* **444** 297–306
[52] Wang C and Diao D 2017 Self-magnetism induced large magnetoresistance at room temperature region in graphene nanocrystallited carbon film *Carbon* **112** 162–8
[53] Yazyev O V and Helm L 2007 Defect-induced magnetism in graphene *Phys. Rev.* B **75** 125408
[54] Santos E J G, Riikonen S, Sánchez-Portal D and Ayuela A 2012 Magnetism of single vacancies in rippled graphene *J. Phys. Chem.* C **116** 7602–6
[55] Mizes H A and Foster J S 1989 Long-range electronic perturbations caused by defects using scanning tunneling microscopy *Science* **244** 559–62
[56] Ney A, Papakonstantinou P, Kumar A, Shang N-G and Peng N 2011 Irradiation enhanced paramagnetism on graphene nanoflakes *Appl. Phys. Lett.* **99** 102504
[57] Kollu P *et al* 2014 Anomalous magnetic behavior in nanocomposite materials of reduced graphene oxide-Ni/$NiFe_2O_4$ *Appl. Phys. Lett.* **105** 52412
[58] Zhou T J, Yu Z, Zhong W, Xu X N, Zhang H H and Du Y W 1999 Larger magnetocaloric effect in two-layered $La_{1.6}Ca_{1.4}Mn_2O_7$ polycrystal *J. Appl. Phys.* **85** 7975
[59] Chaudhary V and Ramanujan R V 2016 Magnetocaloric properties of Fe–Ni–Cr nanoparticles for active cooling *Sci. Rep.* **6** 35156
[60] Das K and Das I 2016 Giant enhancement of magnetocaloric effect at room temperature by the formation of nanoparticle of $La_{0.48}Ca_{0.52}MnO_3$ compound *J. Appl. Phys.* **119** 093903
[61] Pang Y L, Lim S, Ong H C and Chong W T 2016 Research progress on iron oxide-based magnetic materials: synthesis techniques and photocatalytic applications *Ceram. Int.* **42** 9–34

Chapter 11

Degradation of antibiotic pollutants and green hydrogen production from wastewater through a photocatalytic reaction

Saeed Farhang Sahlevani, Héctor Valdés Morales and R V Mangalaraja

In the past decade, the degradation of organic pollutants from wastewater has received increasing attention. Among the organic pollutants, antibiotics have been the center of attention because they are used for the treatment of infections in humans and animals, and even as a feed additive in the agriculture sector. Antibiotics are used extensively as growth promoters in swine, poultry, and fish husbandry. Although the concentrations of antibiotics released into the aquatic system are low, even a low amount of antibiotic in the water could cause serious risks to human health and ecosystems. A large number of scientists have proposed methods to remove antibiotics form wastewater. The advanced oxidation processes based on photocatalysis are able to degrade a large number of pollutants, even antibiotics. In this mini-review, the recent progress in the photocatalysis of antibiotic wastewater is summarized. The photocatalysts commonly used are also discussed. They can be classified mainly as pure ZnO, TiO_2, and their heterojunction composites, as well carbon based materials.

11.1 Introduction

In recent years, public concern has arisen around environmental pollution caused by so-called emerging organic pollutants (EOPs) in water sources and factory effluents, because of their potentially adverse effects on human health and aquatic ecosystems. A large number of physical and biological processes have been proposed to remove pollutants from water, but they do not completely remove the majority of EOPs.

doi:10.1088/978-0-7503-5143-0ch11

Advanced oxidation processes (AOPs) based on catalytic, electrochemical, and photochemical reactions are currently available and are able to degrade a large number of pollutants. An important group of AOPs is represented by photocatalytic processes, in which luminous radiation produces the electronic activation of the semiconductor material that constitutes the photocatalyst. The photocatalytic degradation, based on the generation of extremely reactive hydroxyl radicals, is one of the most attractive methodologies to oxidize antibiotics. However, the photocatalytic oxidation technique often suffers from the photocatalytic material's performance efficacy due to low quantum efficiency, poor utilization of solar energy, and less surface adsorption due to low surface area and porosity.

With the rapid development of the drug and pharmaceuticals industry, the consumption of antibiotics increased among humans and animals. Antibiotics have cured people and animals from bacterial infections, therefore saving many lives. However, for the health of ecosystems and the natural environment, the consequent antibiotic pollution also needs to be addressed. The discharge of industrial and pharmaceutical wastewater has become a major threat to the sustainable development of society [1]. The persistence and difficult degradation properties of antibiotics can cause some serious environmental problems in addition to severe water pollution, such as the emergence of drug resistance. The existence of antibiotics in the water will be favorable for breeding antibiotic-resistant bacteria, posing a threat to the health of people and the effectiveness of antibiotic drugs.

Antibiotics have been used not only for human and animal therapy but also as growth promoters in livestock. Due to being only partly metabolized, the remaining antibiotics are discharged as original substances or metabolites into the environment. Consequently, they have been found in many resources such as wastewater, surface water, and even in drinking water [2]. In the past 5 years, there has been a rapid increase in the number of research publications and citations related to antibiotic degradation (figure 11.1). The aim of the current review is to provide a comprehensive summary of the degradation process of antibiotics (figure 11.2).

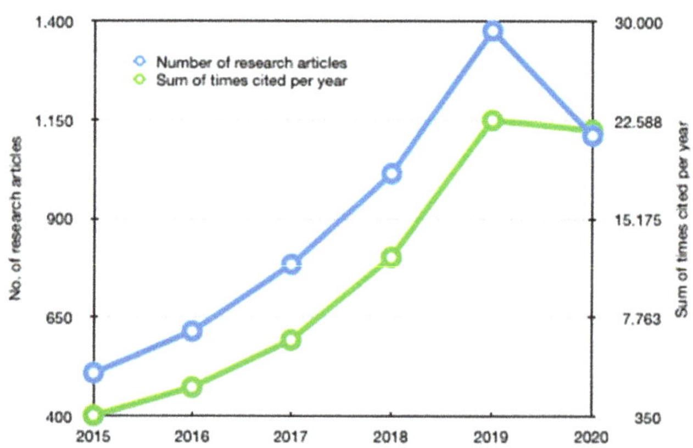

Figure 11.1. The number of research articles and citations on the topic of antibiotic degradation (data obtained from Web of Science in August 2020).

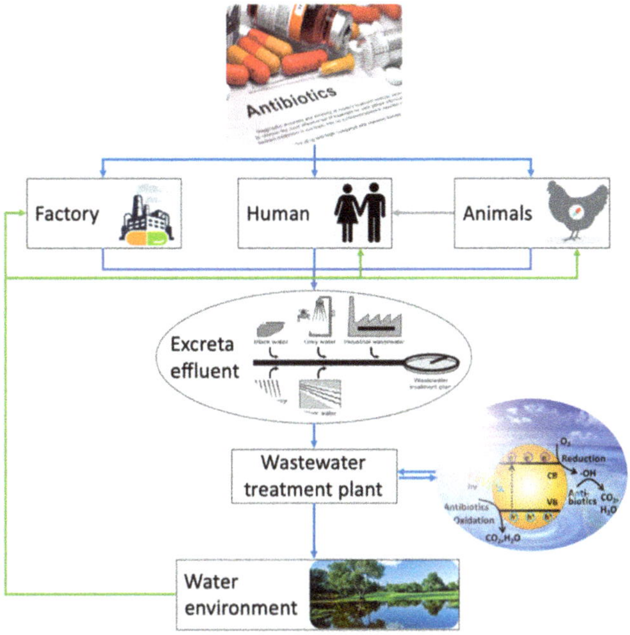

Figure 11.2. Circulation of antibiotics in the environment.

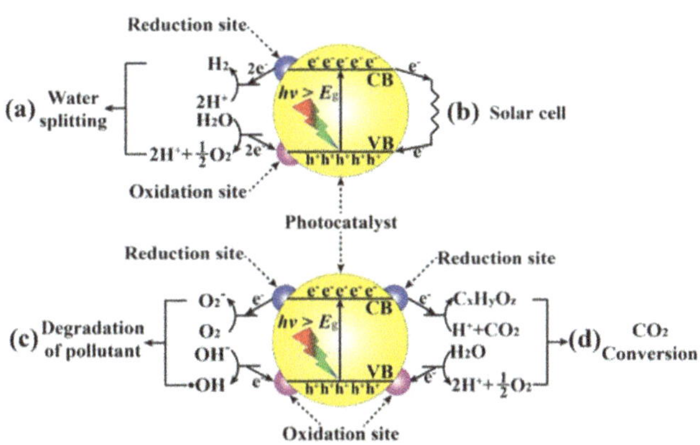

Figure 11.3. Fundamental basics of photocatalysis. A photocatalyst that can (a) produce hydrogen from water, (b) convert solar energy into electric energy, (c) degrade antibiotic pollutants, and (d) reduce CO_2 into organic fuels. (Reproduced with permission from [4]. Copyright 2020 Elsevier.)

11.2 Fundamentals of photocatalytics

The fundamental procedures of photocatalytic hydrogen and the degradation of antibiotic pollution are illustrated in figure 11.3. The whole process includes the following; (i) when the energy of the photons higher than the bandgap of the semiconductors is absorbed by the photocatalyst, then electrons in the valence band can be excited and jump up into the conduction band; (ii) the photogenerated electrons and holes will migrate to the surface of the photocatalysts, and part of the carriers will be recombined in the process; (iii) following migration of the electrons to the surface of the photocatalyst, a reductive pathway is initiated, wherein O_2 is reduced to $\cdot O_2^-$; (iv) conversely, an oxidative pathway is initiated when the holes migrate to the surface of the photocatalyst and **•OH** is generated upon oxidation of $\mathbf{H_2O}$ or the antibiotic molecules. Then, the antibiotic molecules are degraded by the photocatalytically active species $\mathbf{\bullet O_2^-}$ or **•OH**. All these processes influence the final rate and efficiency of the photocatalytic degradation of antibiotics over the photocatalyst system [3, 4].

Four main strategies have been employed in order to design and fabricate visible-light-driven photocatalysts and diverse advanced photocatalytic materials: (i) doping of UV light-driven photocatalysts with metal or non-metal; (ii) the development of new visible-light-driven semiconductor photocatalysts; (iii) the design and structure of heterojunction photocatalysts; and (iv) the design and construction of surface plasmon-resonance enhanced photocatalytic system. The formation of the heterojunction in the desired photocatalyst presented an effective route to achieve charge separation and thereby enhanced photocatalytic efficiency. The valence state heterojunction proved to be a kind of novel heterojunction that could realize the facile transfer of photoinduced carriers and energy without noticeable loss, dramatically improving the photocatalytic activity. However, the valance of the transition metal in the photocatalyst could hardly be controlled and the process was often complex during valence state heterojunction preparation, which became major obstacles for its further development and application. Thus it is important and significant to explore new applications for the fabrication of valence state heterojunctions. The critical aspect to attain high efficiency of the photocatalysis of pollutant degradation is the bandgap value. The energy of the irradiation light should be higher than the bandgap of the photocatalyst. Otherwise, the photocatalyst will not absorb the photons of irradiated light. Furthermore, the position of the valence band should be higher than the oxidation position of the pollutant. On the other hand, the hydroxyl radical ($^{\cdot}OH$) plays an important role in contaminant degradation and the production of $^{\cdot}OH$ is mainly created using two pathways: the first is from multistep reduction by photoelectrons, as mentioned above, and the other is the water oxidized by holes. Moreover, if the photocatalysts are not be able to oxidize water to produce $^{\cdot}OH$, the multistep reduction by photoelectrons to generate superoxide radicals will provide another possible method to produce hydroxyl radicals. It will first initiate through electron reactions with O_2 to produce $^{\cdot}O_2^-$, then the $^{\cdot}O_2^-$ will combine with H^+ to generated $^{\cdot}HO_2^-$. Then, the H_2O_2 as the intermediate product will be produced. Eventually the hydroxyl radicals can be evolved to the dissociation of the H_2O_2 [4].

11.3 Photocatalytic degradation of antibiotic wastewater

Different materials have different band conductive (BC) and valance band (VB) positions, which will exhibit different performance in photocatalysis. In the past decade photocatalyst materials have experienced a crucial development process, and furthermore, have been applied the exploration of photocatalytic antibiotic wastewater degradation, simultaneously the materials also experienced a crucial development. In this chapter, the photocatalysts can be classified into three groups: ZnO, TiO_2, and their hybrid structures, such as core–shell, conjunction, heterojunction, and etc. Furthermore, tetracycline, oxytetracycline, and moxifloxacion were studied as the model antibiotics in wastewater.

11.3.1 ZnO

ZnO is a common photocatalyst that has received a huge amount of attention in past decades. ZnO is indeed one of the most widely investigated oxide photocatalysts due its excellent properties, such as stability, non-toxicity, and biocompatibility [5, 6]. It has a great properties for the absorption of UV irradiation as well as visible light, that make it useful in several fields, such as photocatalysis [5], photo-detectors [7], solar cells, and sensors [8]. However, the photocatalytic activity of ZnO is limited by the high recombination rates of electrons and holes. Hence, extending the optical response and simultaneously suppressing the charge recombination rate are prerequisites for improving the photocatalytic efficiency of metal oxides. There are many reported studies using ZnO as a photocatalyst in the photocatalytic degradation of organic pollutants. Recently Makropoulou *et al* investigated whether the photocatalytic oxidation of the sulfamethoxazole (SMX) can be fulfilled via ZnO under UVA irradiation. They reported that the pH of the ZnO synthesis solution significantly affects the morphology of the catalysts and consequently their photocatalytic activity [9]. The photocatalytic efficiency increases when the initial antibiotic is decreased. In their photodegradation experiment, 200 mg of ZnO was applied for the degradation of 5 mg l^{-1} of SMX. The degradation efficiency was 90%. An increase of the initial antibiotic concentration leads to an excessive coverage of the photocatalyst surface by antibiotic pollutant molecules and reactions by products, which compete for the active sites of the catalyst, as well as for the photogenerated hydroxyl radicals, affecting the degradation rates. SEM images of ZnO nanorod (figure 11.4) catalysts before and after photocatalytic treatment confirm the extent of surface coverage and further reveal the transformation of the surface morphology after contact with organic pollutant molecules from nanorods to nanoparticles [9].

In an earlier study, Elmolla and Chaudhuri [10] investigated the ZnO used for the degradation of antibiotics under UV irradiation. In their photodegradation test, studies on different variables, such as UV irradiation, time irradiation, pH, photocatalyst loading amount, and pollutant concentration. The optimum operating conditions for completed degradation of antibiotic pollutants in an aqueous solution were a concentration of 0.5 g l^{-1}, an irradiation time of 180 min, and pH 11. pH is one of the most affecting parameters on photocatalytic process. It could be the influencing factor on the surface charge of the nanoparticles. In acidic pH, both ZnO as the photocatalyst and the antibiotic are positively charged, hence the adsorption on the surface of ZnO is limited [11]. In other words, at acidic pH values, ZnO could

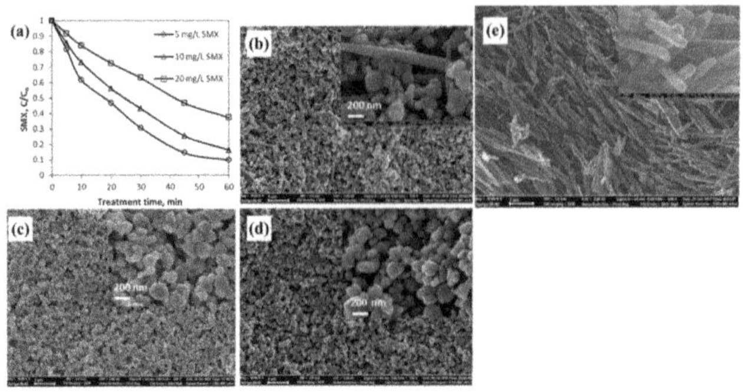

Figure 11.4. (a) The effect of the initial SMX concentration on the photocatalytic degradation of ZnO. (b) SEM images of ZnO after photocatalysis with a concentration of 5 mg l^{-1}, some nanoparticles are still not occupied by SMX molecules, (c) 10 mg l^{-1}, and (d) 20 mg l^{-1}, ZnO nanorods completely dominated by nanoparticles. (e) SEM image of the same ZnO before photocatalysis. (Reproduced with permission from [9]. Copyright 2020 Elsevier.)

lose its oxygen in reactions with H^+, tending to be the ionic form of Zn^{2+}. In the neutral and alkaline pH, the antibiotic has negative charge, whereas ZnO has positive charge. Furthermore, in alkaline pH, the released electrons have the ability to react with oxygen as an electron receptor, which causes the oxygen atom to convert to radicals. The created electron holes are electrons separated from the molecules of organic materials and convert them to the form of R^+ or react with water molecules or hydroxyl ions, therefore producing hydroxyl free radicals which may eventually lead to the decomposition of organic pollutants [11]. There are many reported studies using ZnO as a catalyst in the photocatalytic degradation of antibiotics, such as sulfamethoxazole [12], cephalexin [13], and cefotaxime [14].

11.3.2 TiO_2

TiO_2 semiconductors have a prominent role in the photocatalytic oxidative degradation of antibiotics, which can be ascribed to their efficiency, non-toxicity, and low-cost use in water treatments. TiO_2, with its wide bandgap energy and light absorption in the ultraviolet region, is not considered as a feasible photocatalyst under visible or solar light. Furthermore, its superior redox ability made it the most attractive photocatalyst in the early days and even now. The focus in the research application field could be the oxidation of pollutants to convert solar energy into pollution-free, hydrogen energy to alleviate the energy crisis. Reyes *et al* [15] reported that TiO_2 photocatalysis was used to degrade tetracycline (TC) as a model antibiotic pollutant in aqueous solution under three different light sources (a UV lamp, a solarium device, and a UVA lamp). In their report, the catalyst was applied for the degradation of TC (40 mg l^{-1}). The oxidation of TC was observed when used in a UV lamp. In another study, Zhu *et al* [16] investigated the mechanism of photocatalytic degradation of TC in an aqueous solution by TiO_2 nanoparticles under UV irradiation. In their photodegradation test, the degradation efficiency was 95% within 40 min. There were also similar studies focused on the photocatalytic oxytetracycline wastewater degradation via TiO_2 nanoparticles. In addition to the tetracycline and oxytetracycline, TiO_2 also exhibited an outstanding performance for cephalexin, moxifloxacin, and sulfamethoxazole antibiotic wastewater degradation [17–21]. For example, Khosravi *et al* used montmorillonite coated with TiO_2 which showed better adsorption of cephalexin compared to the other samples, and above 91% cephalexin degradation could be achieved. Pereira *et al* [22] mentioned that the two kinds of antibiotics most extensively used in aquaculture can be 100% degraded by photocatalysis with TiO_2 under solar irradiation. Recently, the degradation of fluoroquinolones (FQ) has received much attention because the World Health Organization (WHO) reported that the presence of FQ in the general environment causes serious problems even at a low concentrations, and FQ has been used widely in pharmaceuticals for humans and animals in past decades. One of research groups investigated the efficiency of TiO_2 nanoparticles for the photodegradation process of gemifloxacin from wastewater. The first order reaction kinetics had a degradation of 95% in less than 30 min of irradiation time and as a result almost complete loss of antibacterial activity against *Escherichia coli* [23]. However, the wide bandgap semiconductors, such as ZnO and TiO_2, constrained it from absorbing visible light, thus only the UV irradiation was responsible. UV light only accounts for about 4% of the overall sunlight spectrum. This drawback limits the use of wide bandgap semiconductors in future applications for antibiotic wastewater degradation. Visible light occupies in excess of 40% of the overall sunlight spectrum. Instead, the hybrid structure or heterojunction photocatalysis of nanomaterials as well as element doping are efficient ways to achieve visible light-driven photocatalysts.

11.3.3 Hybrid nanomaterials

The major drawbacks of the ZnO and TiO_2 nanoparticles are their broad bandgap, the high recombination rates of e–h pairs, and high photo-corrosion. They have a low amount of light absorption in the visible region. It is essential to modify their limitations in order to enhance their physicochemical and optical properties. Several efforts have been made to improve the efficiency of semiconductors, including doping, phase transfer, band regulation, and carbon composites. The results have revealed that the absorption of the composite photocatalysts were extended into the visible light region and the photocatalytic activities were improved in comparison to the pure photocatalysts. ZnO decorated with TiO_2 nanoparticles exhibits superior photodegradation by decolorizing 90% of oxytetracycline in 8 min. UV–Vis absorption studies have shown that the ZnO–TiO_2 composite has three times higher photocatalytic activity compared to pure ZnO [24]. To date, most studies conducted have focused on the doping of ZnO or TiO_2 through the introduction of metals (Al, Mn, Ni, etc) [25], non-metal (C, S, N, etc) [26, 27], or rare-earth elements (Ce, La, Gd, etc) [28]. Jin *et al* reported the removal of norfloxacin using N-TiO_2 within 30 min with the degradation efficiency reaching 99.53%. Furthermore, an early investigation using Fe-doped ZnO for photocatalytic applications found that it contributes significantly towards the generation of H_2O_2 in the system, which is ultimately detrimental for photocatalytic oxidation [29]. Although the doping leads to a decrease of the conductivity band (CB) position or an increase of the valence band (VB) position of the semiconductors, this is also the reason why wide semiconductors with an enhanced visible light absorption can be achieved. The narrow bandgap semiconductors such as sulfides, V_2O_5, or Fe_3O_4 could be selected as the visible-light response materials in order to the increase the efficiency of antibiotic degradation under visible light or solar irradiation. Many studies have focused on antibiotic degradation via nanocomposites as well as doping. Nitrogen, sulfur, and carbon doped ZnO and TiO_2 are popular in the purification of antibiotic wastewater. Farhadian *et al* [30] reported N, S-doped TiO_2 and Z, S-doped ZnO for photocatalytic degradation of tetracycline under visible light. They observed that the Z, S-doped ZnO, under visible light irradiation, exhibited high photocatalytic elimination over 60 min, which was higher than the rate in pure ZnO. Moreover, the results showed that the doping of N, S into a TiO_2 sample shown higher photocatalytic tetracycline degradation than pure TiO_2 nanoparticles. The high activity of N, S in TiO_2 could be attributed to the increase in the light absorption band after N and S doping and improved quantum efficiency due to preventing photoinduced charge recombination by localizing N and S impurities on the surface of pure nanoparticles. Another researcher reported the photodegradation of the tetracycline antibiotic via C-doped TiO_2 [31]. The results showed that 83% degradation efficiency of antibiotic wastewater pollution after four cycles. The mechanisms and performance of photocatalysis in the degradation of pollutant wastewater as well as ciprofloxacin (CIP) can be seen in figure 11.5. Additionally, the efficiency of the degradation of antibiotics could be enhanced by the a core@shell structure, such as $Cu@Fe_3O_4$ core–shell nanoparticles [32]. The oxidative degradation of the antibiotics could be ascribed to the enhanced $^{\cdot}OH$ radical generation by Cu ions and electron exchange between Cu and Fe_3O_4, which are in close contact in a core–shell structure. In addition to core–shell structures, other hybrid structures such as decoration or composites have also been used for drug photodegradation [33–35]. These results prove that the hybrid structures could act as efficient photocatalysts in the degradation of organic pollutants as well as wastewater containing antibiotics.

To date, scholars have focused on the nanosemiconductors coupled with one of the carbonaceous materials, such as graphene [36], carbon nanotubes (CNTs) [37], carbon nanofibers [38], reduced graphene oxide (rGO) [39], or polymeric carbon

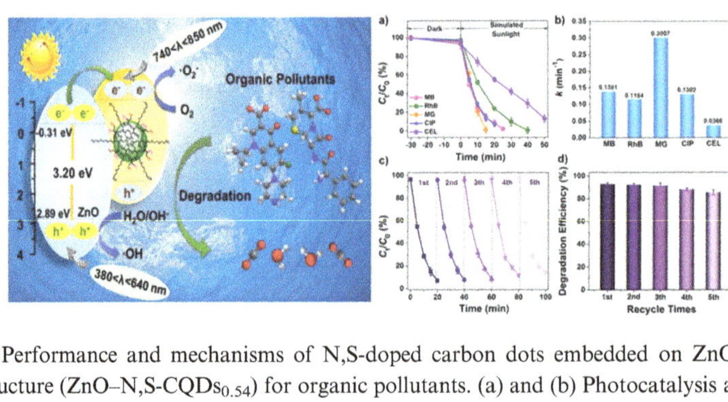

Figure 11.5. >Performance and mechanisms of N,S-doped carbon dots embedded on ZnO with a hybrid nanoflower structure (ZnO–N,S-$CQDs_{0.54}$) for organic pollutants. (a) and (b) Photocatalysis activity of ZnO–N,S-$CQDs_{0.54}$ for the photodegradation of different organic pollutants. (c) and (d) Photocatalytic stability of ZnO–N,S-$CQDs_{0.54}$ for the degradation of CIP within five cycles under sunlight irradiation. (Reproduced with permission from [26]. Copyright 2020 Elsevier.)

nitride (g-C_3N_4) [40] for improving the properties of photocatalysts for visible light-driven applications. For example, Guo and co-workers reported that rGO based semiconductor composites could facilitate the effective photodegradation of drugs during photocatalysis because of the reduced bandgap and accelerative electron transparency. Recently, hybrid semiconductors have received more and more attention due to their use in photocatalytic hydrogen evolution [41–44]. Tang *et al* indicated that the GO affected catalyst properties such as conductivity, improved the activity through minimizing the recombination of photoinduced e–h pairs, and enhanced pollutant removal and hydrogen evolution performance. Furthermore, the photogenerated electron mechanism can be ascribed to the excellent carrier mobility and Fermi energy levels of graphene. The g-C_3N_4 was able to be activated under visible light irradiation and higher efficiency than pure TiO_2 or ZnO in photocatalytic degradation. Hence, the noble materials can usually be coupled with g-C_3N_4 for enhanced performance of photocatalysis through the surface plasmon resonance effect. The bandgap value of g-C_3N_4 is 2.7, with both CB and VB positions being favorable for antibiotic degradation. Wang and his group synthesized g-bridged 2D–2D $Bi_5FeTi_3O_{15}$–ultrathin g-C_3N_4 Z-scheme heterojunction photocatalysts with powerful interfacial charge transfer that proved to have good photoreduction for efficient photocatalytic degradation of antibiotics [45]. The ternary heterostructure nanocomposite exhibited the best tetracycline degradation performance under visible light or solar irradiation. Recently, Yapi *et al* [46] reported improved photocatalytic efficiency of photocatalysis for the degradation of pollutants based on the ternary composite Ag–TiO_2–rGO. The addition of reduced graphene oxide and silver nanoparticles can extend the absorption range to visible light and the absorption edge of the heterojunction composite has a tight shift compared to the pure TiO_2 (figure 11.6). This shift can be attributed to the

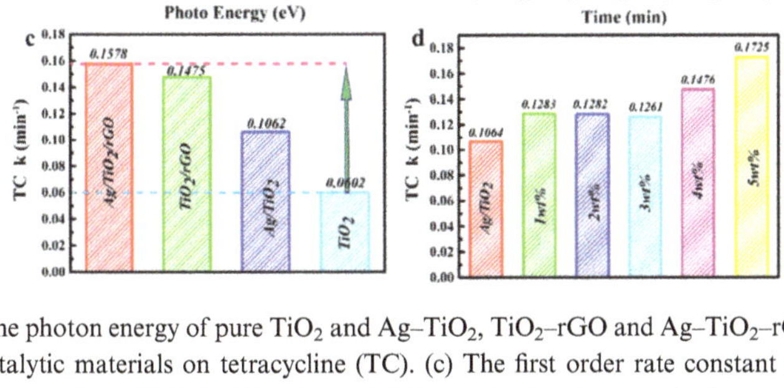

Figure 11.6. (a) The photon energy of pure TiO_2 and Ag–TiO_2, TiO_2–rGO and Ag–TiO_2–rGO composites. (b) Different photocatalytic materials on tetracycline (TC). (c) The first order rate constant of different photocatalysts for the degradation of TC. (d) The effect of different GO loadings on the first order rate constant of photocatalysis. (Reproduced with permission from [46]. Copyright 2020 Elsevier.)

Table 11.1. Report on photocatalytic antibiotic degradation.

Photocatalysts	Antibiotic pollutant	Experimental conditions	Light source/ irradiation time	Percentage of degradation and findings	Reference
ZnO with nanorod morphology	Sulfamethoxazole (SMX)	Potassium ferrioxalate actinometry was used for the estimation of the incident photon flux received by the reactant solution. The solution was stirred in the dark for 30 min for a well adsorbed substrate. All experiments were performed at room temperature. The effect of initial SMX concentration on photocatalytic performance was evaluated.	UVA irradiation (λ = 356nm) 60 min	The photocatalytic efficiency increases when the initial SMX concentration is decreased.	[9]
ZnO nanowires	Cephalexin	Room temperature Light intensity: 7001 ± 5 W m^{-2} Catalyst concentration: 100 μgL^{-1} Cephalexin concentration: 20, 40, 80, 100 μgL^{-1} Background solution ionic strength: 5 mM NaCl Solution pH initial: 7.2	Simulated sunlight Cut off $\lambda < 290$ nm 180 min	The ZnO nanowires have excellent photocatalytic activity for cephalexin. The degradation efficiency of cephalexin was substantially increased by increasing solution pH (alkaline conditions).	[13]
ZnO in the presence of fluoride ions (F-ZnO)	Sulfamethoxazole (SMX)	Intensity of UVC radiation: 3.5 mJ cm^{-2} Airflow rate: 2.5 l min^{-1} T: 21 ° C Concentration of NH_4F: 2.505 mM Photocatalysis dosage: 1.48 g L^{-1} pH: 4.72 Airflow rate: 2.5 l min^{-1} Fluoride ion concentration: 2.505 mM	UVC irradiation 30 min	Experimental SEM showed 96% removal. The effects of F-ZnO on the SMX removal efficiency, kinetics, energy consumption, toxicity removal, and stability of photocatalyst were successful.	[50]

N-doped TiO_2 ($N\text{-}TiO_2$)	Ciprofloxacin (CPX)	CPX concentration: 10 mg l^{-1} pH: 7 Catalyst dose: 0.5 g l^{-1}	Visible light 180 min	Maximum CPX degradation was 55% under optimum conditions. The photocatalytic activity of $N\text{-}TiO_2$ was much higher than for TiO_2. The nitrogen doping helped in narrowing the bandgap of $N\text{-}TiO_2$ to extend its activity under blue visible light.	[51]
N-doped TiO_2 Spherical morphology	Norfloxacin (NOR)	Light intensity: 150 mW cm^{-2} NOR concentration: 6.03 mg l^{-1} $N\text{-}TiO_2$ dose: 0.54 g l^{-1} pH: 6.37	Visible light irradiation 30 min	The light absorption ranged from the UV region to the visible light region since the bandgap was reduced to 2.92 eV. NOR degradation efficiency could reach to 99.53% within 30 min.	[27]
Mn^{2+},Co^{2+} co-doped TiO_2 ($Mn\text{–}Co\text{–}TiO_2$)	Enoxacin (ENX)	Light intensity: 45.2 mW cm^{-2} The flow rate of composition of acetonitrile and methanol (2:1): 1.2 ml min^{-1} ENX concentration: 10 mg L^{-1}	Solar light irradiation	Mn^{2+}, Co^{2+} reduced the bandgap of TiO_2. The fruitful photocatalytic and non-toxic degradation of ENX in synthetic and real wastewater.	[29]
N,S-doped TiO_2 (NST)/N,S-doped ZnO (NSZ)	Tetracycline (TC)	Room temperature pH: 7 Photocatalyst dose: 1 g l^{-1} TC concentration: 25 mg l^{-1}	Visible light 60 min	The adsorption rate was very high in the first 30 min. This is reflects the physical absorption process. The NSZ has 80% absorbance, that is attributed to the large porosity as well as the high surface area. The NST shown 68% TC degradation while un-doped TiO_2 exhibited 37% degradation. This increase was attributed to the light absorption band of TiO_2 after N, S doping and prevention of photoinduced charge recombination by localized N and S impurities.	[30]

(Continued)

Table 11.1. (*Continued*)

Photocatalysts	Antibiotic pollutant	Experimental conditions	Light source/ irradiation time	Percentage of degradation and findings	Reference
$Ag/TiO_2/rGO$ ternary composites	Tetracycline (TC)	TC concentration: 20 mg l^{-1} The initial pH solution was monitored to be 3, 5, 7, 9, 11	300 W mercury lamp ($\lambda < 400$) 60 min	This ternary composite shown excellent photoreduction properties, with more than 80% degradation within 10 min. The electronic coupling system exhibited excellent long service life of the photocatylist. The degradation efficiency of the ternary composite can reach over 97% within 40 min at pH 5 (figure 11.6).	[46]
Ag doped ZnO nanoparticles with graphite layer	Metronidazole (MNZ)	UV intensity: 6.5 mW cm^{-2} The temperature solution was constant MNZ concentration: 30 mg L^{-1} Catalyst concentration: 0.5 g l^{-1} pH: 9	UV radiation/ 60 min Solar irradiation/ 180 min	88.5% of MNZ was removed under UV. 97.3% of MNZ was removed under solar. Ag reduced the rate of the recombination of e–h pairs and enhanced the photocatalytic activity in a wide range of light wavelengths. The graphite surface also acted as an electron sink to efficiently inhibit the photo-corrosion of ZnO, and hence improved the photostability of the composites. The combination of those components improved the degradation of MNZ under UV and simulated solar irradiation.	[2]

plasma effect of the metal surface, thus enhancing the light absorption range of the photocatalysts. Furthermore rGO could move the Fermi energy level of the composites in a favorable direction, thereby enhancing the absorption of visible light or solar irradiation by the heterojunction composite. Evidently, the amount of graphene oxide loading enhances the adsorption activity while increasing the activity of the photocatalytic reaction as an excellent conductor of photogenerated electrons. In photocatalytic organic degradation, the Ag–TiO_2–rGO exhibited excellent tetracycline removal in 10 min and better efficiency within 60 min compared to pure TiO_2, the heterojunction nanocomposite showing higher photocatalytic activity for the degradation of tetracycline. However, the practical application of these photocatalysts is still facing limitations, such as degree of photocatalysis and reusability (table 11.1).

11.4 Conclusions and outlook

We have presented a review of the current literature reporting the degradation of antibiotics using various photocatalysts. This mini-review finds that the application of extensive photocatalysis in organic waste polluted waters should be extended from laboratory scale experiments to real wastewater treatment plants. ZnO, TiO_2, and their composite structures are the materials applied most widely as photocatalysts in water decontamination. The valence band photocatalysis and oxidation potentials of antibiotics could be an important factor for pollution degradation. The results obtained show that the valence bands in photocatalysis have an important relation to the organic impurity oxidization potential. Additionally, the heterojunction mechanism can be used as a strategy to enhance the photocatalytic performance of photocatalysts. The photogenerated electron–hole pair in a heterojunction system can be separated spatially. It was noted that a higher loading amount of graphene oxide also increased the photocatalytic activity. Overall, the morphology of photocatalysis, specific surface area, pH, and co-catalysis are effective parameters to increase the high activity of photocatalysis through harnessing photon energy from the UV or visible region. However, the use of this semiconductor material presents advantages and known limitations. Note that the practical application and commercialization of these semiconductor materials demonstrate that this research field is important and exciting. Therefore, it is very important to continue working on the search for new, more efficient photocatalysts with better performance at the large scale for practical applications.

References

[1] Liu W, Zhou J and Yao J 2020 Shuttle-like CeO_2/g-C_3N_4 composite combined with persulfate for the enhanced photocatalytic degradation of norfloxacin under visible light *Ecotoxicol. Environ. Saf.* **190** 110062

[2] Tran M L, Nguyen C H, Fu C C and Juang R S 2019 Hybridizing Ag-doped ZnO nanoparticles with graphite as potential photocatalysts for enhanced removal of metronidazole antibiotic from water *J. Environ. Manage.* **252** 109611

[3] Li D and Shi W 2016 Recent developments in visible-light photocatalytic degradation of antibiotics *Cuihua Xuebao/Chinese J. Catal.* **37** 792–9
[4] Wei Z, Liu J and Shangguan W 2020 A review on photocatalysis in antibiotic wastewater: pollutant degradation and hydrogen production *Chinese J. Catal.* **41** 1440–50
[5] Farhang Sahlevani S *et al* 2019 Influence of refluxing time and HMTA on structural and optical properties of rod, prism like ZnO nanostructures *J. Mater. Sci., Mater. Electron.* **30** 5670–80
[6] Farhang-Sahlevani S *et al* 2019 A facile hydrothermal synthesis of CeO_2 nanocubes decorated ZnO nanostructures: optical and enhanced photocatalytic properties *J. Mater. Sci., Mater. Electron.* **30** 11643–51
[7] Deka Boruah B 2019 Zinc oxide ultraviolet photodetectors: rapid progress from conventional to self-powered photodetectors *Nanoscale Adv.* **1** 2059–85
[8] Sharma P, Kumar N, Chauhan R, Singh V, Srivastava V C and Bhatnagar R 2020 Growth of hierarchical ZnO nano flower on large functionalized rGO sheet for superior photocatalytic mineralization of antibiotic *Chem. Eng. J.* **392** 123746
[9] Makropoulou T, Kortidis I, Davididou K, Motaung D E and Chatzisymeon E 2020 Photocatalytic facile ZnO nanostructures for the elimination of the antibiotic sulfamethoxazole in water *J. Water Process Eng.* **36** 101299
[10] Elmolla E S and Chaudhuri M 2010 Degradation of amoxicillin, ampicillin and cloxacillin antibiotics in aqueous solution by the UV/ZnO photocatalytic process *J. Hazard. Mater.* **173** 445–9
[11] Farzadkia M, Rahmani K, Gholami M, Esrafili A, Rahmani A and Rahmani H 2014 Investigation of photocatalytic degradation of clindamycin antibiotic by using nano-ZnO catalysts *Korean J. Chem. Eng.* **31** 2014–9
[12] Goncalves M G *et al* 2020 Relationship of the physicochemical properties of novel ZnO/biochar composites to their efficiencies in the degradation of sulfamethoxazole and methyl orange *Sci. Total Environ.* **748** 141381
[13] He J *et al* 2019 Photocatalytic degradation of cephalexin by ZnO nanowires under simulated sunlight: kinetics, influencing factors, and mechanisms *Environ. Int.* **132** 105105
[14] Mohammed R, Ali M E M, Gomaa E and Mohsen M 2020 Green ZnO nanorod material for dye degradation and detoxification of pharmaceutical wastes in water *J. Environ. Chem. Eng.* **8** 104295
[15] Reyes C *et al* 2006 Degradation and inactivation of tetracycline by TiO_2 photocatalysis *J. Photochem. Photobiol.* A **184** 141–6
[16] Zhu X, Wang Y, Sun R and Zhou D 2013 Chemosphere photocatalytic degradation of tetracycline in aqueous solution by nanosized TiO_2 *Chemosphere* **92** 925–32
[17] Khosravi R, Zarei A, Heidari M, Ahmadfazeli A, Vosughi M and Fazlzadeh M 2018 Application of ZnO and TiO_2 nanoparticles coated onto montmorillonite in the presence of H_2O_2 for efficient removal of cephalexin from aqueous solutions *Korean J. Chem. Eng.* **35** 1000–8
[18] Bansal P, Verma A, Aggarwal K, Singh A and Gupta S 2016 Investigations on the degradation of an antibiotic cephalexin using suspended and supported TiO_2: mineralization and durability studies, *Can. J. Chem. Eng.* **94** 1269–76
[19] Van Doorslaer X, Heynderickx P M, Demeestere K, Debevere K, Van Langenhove H and Dewulf J 2012 TiO_2 mediated heterogeneous photocatalytic degradation of moxifloxacin: operational variables and scavenger study *Appl. Catal.* B **111–112** 150–6

[20] Van Doorslaer X, Haylamicheal I D, Dewulf J, Van Langenhove H, Janssen C R and Demeestere K 2015 Heterogeneous photocatalysis of moxifloxacin in water: chemical transformation and ecotoxicity *Chemosphere* **119** S75–80
[21] Xekoukoulotakis N P *et al* 2010 Kinetics of UV-A/TiO_2 photocatalytic degradation and mineralization of the antibiotic sulfamethoxazole in aqueous matrices *Catal. Today* **161** 163–8
[22] Elmolla E S and Chaudhuri M 2010 Photocatalytic degradation of amoxicillin, ampicillin and cloxacillin antibiotics in aqueous solution using UV/TiO_2 and UV/H_2O_2/TiO_2 photocatalysis *DES* **252** 46–52
[23] Ibrahim F A, Al-Ghobashy M A, Abd El-Rahman M K and Abo-Elmagd I F 2017 Optimization and in line potentiometric monitoring of enhanced photocatalytic degradation kinetics of gemifloxacin using TiO_2 nanoparticles/H_2O_2 *Environ. Sci. Pollut. Res.* **24** 23880–92
[24] Singh J, Kumar S, Rishikesh, Manna A K and Soni R K 2020 Fabrication of ZnO–TiO_2 nanohybrids for rapid sunlight driven photodegradation of textile dyes and antibiotic residue molecules *Opt. Mater. (Amst).* **107** 110138
[25] Hosseini M, Esrafili A, Farzadkia M, Kermani M and Gholami M 2020 Degradation of ciprofloxacin antibiotic using photo-electrocatalyst process of Ni-doped ZnO deposited by RF sputtering on FTO as an anode electrode from aquatic environments: synthesis, kinetics, and ecotoxicity study *Microchem. J.* **154** 104663
[26] Qu Y, Xu X, Huang R, Qi W, Su R and He Z 2020 Enhanced photocatalytic degradation of antibiotics in water over functionalized N,S-doped carbon quantum dots embedded ZnO nanoflowers under sunlight irradiation *Chem. Eng. J.* **382** 123016
[27] Jin X *et al* 2019 Photocatalytic degradation of norfloxacin using N-doped TiO_2: optimization, mechanism, identification of intermediates and toxicity evaluation *Chemosphere* **237** 124433
[28] Zammit I, Vaiano V, Ribeiro A R, Silva A M T, Manaia C M and Rizzo L 2019 Immobilised cerium-doped zinc oxide as a photocatalyst for the degradation of antibiotics and the inactivation of antibiotic-resistant bacteria *Catalysts* **9** 222
[29] Sayed M *et al* 2018 Narrowing the band gap of TiO_2 by co-doping with Mn^{2+} and Co^{2+} for efficient photocatalytic degradation of enoxacin and its additional peroxidase like activity: a mechanistic approach *J. Mol. Liq.* **272** 403–12
[30] Farhadian N, Akbarzadeh R, Pirsaheb M, Jen T C, Fakhri Y and Asadi A 2019 Chitosan modified N, S-doped TiO_2 and N, S-doped ZnO for visible light photocatalytic degradation of tetracycline *Int. J. Biol. Macromol.* **132** 360–73
[31] Oseghe E O and Ofomaja A E 2018 Facile microwave synthesis of pine cone derived C-doped TiO_2 for the photodegradation of tetracycline hydrochloride under visible-LED light *J. Environ. Manage.* **223** 860–7
[32] Pham V L, Kim D G and Ko S O 2018 Oxidative degradation of the antibiotic oxytetracycline by Cu@Fe_3O_4 core–shell nanoparticles *Sci. Total Environ.* **631–632** 608–18
[33] Elfeky A S *et al* 2020 Multifunctional cellulose nanocrystal /metal oxide hybrid, photodegradation, antibacterial and larvicidal activities *Carbohydr. Polym.* **230** 115711
[34] Lu X *et al* 2016 Controllable synthesis of graphitic C_3N_4/ultrathin MoS_2 nanosheet hybrid nanostructures with enhanced photocatalytic performance *Dalt. Trans.* **45** 15406–14
[35] Changanaqui K, Brillas E, Alarcón H and Sirés I 2020 ZnO/TiO_2/Ag_2Se nanostructures as photoelectrocatalysts for the degradation of oxytetracycline in water *Electrochim. Acta* **331** 135194

[36] Yan M *et al* 2017 Fabrication of nitrogen doped graphene quantum dots-BiOI/$MnNb_2O_6$ p–n junction photocatalysts with enhanced visible light efficiency in photocatalytic degradation of antibiotics *Appl. Catal.* B **202** 518–27
[37] Liu S *et al* 2018 Anchoring Fe_3O_4 nanoparticles on carbon nanotubes for microwave-induced catalytic degradation of antibiotics *ACS Appl. Mater. Interfaces* **10** 29467–75
[38] Zhang Y, Zhang B-T, Teng Y, Zhao J, Kuang L and Sun X 2020 Carbon nanofibers supported Co/Ag bimetallic nanoparticles for heterogeneous activation of peroxymonosulfate and efficient oxidation of amoxicillin *J. Hazard. Mater.* **400** 123290
[39] Guo H *et al* 2019 Enhanced catalytic performance of graphene–TiO_2 nanocomposites for synergetic degradation of fluoroquinolone antibiotic in pulsed discharge plasma system *Appl. Catal.* B **248** 552–66
[40] Hu K *et al* 2020 Facile synthesis of Z-scheme composite of TiO_2 nanorod/g-C_3N_4 nanosheet efficient for photocatalytic degradation of ciprofloxacin *J. Clean. Prod.* **253** 120055
[41] Tang L, Liu L and Yang F 2019 $FeMoO_4$-graphene oxide photo-electro-catalyst for berberine removal and hydrogen evolution *Int. J. Hydrogen Energy* **44** 19755–61
[42] Dong H *et al* 2019 High-throughput production of ZnO-MoS_2-graphene heterostructures for highly efficient photocatalytic hydrogen evolution *Materials* **12** 2233
[43] Samsudin M F R and Sufian S 2020 Hybrid 2D/3D g-C_3N_4/$BiVO_4$ photocatalyst decorated with RGO for boosted photoelectrocatalytic hydrogen production from natural lake water and photocatalytic degradation of antibiotics *J. Mol. Liq.* **314** 113035
[44] Wei Z *et al* 2019 Photocatalytic hydrogen evolution with simultaneous antibiotic wastewater degradation via the visible-light-responsive bismuth spheres-g-C_3N_4 nanohybrid: waste to energy insight *Chem. Eng. J.* **358** 944–54
[45] Wang K, Li J and Zhang G 2019 Ag-bridged Z-Scheme 2D/2D $Bi_5FeTi_3O_{15}$/g-C_3N_4 heterojunction for enhanced photocatalysis: mediator-induced interfacial charge transfer and mechanism insights *ACS Appl. Mater. Interfaces* **11** 27686–96
[46] Hou Y *et al* 2019 Ultrasonic impregnation assisted *in situ* photoreduction deposition synthesis of Ag/TiO_2/rGO ternary composites with synergistic enhanced photocatalytic activity *J. Taiwan Inst. Chem. Eng.* **104** 139–50
[47] Chen Q, Zhang M, Li J, Zhang G, Xin Y and Chai C 2020 Construction of immobilized 0D/1D heterostructure photocatalyst Au/CuS/CdS/TiO_2 NBs with enhanced photocatalytic activity towards moxifloxacin degradation *Chem. Eng. J.* **389** 124476
[48] Palominos R A, Mondaca M A, Giraldo A, Peñuela G, Pérez-Moya M and Mansilla H D 2009 Photocatalytic oxidation of the antibiotic tetracycline on TiO_2 and ZnO suspensions *Catal. Today* **144** 100–5
[49] Topkaya E, Konyar M, Yatmaz H C and Öztürk K 2014 Pure ZnO and composite ZnO/TiO_2 catalyst plates: a comparative study for the degradation of azo dye, pesticide and antibiotic in aqueous solutions *J. Colloid Interface Sci.* **430** 6–11
[50] Mirzaei A, Yerushalmi L, Chen Z, Haghighat F and Guo J 2018 Enhanced photocatalytic degradation of sulfamethoxazole by zinc oxide photocatalyst in the presence of fluoride ions: optimization of parameters and toxicological evaluation *Water Res.* **132** 241–51
[51] Karim A V and Shriwastav A 2020 Degradation of ciprofloxacin using photo, sono, and sonophotocatalytic oxidation with visible light and low-frequency ultrasound: degradation kinetics and pathways *Chem. Eng. J.* **392** 124853

IOP Publishing

Nanostructured Materials for Sustainable Energy and Environmental Remediation

R V Mangalaraja, R Udayabhaskar, P Sathishkumar and Joydeep Dutta

Chapter 12

Fluorescent nanoclusters used as a probe for sensing of toxic chemicals and biomolecules

Krishnamoorthy Shanmugaraj, R V Mangalaraja, Radhamanohar Aepuru, Cristian H Campos, Malaichamy Ilanchelian and Mohamad Hafiz Bin Mamat

12.1 Introduction

Toxic chemicals and biomolecules play an important role in human health, forensic, biomedical, and environmental science [1]. Thus the sensing of these toxic chemicals and biomolecules is essential, and can be achieved via various smart sensing devices, which are lower cost than traditional methods and have the potential for monitoring environments and food, and in other important industrial applications. Furthermore, the advancement of cost-effective and highly sensitive sensor techniques requires superior technology coupled with basic knowledge in materials science, biology, and chemistry [1]. Generally, sensors contain two important mechanisms: (i) a recognition element to provide selective/specific binding with the target molecules and (ii) a transducer component for signalling the binding event. An effective sensor relies heavily on the authentication process for both components, based on selectivity, detection range (limit of detection (LOD)), signal-to-noise ratio (S/N), and response time [2, 3] As a result, developing sensors with higher efficiency depends on the development of new materials to improve both the transduction and recognition processes. Currently, various methods have been developed to meet the demand for the detection of toxic chemicals and biomolecules, such as electroanalysis, [4] potentiometry, [5] spectrophotometry, [6] chemiluminescence, [7] catalyst-assisted flow-injection [8], and iodometric methods [9]. On the other hand, all of these techniques are relatively expensive and time-consuming, and require sophisticated instruments and complex sample treatment.

doi:10.1088/978-0-7503-5143-0ch12

In contrast, spectrofluorometric detection has obtained significant attention due to its high sensitivity, selectivity, lower time requirements, ease of operation, and reproducibility. The fluorescence based detection methods are 1000-fold more sensitive than absorbance based optical sensors [10, 11]. The important challenges in fluorescence methods are the selection of the right fluorescent compound and the transformation of the sensing events into changes in the fluorescence intensity. Fluorescence based sensing techniques using photoluminescent polymers, small organic molecules, nanomaterials, and semiconductor quantum dots (QDs) have received considerable attention owing to their sensitivity, convenient visual detection, and short response times [12–16] Unfortunately, the above mentioned methods suffer from low photostability and difficult synthesis processes, while the application of QDs (CdS, CdSe, and ZnS) is restricted owning to the serious health concerns and environmental issues caused by heavy metals. These fluorescence probes that can overcome the above limitations with high rapidity and sensitivity are highly desirable [17].

In the last few decades, considerable development in the synthesis of highly fluorescent metal nanoclusters (NCs) (such as Au, Ag, and Cu) has provided a novel class of fluorescent materials with an large Stokes shift, ultrasmall size, good biocompatibility, and apparent luminescence. These NCs possess very low toxicity compared to QDs [18]. Owing to the lower toxicity of NCs, they have been successfully applied in chemical sensors, electronic devices catalysis, and biological imaging [19, 20]

Fluorescent metal NCs have attracted great interest from scientists due to their excellent features, including photostability, biocompatibility, ease of synthesis, and subnanometer size [19, 20]. These metal NCs display the link between nanoparticles (NPs) (exhibiting plasmons) and metal atoms (exhibiting distinct optical properties) and exhibit molecule-like behaviour. The optical and electronic behaviour of metals mainly depends on their size, particularly in the nanometre range (figure 12.1). There is no energy gap separation between the conduction band and valance band for bulk metals, and as a result electrons can move freely without any barriers. The electron mean free path is also determined from the scattering of these electrons. In the case of metal nanoparticles, the size is smaller than or comparable to the mean free path of the electrons, where the movement of electrons is hindered by the NPs' size and surface interaction is mainly expected. This behaviour gives rise to surface plasmon resonance (SPR) [21]. The size of the metals is again reduced to less than 1 nm or down to a few atoms in metal NCs, which leads to discontinuation of the continuous band structure, which is separated into discrete energy levels. Therefore, these NCs are not plasmonic or conduction, as the interaction with light still exists, but are identical to dye molecules, which via electronic transitions between the energy levels, produce light emission and absorption. This kind of optical behaviour gives them potential as fluorescence probes in bioimaging and sensing as, for example, fluorophores and QDs [20].

Noble metal NCs contain only a few to several hundred metal atoms with a size of < 2 nm, which is comparable to the electron Fermi wavelength of Ag or Au. Gold nanoclusters (AuNCs) do not show the SPR band in the visible region but have

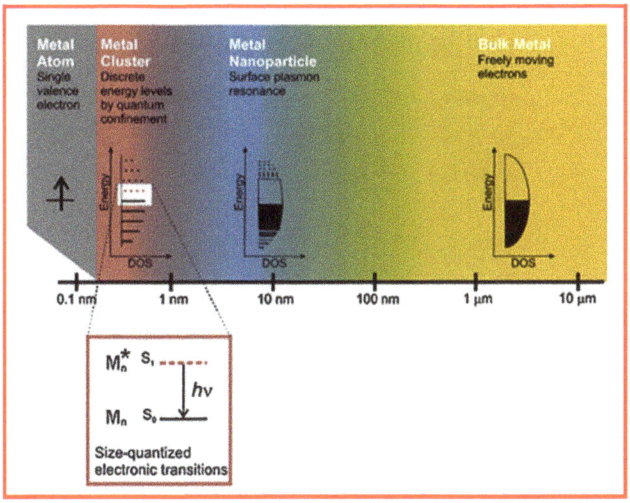

Figure 12.1. Schematic representation of various energy levels of bulk metal, metal NPs, metal NCs, and single atoms. (Reproduced with permission from [36]. Copyright 2020 The Royal Society of Chemistry.)

fluorescence in the visible to near-infrared regions, unlike the large spherical AuNPs. Among the various NCs (including Au, Ag, Pt, and Cu), AuNCs have gained much interest due to the advantages of the Au element itself, and further due to the various available synthesis strategies and their extraordinary stability, biocompatibility, and potential for applications [22, 23]. However, the facile synthesis of fluorescent NCs with various ligands and tunable fluorescence for applications as optoelectronic emitters or biological labels has gained importance in recent years [24–26].

Many fluorescent AuNCs have been synthesized using different approaches, such as photoreduction, microwave heating, sonochemical, thermal reduction, template-assisted, and ligand-induced etching synthesis. AuNCs are synthesized in the presence of suitable ligands or templates and have a metal core–ligand shell structure. In addition to the metal core, it contains only few to several hundred Au atoms, with the ligand shell consisting of DNA, dendrimers, polymers, proteins, small molecules, or thiolates [27–32]. The protecting templates or ligands are considered to be not only necessary for stabilization, but also related to the origin of the fluorescence of the AuNCs [33, 34]. Both the ligand shell and Au core present in AuNCs can interact specifically with various analytes as the recognition components in AuNC based fluorescence applications [35].

Therefore, the synthesis of various metal NCs with appropriate surface ligands to develop portable fluorescence sensors for the detection of various analytes is required. In this chapter we summarize the various methods of NC synthesis and we review the recent fluorescence detection applications for a wide range of analytes, including contaminants, biomolecules, and toxic chemicals with an emphasis on the sensing mechanism.

12.2 Synthesis methods of nanoclusters

Several synthesis methods for NCs have been reported, including both bottom-up and top-down approaches [19, 20]. Over the past two decades, substantial efforts

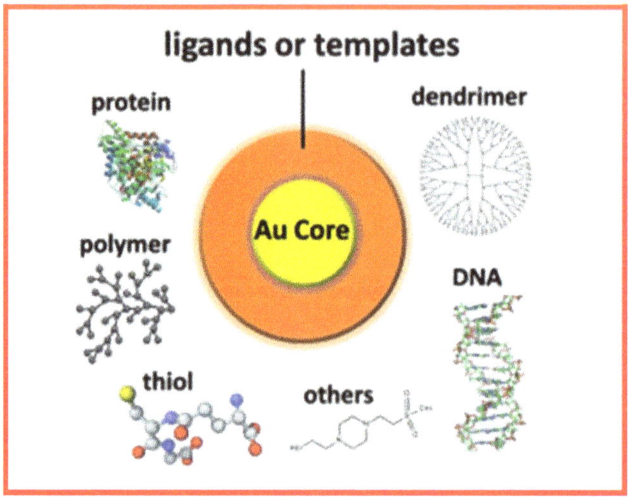

Figure 12.2. Schematic representation of fluorescent AuNCs protected by different ligands. (Reproduced with permission from [19]. Copyright 2014 The Royal Society of Chemistry.)

have been made toward the preparation of fluorescent NCs, focusing on controlling their size, functionality, and stability.

The size dependent preparation of NCs with good quality is of great importance. In the obtaining high-quality metal NCs, some important factors must be considered: (i) the reducing condition, usually light irradiation, sonication, or strong reducing agents need to be employed to improve the quantum yield of NCs; (ii) a long aging time; and (iii) the ligand should have strong interaction with metal NCs. Considering these important factors, various techniques have been used to prepare high-quality NCs with different scaffolds that can be controlled. These scaffolds contain several groups that allow strong interaction with metal ions to form high-quality clusters and also prevent aggregation of NCs into larger NPs [37–39]. These significantly change the electronic structure and also influence their properties. The scaffolds used for the synthesis of clusters include DNA, proteins, small thiol molecules, dendrimers, polymers, and peptides (figure 12.2). Recently, various studies on copper, silver, and other NCs have also been reported [18, 19, 39].

12.3 Applications of fluorescent metal nanoclusters

Metal NCs have been used widely as probes for the sensing of various analytes such as cations, anions, toxic chemicals, and biomolecules, etc. Metallic NCs are highly reactive, giving rise to improved selectivity and sensitivity due to their higher surface area, ultrasmall size, and functionalization with different capping ligands. Therefore, in this section we aim to explore the recent advancements and developments of NCs for the detection of various analytes, including anions, cations, small molecules, toxic compounds, and biomolecules.

12.3.1 Detection of cations and anions

Fluorescence quenching or enhancing of metal NCs by cations and anions are direct methods for the detection of metal ions (e.g. Cr^{3+}, Pb^{2+}, Hg^{2+}, and Cu^{2+}) and anions (CN^-, S^{2-}, NO_2^-, Cl^-, and I^-), which could quench the fluorescence of AgNCs, AuNCs, and CuNCs [40]. These heavy metal ions and anions are also considered to be important environmental pollutants because they can causes various health issues when to living beings are exposed. The US Environmental Protection Agency (US EPA) recommends a maximum accepted level of these ions in drinking water [10, 40, 41], which are shown in table 12.1, and related diseases are caused by exceeded metal ion levels. Therefore, the detection of these metal ions in the environment and biological samples is of great importance owing to their toxic effects on human health and ecosystems. Recently, various types of fluorescent metal NCs have been used for the sensing of these heavy metal ions.

12.3.1.1 Metal ion sensors

Recently, Sahu *et al* developed a method for the sensing of Hg^{2+} ions using L-cysteine assisted chitosan reduced gold nanoclusters (L-cys–CS-AuNCs) as a fluorescence probe [42]. The L-cys–CS-AuNCs were found to be effective fluorescence probes for the sensitive and selective detection of Hg^{2+} ions by 'turning off' the fluorescence quenching mechanism (figure 12.3). Upon Hg^{2+} ion addition to the L-cys–CS-AuNCs, the orange fluorescence of the L-cys–CS-AuNCs was decreased significantly, which is due to the strong metallophilic interactions of the $5d^{10}$ of Hg^{2+} and $5d^{10}$ of Au^+ on the gold surface. This method was selective, sensitive, and environmentally friendly with a limit of detection (LOD) of 3 nM. Moreover, the

Table 12.1. The acceptable level of toxic ions in drinking water according to the US EPA and their potential health effects [40].

Toxic ions	US EPA permissible level	Potential health effects
Cd^{2+}	45.29 nM	Kidney disease
Hg^{2+}	10.07 nM	Kidney disease
As^{3+}	0.14 M	Skin allergy and potential increased risk of cancer
Pd^{2+}	73.09 nM	Kidney disease and high blood pressure
Ag+	0.94 M	Skin discoloration and cornea disease in eyes
Cr^{3+}	2.00 M	Allergic dermatitis
Cu^{2+}	21.13 M	Gastrointestinal distress and liver or kidney disease
NO_2^-	27.39 M	Neurological problems and anaemia
CN^-	7.40 M	Thyroid and neurological problems
S^{2-}	0.21 mM	Blue baby syndrome and shortness of breath
Cl^-	27.39 μM	Respiratory paralysis and irritation of the mucous membranes

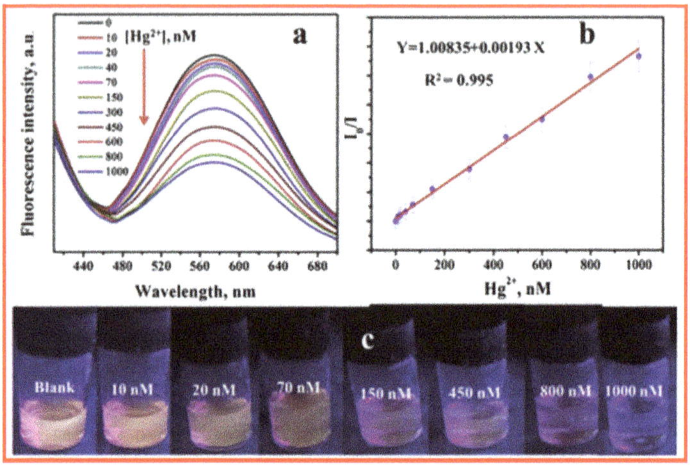

Figure 12.3. (a) Fluorescence emission spectra, (b) relative intensity, and (c) photographs under UV light of L-cys–CS-AuNCs upon the addition of Hg^{2+} ions. (Reproduced with permission from [42]. Copyright 2020 Elsevier.)

L-cys–CS-AuNCs were also used in the detection of Hg^{2+} ions in real water samples such as pond, river, and tap water.

Recently, red fluorescent glutathione-lactose functionalized gold nanoclusters (GSH-lactose-AuNCs) was prepared for the detection of Cu^{2+} and Al^{3+} ions in real water samples [43]. The fluorescence spectra of the GSH-lactose-AuNCs at 635 nm showed independent behaviour with fluorescence quenching upon the addition of Al^{3+} ions and fluorescence quenching for Cu^{2+} ions. As a result, good linear relationships between the fluorescence intensity and an Al^{3+} ion concentration in the range of 0.075–100 μM with an LOD of 12 nM and a Cu^{2+} ion concentration in the range of 0.050–100 μM with an LOD of 52 nM were obtained, respectively. Further, the size of the GSH-lactose-AuNCs increased from 1.03 nm $\pm$ 0.37 nm to 6.45 nm $\pm$ 0.28 nm and 5.05 nm $\pm$ 0.18 nm upon the addition of Al^{3+} and Cu^{2+} ions, respectively, confirming the GSH-lactose-AuNC aggregation (figure 12.4). In addition, a real sample analysis for Al^{3+} ions was also carried out using an antacid drug. Further, GSH-lactose-AuNCs was also used for the detection of Al^{3+} ion in industrial water collected from aluminium manufacturing industries.

In another study, Akshath *et al* prepared protein modified gold nanoclusters (FGNCs), which can be used as a sensor for the detection of cobalt and cadmium ions with LODs in the range of 5–165 ng ml^{-1} and 20–1000 ng ml^{-1}, respectively [44]. The increase in the fluorescence intensity of the FGNCs in the presence of cadmium ions was due to the fluorescence resonance energy transfer (FRET) taking place between the tryptophan and NCs. Further, the proposed method was also used for the detection of vitamin B_{12} in the ampoules and validated.

In another report, cobalt (Co^{2+}) was detected by using glutathione stabilized Au nanoclusters (GSH-AuNCs) [45]. GSH-AuNCs showed green emission with a maximum emission wavelength at 500 nm under UV light illumination. With the

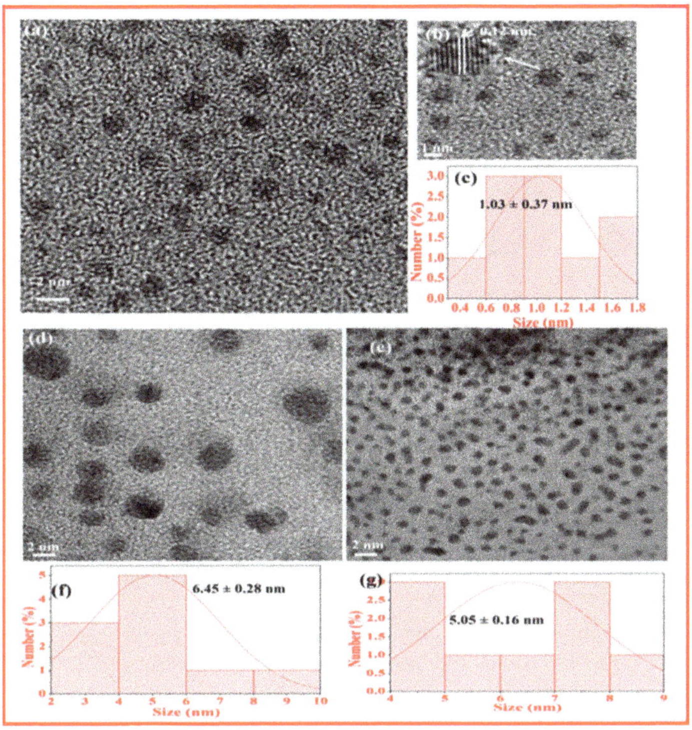

Figure 12.4. (a) HR-TEM images, (b) lattice fringes, and (c) histogram of GSH-lactose-AuNCs. (d) and (e) HR-TEM images and (f) and (g) histograms of GSH-lactose-AuNCs upon the addition of Al^{3+} ions and Cu^{2+} ions, respectively. (Reproduced with permission from [43]. Copyright 2020 Elsevier.)

addition of Co^{2+} ions, the fluorescence intensity of the GSH-AuNCs was decreased with a linear range of 2.0 μM to 50.0 μM and the LOD was found to be 0.124 μM (figure 12.5). This method was also utilized for the successful detection of Co^{2+} ions in various environmental water samples.

12.3.1.2 Anion sensors

Iodine is a fundamental micronutrient for living organisms insufficiency of might cause congenital iodine deficiency syndrome, hypothyroidism, mental disability, and goitre [46, 47]. Pereira *et al* reported poly(vinylpyrrolidone)-supported copper nanoclusters (CuNCs) for the selective detection of iodide with an LOD of 1.0 ng ml^{-1} [48]. The fluorescence quenching of CuNCs upon the addition of iodine was ascribed to a sphere of action static quenching model. In addition, this paper based sensor using CuNCs has also been used for the *in situ* detection of iodine generation in various samples.

Yang *et al* developed polyethyleneimine-assisted dual-emissive gold nanoclusters (DE-AuNCs) for the detection of CN^- ions (figure 12.6) [49]. The ratiometric fluorescence intensity change of DE-AuNCs was based on the mechanism of etching of the Au(I) surface by CN^- ions. The practical application of the DE-AuNCs probe was also demonstrated in the presence of urine and river samples.

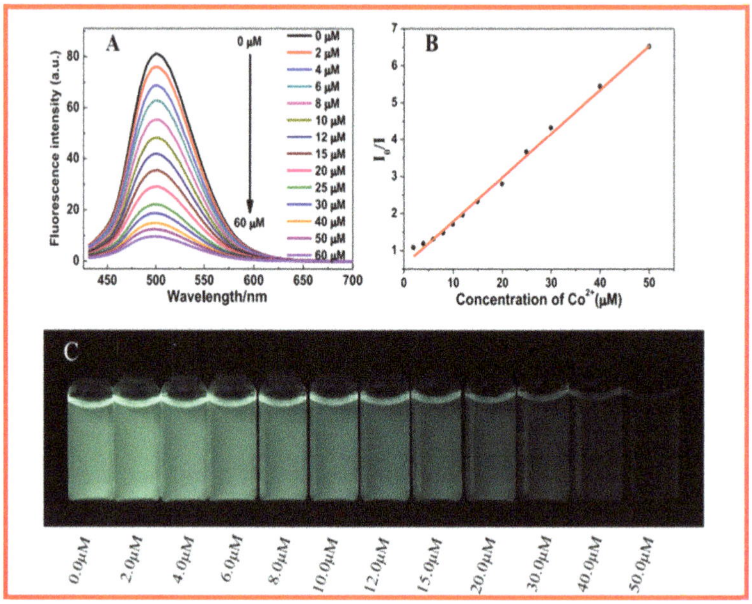

Figure 12.5. The fluorescence (a) quenching, (b) quenching ratio, and (c) photographs of GSH-Au NCs upon the addition of Co^{2+} ions. (Reproduced with permission from [45]. Copyright 2021 Elsevier.)

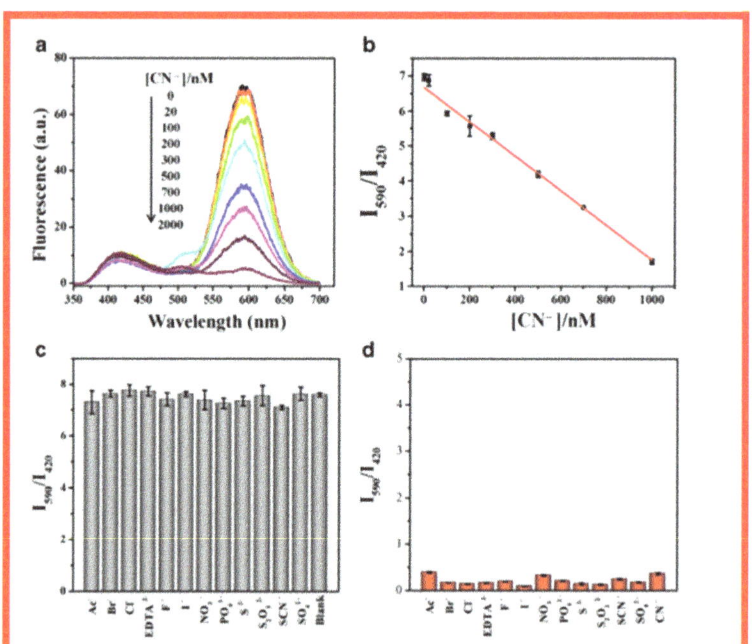

Figure 12.6. (a) Fluorescence spectra and (b) plot of intensity ratio (I_{590}/I_{420}) of DE-AuNCs upon adding different concentrations of CN^-. The fluorescence intensity ratio (I_{590}/I_{420}) of DE-AuNCs with 10 μM anions in the (c) absence and (d) presence of 2 μM CN^-. (Reproduced with permission from [49]. Copyright 2020 Springer.)

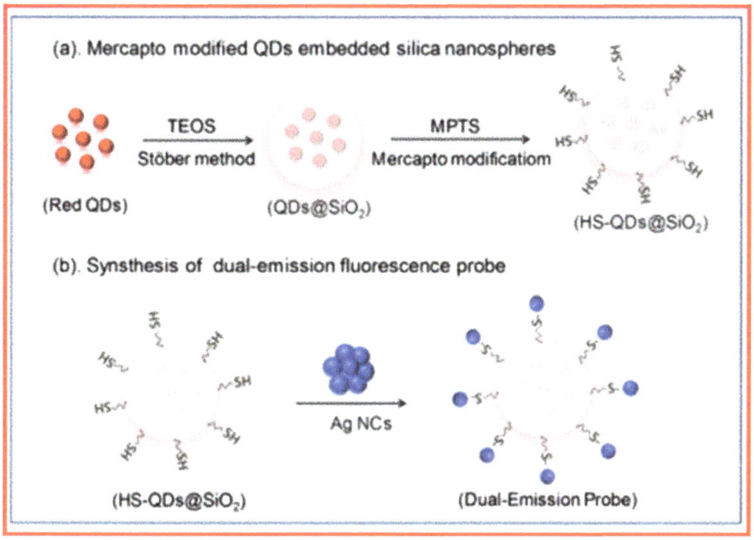

Figure 12.7. (a) The synthesis of mercapto functionalized QDs@SiO_2 nanospheres and (b) the surface assembly of AgNCs onto SiO_2 nanospheres. (Reproduced with permission from [50]. Copyright 2017 Elsevier.)

Yan *et al* developed silver nanocluster stabilized silica nanospheres for the detection of S^{2-}, with an LOD of 32 nM (figure 12.7) [50]. This nanohybrid was used for on-site detection of gaseous hydrogen sulfide gas using this nanohybrid. Further, the selectivity of this nanohybrid in the presence of other interference (Cl^-, F^-, I^-, Br^-, CO_3^{2-}, HPO_4^{2-}, SO_3^{2-}, SO_4^{2-}, $H_2PO_4^-$, HCO_3^-, HSO_3^-) was studied and it showed superior selectivity towards the detection of S^{2-} only, other anions did not show significant changes in the fluorescence intensity.

Lu *et al* reported the synthesis of glutathione onto 11-mercaptoundecanoic acid capped gold nanoclusters (GSH–MUA-AuNCs) for the sensing of hydrogen sulfide (H_2S). The fluorescence intensity of GSH–MUA-Au NCs was increased gradually upon the addition of H_2S (from 0.5 μM to 4 μM) with an LOD of 35 nM. Under optimized conditions, other anions did not interfere with the determination of hydrogen sulfide. In a practical application of the GSH–MUA-AuNCs were also used for the detection of H_2S in human serum samples [51].

Thiosalicylic acid capped-copper nanoclusters (TA-CuNCs) were utilized for the selective and sensitive detection of NO_2^-, with an LOD of 5 μM [52]. The fluorescence quenching of TA-CuNCs upon the addition of NO_2^- was through a redox reaction between TA and NO_2^-. Moreover, this approach was also used to determine the NO_2^- concentration in lake water samples.

Zheng *et al* developed a simple and sensitive fluorescent sensor for the determination of NO_2^- using fluorescent CuNCs [53]. The fluorescence of the CuNCs was effectively and selectively quenched by the NO_2^- through the oxidation of Cu (0) to Cu (I) in the core of the CuNCs. Under optimal conditions, the fluorescence intensity was decreased in the concentration ranges of 0.0125–125 M and 125–5000 M with an

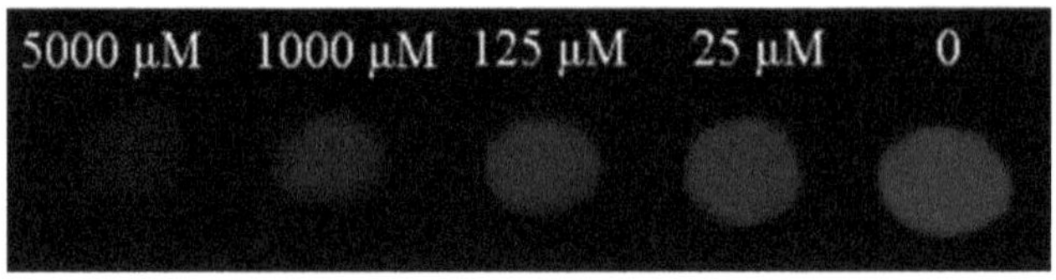

Figure 12.8. Photographic image of CuNCs on cellulose paper upon the addition of different concentrations of NO_2^- ions from 5000, 1000, 125, 25, and 0 μM under UV lamp illumination. (Reproduced with permission from [53]. Copyright 2016 Elsevier.)

LOD of 3.6 nM. Further, a paper based sensor was also developed using this CuNCs for the detection of NO_2^- ions under UV illumination (figure 12.8).

12.3.2 Detection of small molecules

Dopamine (DA) (a monoamine neurotransmitter) is used to transmit signals from one nerve cell to another. This brain endocrine is essentially responsible for the transmission of data on feelings of excitement and cheerfulness in the brain, and is likewise identified with addiction [54–57]. Ling *et al* developed glutathione and dihydrolipoic acid protected silver nanoclusters (GSH-DHLA-Ag nanoclusters) for sensing of DA [58]. The fluorescence intensity of the GSH-DHLA-Ag NCs was increased in the presence of DA owing to the aggregation-inducing emission enhancement effect. The linear range of DA concentration was 0–800 nM, with the LOD found to be 10 nM. A schematic illustration of DA sensing is shown in figure 12.9.

Cysteine capped-copper/molybdenum bimetallic nanoclusters (Cu–MoNCs) used for the fluorescence detection of methotrexate (MTX) via an inner filter effect (IFE) were proposed by Nerthigan *et al* [59]. Methotrexate (MTX) is an anti-neoplastic drug used for the treatment of various cancers. The Cu–MoNCs display bluish-green fluorescence and a relative quantum yield of 26% (figure 12.10).

The fluorescence was linear with the MTX concentration range of 50 nM to 100 μM with an LOD of 13.7 nM. This proposed method was successfully applied in the detection of the MTX drug in human urine sample with recovery of 99%.

4-nitrophenol (4-NP) is used widely in the production of pharmaceuticals and fine chemicals. Green-emitting gold NCs (AuNCs) have been synthesized using thiol-β-cyclodextrin as a ligand for sensing of 4-NP in water samples [60]. The fluorescence intensity at 502 nm of the AuNCs showed a significant decrease in the presence of an increasing concentration of 4-NP from 0.1 μM to 100 μM with the LOD found to be 90 nM. The fluorescence detection mechanism for the detection of 4-NP is due to the host–guest interaction between β-cyclodextrin and 4-NP molecules.

Receptive oxygen species (ROSs) are a progression of intermediate products of molecular oxygen created in living creatures during reduction processes [61]. Hydrogen peroxide (H_2O_2), as a significant ROS, assumes a huge part in different physiological processes [62]. Recently, Zhou *et al* developed polyetheleneimine-capped silver nanoclusters (AgNCs) as a probe for sensing of H_2O_2 [63]. The fluorescence of the AgNCs was reduced significantly by the reduction of glutathione

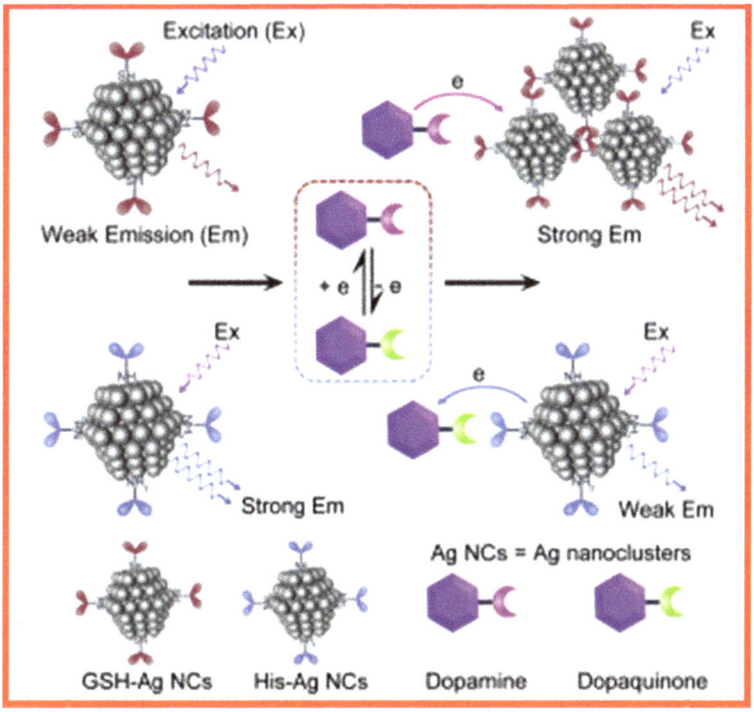

Figure 12.9. Schematic representation of the mechanism of GSH-DHLA-Ag and His-Ag nanoclusters for the detection of DA. (Reproduced with permission from [58]. Copyright 2020 Elsevier.)

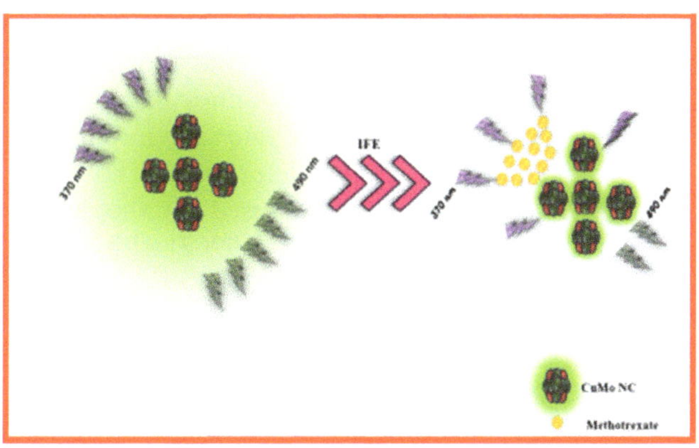

Figure 12.10. Sensing mechanism for the detection of the MTX drug using a Cu–MoNC probe. (Reproduced with permission from [59]. Copyright 2019 Springer.)

through the binding of its thiol groups. Usually, H_2O_2 converts GSH into an oxidized form with free-thiol elimination and prohibits the quenching interaction. The fluorescence response was increasing in a H_2O_2 concentration range from 0.1 μM to 20 μM with an LOD of 35 nM. Further, this method was used for the detection of glucose in human serum samples.

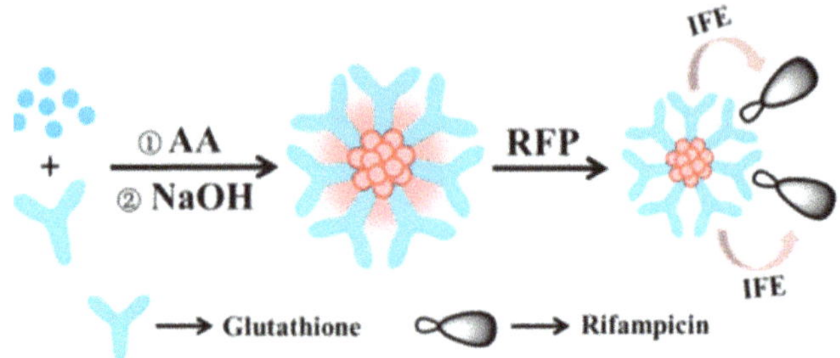

Figure 12.11. A schematic representation of the detection of rifampicin using GSH-CuNCs based on the IFE of fluorescence. (Reproduced with permission from [68]. Copyright 2020 The Royal Society of Chemistry.)

Rifampicin (RFP) is family of the rifamycin group and is used as an excellent antibacterial drug [64, 65]. An overdose of rifampicin in patients can develop resistance to medication in pathogenic microbes in the human body, which makes it more difficult to destroy micro-organisms and treat infections [66, 67]. RFP has been detected using GSH-CuNCs as a probe (figure 12.11) [68]. In this work, GSH-CuNC fluorescence was decreased significantly by rifampicin owing to the IFE of the fluorescence mechanism. GSH-CuNCs were used for the detection of rifampicin in the linear range of 50 pM to 10 000 pM with an LOD of 16 pM. Moreover, the GSH-CuNCs were used to detect rifampicin in different real samples (eye drops, capsules, serum samples) with quantitative spike recoveries ranging from 97% to 105%.

L-histidine (His) is one of the amino acids and plays a vital part in biological systems [69]. An abnormal level of His may lead to various diseases such as Parkinson's disease, erythropoiesis, and epilepsy, and a lower level of His may also lead to lung disease, Alzheimer's disease, and kidney disease [70]. Therefore, the detection of His in biological samples has become very important in recent years. Recently, Au–CuNCs were used for the detection of His with a nanomolar detection level [71]. The calibration plot was used for the detection of His in a concentration range from 3 to 10 000 nM ($R^2 = 0.9962$) with the LOD estimated as 0.9 nM.

Adenosine triphosphate (ATP) is an energy-carrying molecule in living beings and plays a significant role in enzymatic activities and an abnormal concentration leads to many diseases [72]. Dong's group developed a convenient and simple method for the detection of ATP using DNA-stabilized Cu–AgNCs as fluorescence probes and an LOD of 7.0 μM was obtained (figure 12.12) [72]. The fluorescence of DNA-Cu–AgNCs increases appreciably with ATP concentration with the interaction between the ATP and its aptamer causing two darkish Cu–AgNCs to be situated at the 5′ and 3′ termini close to each other. Further, the DNA-Cu–AgNCs–ATP system was also used for the detection of adenosine deaminase (ADA) with a linear response of 5–50 U L^{-1} and an LOD of 5 U L^{-1}. The method was also used in the detection of ADA and ATP in a solution of foetal bovine serum.

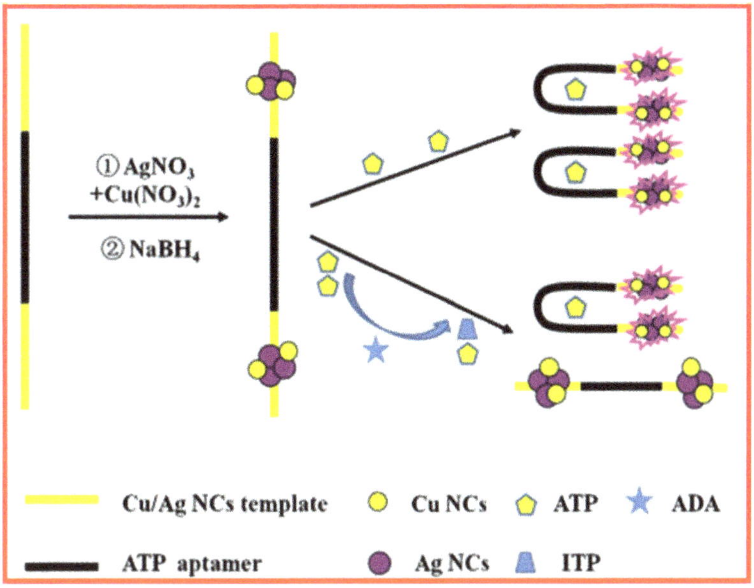

Figure 12.12. Schematic representation for the detection of ATP and ADA using DNA-Cu–Ag NCs. (Reproduced with permission from [72]. Copyright 2020 Springer.)

12.3.3 Pesticide detection

Pesticides have consistently been a topic of central concern in food safety and contemporary agricultural production. Two subclasses of pesticides, organochlorine pesticides (OCPs) and organophosphorus pesticides (OPPs), are used widely all over the world due to their low-cost, easy preparation, and outstanding insecticidal efficiency [73]. However, the inappropriate use of OCPs and OPPs has caused harm to the soil, atmosphere, and water systems, and also severe human health problems and food safety issues [74]. Recently, Nazir *et al* reported the use of fluorescent AuNCs for the detection of 2-methyl-4-chlorophenoxyacetic acid (MCPA) (figure 12.13) [75]. AuNCs stabilized with lysozyme (Lys-AuNCs), histidine (His-AuNCs), and RNase (RNase-AuNCs) were synthesized for the detection of MCPA using a fluorescence turn-on mechanism. Initially, the fluorescence of the AuNCs was quenched upon the addition of Cu^{2+} ions followed by fluorescence recovery in the presence of MCPA. These NCs had LODs for the detection of MCPA of, from highest to lowest, His-AuNCs (21.87 nM), Lys-AuNCs (12.97 nM), and RNase-AuNCs (9.26 nM)).

Sharma *et al* developed an ultrasensitive and cost effective method for the detection of ethyl parathion using bimetallic BSA@AuAgNCs [76]. This sensing method is based on fluorescence quenching with Cu^{2+} ion addition and an enhanced phenomenon owing to the generation of thiocholine (TCh), a catalytic product of the enzymatic reaction of acetylthiocholine (ATCh) using the acetylcholinesterase (AChE) enzyme. The formation of TCh desirably grabs Cu^{2+} from the BSA@AuAgNCs-Cu^{2+} ensemble and recovers the fluorescence intensity of the BSA@AuAgNCs. The inhibitory action of the

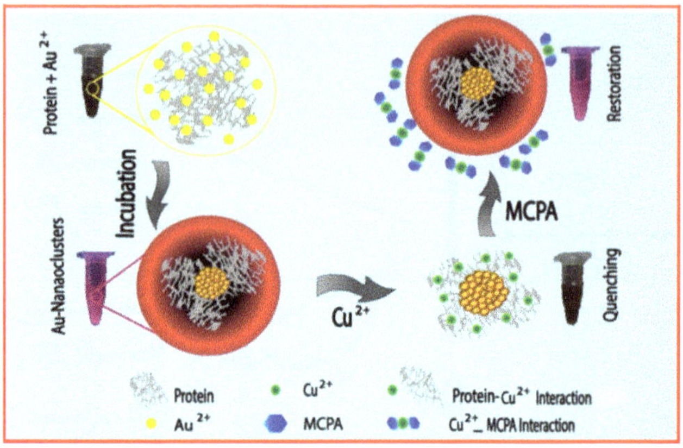

Figure 12.13. Schematic representation of the detection of MCPA using NCs through a fluorescence quenching and restoration method. (Reproduced with permission from [75]. Copyright 2020 Springer.)

AChE enzyme in the presence of ethyl parathion may leading to a decrease in the formation of TCh along with a decrease in TCh-Cu^{2+} that finally retrieved quenched of bimetallic NCs. This biosensor was suitable for the sensing of ethyl parathion in pM range with an LOD of 2.40 pM, which is the lowest known so far.

Dithiocarbamates (DTCs) belong to the family of fungicides and are used widely across the globe. Therefore, the monitoring of the DTC concentration in environmental samples has been very important in recent years. Recently, CTAB-encapsulated fluorescent CuNCs were developed by Chen *et al* for the specific recognition of DTCs (figure 12.14) [77]. The synthesized CuNCs were used for the highly selective and sensitive detection of DTCs using both fluorometric and colorimetric methods. In addition, this method was also used for the sensing of DTCs in real samples.

12.3.4 Nucleic-acid detection

MicroRNAs (miRNAs) are considered important biomarkers for disease therapeutics and diagnostics. Recently, Gu *et al* developed an enzyme-free amplification platform along with a catalytic hairpin assembly (CHA) using DNA-templated silver nanoclusters (DNA–AgNCs), for the detection of miRNA [78]. Here, two DNA hairpins (H1 and H2) were developed for the target miR-21 induced CHA, then the emission quenching of DNA–AgNCs by BHQ1 was used to construct an activatable probe. This method for the detection of miR-21 showed a linear ratio for different concentrations from 200 pM to 20 nM with the LOD found to be 200 pM (figure 12.15). This cost-effective and simple method holds huge potential for applications in clinical diagnostics and biomedical research. The proposed method has several advantages, such as simple operation, no cleaning requirement, being enzyme-free, and an optical performance, and provides a new plan for the design of novel fluorescent probes, with highly selective and sensitive detection of miRNA.

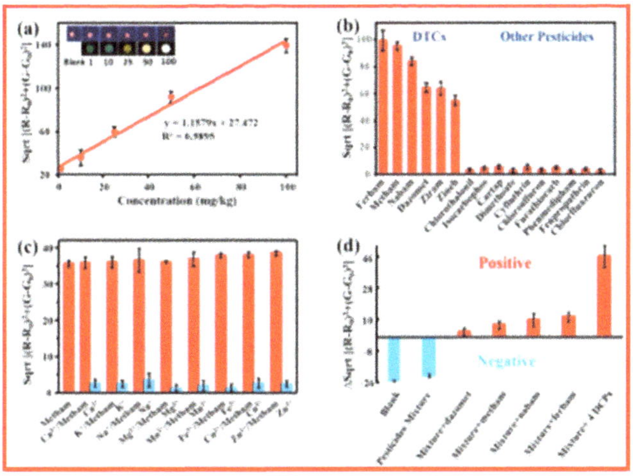

Figure 12.14. Sensing of DTCs. (a) Linear response between the concentration of metham and the response signal of paper based sensors. (b) Selectivity of CuNCs in the presence of other interferences. (c) Competitive assay for the detection of DTCs in the presence of other interferences. (d) The pesticide mixture influences the experiments. (Adapted from [77]. Copyright 2020 Elsevier.)

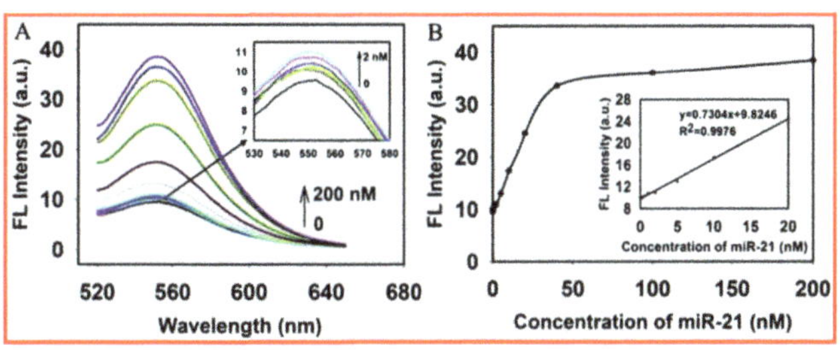

Figure 12.15. (a) The fluorescence intensity of an assay against various concentrations of miR-21 (inset: a low concentration of miR-21). (b) Fluorescence intensities against concentration for R-21 concentration (inset: linear behaviour between the fluorescence peak intensity and concentration of miR-21). (Reproduced with permission from [78]. Copyright 2020 The Royal Society of Chemistry.)

Tan *et al* developed a rapid and simple method for the detection of miR-122 using DNA-templated AgNCs [79]. Exo III was used to break the Ag cluster template and help in the RNA–DNA complex cycle. Without the target, the Ag clusters remain intact and DNA–AgNCs are synthesized under the action of silver nitrate ($AgNO_3$) and sodium borohydride ($NaBH_4$), producing an intense fluorescence signal. By comparing the changes in the fluorescence peak, it was found that this strategy has good selectivity with the LOD found to be 84.0 pM. This method can provide an important indication for the early diagnosis of hepatitis and liver cancer.

Recently, a novel oligonucleotide-stabilized AgNC beacon (NCB) fluorescence probe was developed by Wen *et al* for sensing of oligonucleotide sequence targets [80]. This assay contained two tailored DNA strands a signal probe stand along with a fluorescence enhancing strand. Using this, two NCBs were developed for the

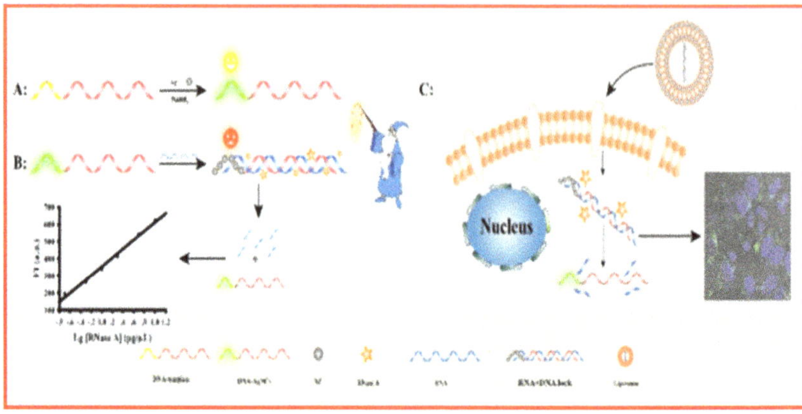

Figure 12.16. Ultra-selective and sensitive detection of RNase A using AgNCs. (a) Preparation of AgNCs. (b) Activity assay for RNase A. (c) Living cell RNase imaging *in vitro*. (Reproduced with permission from [81]. Copyright 2020 Elsevier.)

determination of two disease-related oligonucleotide sequences, and the outcome confirmed the two target oligonucleotide sequences in the ranges 50–200 and 50–600 nM along with an LOD of 25 and 20 nM, respectively.

Dang *et al* developed an ultrasensitive, label-free, and fluorescence-on technique for the determination of RNase using DNA–AgNCs as the fluorescence probe (figure 12.16) [81]. The fluorescence intensity of the AgNCs was retained owing to the RNA degradation by RNase A. From the fluorescence intensity changes caused by RNase A, there was found a linear relationship with the RNase A concentration in the range from 0.2 pg μL^{-1} to 10 pg μL^{-1} with an LOD of 0.098 pg/μL. Further, this method was also used for detecting an RNase A change in living cells.

12.3.5 Protein detection

Sych *et al* developed a straightforward technique for immunoglobulin (Ig) and serum albumin detection using AgNCs as a probe [82]. The luminescent probes of AgNCs with IgG, pentaglobin, and human serum albumins and their mixtures were synthesizing by combining $AgNO_3$ with the other proteins through the addition of $NaBH_4$. Each type of complex showed different fluorescence peaks, which allowed AgNCs to be to applied to sense Ig and HSA selectively in their mixtures. This method could be used for the detection of various immunoglobulins and albumin in biological samples.

Ardekani *et al* designed and synthesized a AgNC based aptasensor for the determination of lysozyme (LYZ) [83]. The method consisted of a complementary DNA sequence and an RNA aptamer with AgNCs as the template. Upon the addition of aptamer, the DNA–AgNCs showed fluorescence quenching owing to the hybridization of the aptamer. In the presence of LYZ, aptamer binding of the DNA–AgNCs was prohibited owing to the stronger interaction between the LYZ and the aptamer. The addition of a DNA template leads to fluorescence enhancement of the DNA–AgNCs which is dependent on the LYZ concentration. The LOD

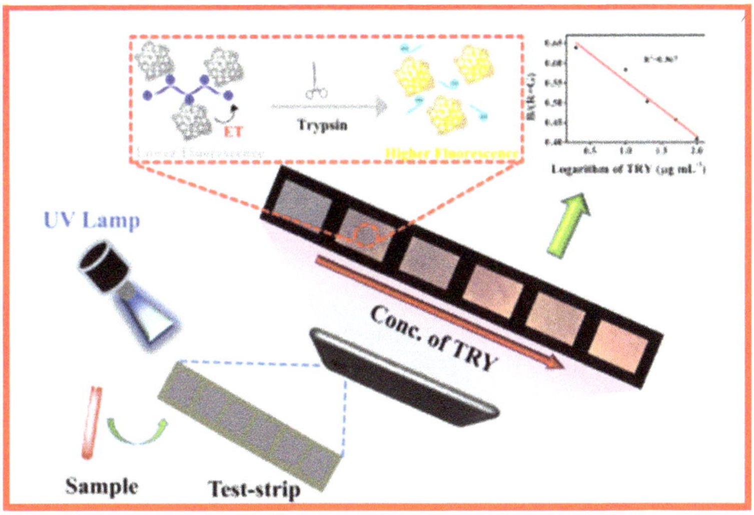

Figure 12.17. A schematic representation of the principles of the detection of TRY. (Reproduced with permission from [84]. Copyright 2019 Elsevier.)

for this method was found to be 10 nM. The proposed method has many advantages such as high sensitivity, low-cost, convenient operation, and easy fabrication for the determination of LYZ.

Recently, fluorescent AuNCs were used as a fluorescence probe for point-of-care determination of trypsin (TRY) (figure 12.17) [84]. The fluorescence signal of the AuNCs was quenched by cytochrome c (Cyt c) via an electron transfer mechanism. Then, the addition of TRY leads to a considerable increase in fluorescence enhancement due to the hydrolysis of Cyt c to produce small fragments. Thus the AuNC system showed a highly selective and sensitive response towards TRY with a linear range of 0.2–100 $\mu g\ mL^{-1}$ and an LOD of 0.08 $\mu g\ mL^{-1}$. Notably, this assay was utilized to construct paper based test strips for the visual and rapid detection of TRY. With a combination of ImageJ software and a smartphone, an image processing algorithm was developed for the determination of TRY.

12.4 Conclusions and trends

Fluorescent metal NCs are a novel class of fluorescent nanomaterials. They have fascinating chemical and physical properties, including higher surface area, higher luminescence biocompatibility, low photobleaching, ultrasmall size, catalytic efficiency, and higher solubility in aqueous media compared to current materials, which lead to the utilization for various sensing applications, such as biochemical and toxic chemical sensing, phototherapy, catalysis, and bioimaging applications. In this chapter, we summarize the advancements in the synthesis of fluorescent NCs and the current progress on fluorescent NCs for the determination of various toxic chemicals and biomolecules. On the whole, major progress has been achieved in the field of metal NCs in recent years. It is our strong belief that more and more selective and sensitive detection and imaging methods using NCs will soon become the gold standard for biological and clinical applications.

Acknowledgements

This work was supported by CONICYT-ANID FONDECYT postdoctoral grant 3180232 and FONDEF ID20I10153.

References

[1] Saha K, Agasti S S, Kim C, Li X and Rotello V M 2012 Gold nanoparticles in chemical and biological sensing *Chem. Rev.* **112** 2739

[2] Anker N, Hall W P, Lyandres O, Shah N C, Zhao J and Van Duyne R P 2008 Biosensing with plasmonic nanosensors *Nat. Mater.* **7** 442

[3] Shipway A N, Katz E and Willner I 2000 Nanoparticle arrays on surfaces for electronic, optical, and sensor applications *Phys. Chem.* **1** 18–52

[4] Ordeig O, Mas R, Gonzalo J, Campo F J D, Muoz F J and de Haroet C 2005 Continuous detection of hypochlorous acid/hypochlorite for water quality monitoring and control *Electroanalysis* **17** 164–8

[5] Soldatkin A P, Gorchkov D V, Martelet C and Renault N J 1997 New enzyme potentiometric sensor for hypochlorite species detection *Sens. Actuators* B **43** 99

[6] Soto N O, Horstkotte B, March J G and Martin V C 2008 An environmental friendly method for the automatic determination of hypochlorite in commercial products using multisyringe flow injection analysis *Anal. Chim. Acta* **611** 182

[7] Claver J B, Miron M and Capitan V 2004 Determination of hypochlorite in water using a chemiluminescent test strip *Anal. Chim. Acta* **522** 267

[8] March J G and Simonet B M 2007 A green method for the determination of hypochlorite in bleaching products based on its native absorbance *Talanta* **73** 232

[9] Vogel A I 1989 *A Text Book of Quantitative Chemical Analysis* (London: Longman)

[10] Deepak F L 2018 *Metal Nanoparticles and Clusters:Advances in Synthesis, Properties and Applications* (London: Springer)

[11] Qu X, Li Y, Li L, Wang Y, Liang J and Liang J 2015 Fluorescent gold nanoclusters: synthesis and recent biological application *J. Nanomater.* **2015** 1–23

[12] Demirel G B, Daglar B and Bayindir M 2013 Extremely fast and highly selective detection of nitroaromatic explosive vapours using fluorescent polymer thin films *Chem. Commun.* **49** 6140

[13] Nagarkar S S, Desai A V and Ghosh S K 2014 A fluorescent metal–organic framework for highly selective detection of nitro explosives in the aqueous phase *Chem. Commun.* **50** 8915

[14] Zhao L M, Ye J W, Li W B, Bogale R F, Wang B, Gong W T and Ning G L 2014 A luminescent cyclic binuclear zinc–pamoate complex for highly selective and sensitive sensing of 2,4,6-trinitrophenol *Inorg. Chem. Commun.* **46** 212

[15] Bhalla V, Gupta A and Kumar M 2012 Fluorescent nanoaggregates of pentacenequinone derivative for selective sensing of picric acid in aqueous media *Org. Lett.* **14** 3112

[16] Pazhanivel T, Nataraj D, Devarajan V P, Mageshwari V, Senthil K and Soundararajan D 2013 Improved sensing performance from methionine capped CdTe and CdTe/ZnS quantum dots for the detection of trace amounts of explosive chemicals in liquid media *Anal. Methods* **5** 910

[17] Shanmugaraj K and John S A 2018 Inner filter effect based selective detection of picric acid in aqueous solution using green luminescent copper nanoclusters *New J. Chem.* **42** 7223

[18] Derfus A M, Chan W C W and Bhatia S N 2004 Probing the cytotoxicity of semiconductor quantum dots *Nano Lett.* **4** 11
[19] Sun J and Jin Y 2014 Fluorescent Au nanoclusters: recent progress and sensing applications *J. Mater. Chem.* C **2** 8000
[20] Li J, Zhu J J and Xu K 2014 Fluorescent metal nanoclusters: from synthesis to applications *Trends Anal. Chem.* **58** 90
[21] Xu H X and Suslick K S 2010 Water-soluble fluorescent silver nanoclusters *Adv. Mater.* **22** 1078
[22] Zheng J, Zhou C, Yu M and Liu J 2012 Different sized luminescent gold nanoparticles *Nanoscale* **4** 4073
[23] Shang L, Yang L X, Stockmar F, Popescu R, Trouillet V, Bruns M, Gerthsen D and Nienhaus G U 2012 Microwave-assisted rapid synthesis of luminescent gold nanoclusters for sensing Hg^{2+} in living cells using fluorescence imaging *Nanoscale* **4** 4155
[24] Muhammed M A H and Pradeep T 2010 *Advanced Fluorescence Reporters in Chemistry and Biology* ed A P Demchenko vol 2 (Berlin: Springer)
[25] Diez I and Ras R 2010 *Advanced Fluorescence Reporters in Chemistry and Biology* vol 2 (Berlin: Springer)
[26] Lin C, Lee C, Hsieh J, Wang H, Li J, Shen J, Chan W, Yeh H and Chang W 2009 Review: synthesis of fluorescent metallic nanoclusters toward biomedical application: recent progress and present challenges *J. Med. Biol. Eng.* **29** 276
[27] Xie J P, Zheng Y G and Ying J Y 2009 Protein-directed synthesis of highly fluorescent gold nanoclusters *J. Am. Chem. Soc.* **131** 888
[28] Kennedy T A, MacLean J L and Liu J 2012 Blue emitting gold nanoclusters templated by poly-cytosine DNA at low pH and poly-adenine DNA at neutral pH *Chem. Commun.* **48** 6845
[29] Duan H W and Nie S M 2007 Etching colloidal gold nanocrystals with hyperbranched and multivalent polymers: a new route to fluorescent and water-soluble atomic clusters *J. Am. Chem. Soc.* **129** 2412
[30] Zheng J, Petty J T and Dickson R M 2003 High quantum yield blue emission from water-soluble Au_8 nanodots *J. Am. Chem. Soc.* **125** 7780
[31] Yu Y, Yao Q F, Luo Z T, Yuan X, Lee J Y and Xie J P 2013 Precursor engineering and controlled conversion for the synthesis of monodisperse thiolate-protected metal nanoclusters *Nanoscale* **5** 4606
[32] Ho J A A, Chang H C and Su W T 2012 DOPA-mediated reduction allows the facile synthesis of fluorescent gold nanoclusters for use as sensing probes for ferric ions *Anal. Chem.* **84** 3246
[33] Wu Z K and Jin R C 2010 On the ligand's role in the fluorescence of gold nanoclusters *Nano Lett.* **10** 2568
[34] Yuan X, Luo Z T, Yu Y, Yao Q F and Xie J P 2013 Luminescent noble metal nanoclusters as an emerging optical probe for sensor development *Chem. Asian J.* **8** 858
[35] Sun J and Jin Y 2014 Fluorescent Au nanoclusters: recent progress and sensing applications *J. Mater. Chem.* C **2** 8000
[36] Diez I and Ras R 2010 *Advanced Fluorescence Reporters in Chemistry and Biology* vol 2 (Berlin: Springer)
[37] Daniel M C and Astruc D 2004 Gold nanoparticles: assembly, supramolecular chemistry, quantum-size-related properties, and applications toward biology, catalysis, and nanotechnology *Chem. Rev.* **104** 293

[38] Muhammed M A H and Pradeep T 2010 *Advanced Fluorescence Reporters in Chemistry and Biology* ed A P Demchenko vol 2 (Berlin: Springer) p 333

[39] Zhang L and Wang E 2014 Metal nanoclusters: new fluorescent probes for sensors and bioimaging *Nano Today* **9** 132

[40] Vasimalai N and Argüelles M T F *Metal Nanoparticles and Clusters* (Cham: Springer)

[41] https://www.epa.gov/ground-water-and-drinking-water/national-primary-drinking-water-regulations

[42] Sahu D, Mohapatra P and Swain S K 2020 Highly orange fluorescence emission by water soluble gold nanoclusters for "turn off" sensing of Hg^{2+} ion *J. Photochem. Photobiol.* A **386** 112098

[43] Kateshiya M R, Malek N I, Murthy Z V P and Kailasa S K 2020 Designing of glutathione-lactose derivative for the fabrication of gold nanoclusters with red fluorescence: Sensing of Al^{3+} and Cu^{2+} ions with two different mechanisms *Opt. Mater.* **100** 109704

[44] Akshath U S, Bhatt P and Singh S A 2020 Differential interaction of metal ions with gold nanoclusters and application in detection of cobalt and cadmium *J. Fluoresc.* **30** 37

[45] Zhao R X, Liu A Y, Wen Q L, Wu B C, Wang J, Hu Y L, Pu Z F, Ling J and Cao Q 2021 Glutathione stabilized green-emission gold nanoclusters for selective detection of cobalt ion *Spectrochim. Acta* A **254** 119628

[46] Andersson M, De Benoist B, Darnton-Hill I and Delange F 2007 *Iodine Deficiency in Europe: A Continuing Public Health Problem* (Geneva: UNICEF)

[47] Zimmermann M B, Jooste P L and Pandav C S 2008 Iodine-deficiency disorders *Lancet* **372** 1251

[48] Pereira F P, Capón N, de la Calle I, Lavilla I and Bendicho C 2019 Fluorescent poly (vinylpyrrolidone)-supported copper nanoclusters in miniaturized analytical systems for iodine sensing *Sens. Actuators* B **299** 126979

[49] Yang H, Yang Y, Liu S, Zhan X, Zhou H, Li X and Yuan Z 2020 Ratiometric and sensitive cyanide sensing using dual-emissive gold nanoclusters *Anal. Bioanal. Chem.* **412** 5819

[50] Yan Y, Zhang K, Yu H, Zhu H, Sun M, Hayat T, Alsaedi A and Wang S 2017 Sensitive detection of sulfide based on the self-assembly of fluorescent silver nanoclusters on the surface of silica nanospheres *Talanta* **174** 387

[51] Lu F, Yang H, Tang Y, Yu C J, Wang G, Yuan Z and Quan H 2020 11-Mercaptoundecanoic acid capped gold nanoclusters with unusual aggregation-enhanced emission for selective fluorometric hydrogen sulfide determination *Microchim. Acta* **187** 200

[52] Cang J, Wang C W, Chen P C, Lin Y J, Lib Y C and Chang H T 2017 Control of pH for separated quantitation of nitrite and cyanide ions using photoluminescent copper nanoclusters *Anal. Methods* **9** 5254

[53] Zheng X J, Liang R P, Li Z J, Zhang L and Qiu J D 2016 One-step, stabilizer-free and green synthesis of Cu nanoclusters as fluorescent probes for sensitive and selective detection of nitrite ions *Sens. Actuators* B **230** 314

[54] Sansuk S, Bitziou E, Joseph M B, Covington J A, Boutelle M G and Unwin P R 2013 Ultrasensitive detection of dopamine using a carbon nanotube network microfluidic flow electrode *Anal. Chem.* **85** 163

[55] Sun C L, Chang C T, Lee H H, Zhou J, Wang J and Sham T K 2011 Microwave-assisted synthesis of a core–shell MWCNT/GONR heterostructure for the electrochemical detection of ascorbic acid, dopamine, and uric acid *ACS Nano* **5** 7788

[56] Lin Y H, Chen C E, Wang C Y, Pu F, Ren J S and Qu X G 2011 Silver nanoprobe for sensitive and selective colorimetric detection of dopamine via robust Ag–catechol interaction *Chem. Commun.* **47** 1181

[57] Ji X, Palui G, Avellini T, Na H B, Yi C Y and Knappenberger K L 2012 On the pH-dependent quenching of quantum dot photoluminescence by redox active dopamine *J. Am. Chem. Soc.* **134** 6006

[58] Ling Y, Wang L, Zhang X Y, Wang X H, Zhou J, Sun Z, Li N B and Luo H Q 2020 Ratiometric fluorescence detection of dopamine based on effect of ligand on the emission of Ag nanoclusters and aggregation-induced emission enhancement *Sens. Actuators* B **310** 127858

[59] Nerthigan Y, Sharma A K, Pandey S and Wu H F 2019 Cysteine capped copper/molybdenum bimetallic nanoclusters for fluorometric determination of methotrexate via the inner filter effect *Microchim. Acta* **186** 130

[60] Li Y, Wen Q L, Liu A Y, Long Y, Liu P, Ling J, Ding Z T and Cao Q E 2020 One-pot synthesis of green-emitting gold nanoclusters as a fluorescent probe for determination of 4-nitrophenol *Microchim. Acta* **187** 106

[61] Muhr V, Buchner M, Hirsch T, Jovanović D J, Dolić S D and Dramićanin M D 2017 Europium-doped $GdVO_4$ nanocrystals as a luminescent probe for hydrogen peroxide and for enzymatic sensing of glucose *Sens. Actuators* B **241** 349

[62] Zhou Z L, Li Y Q, Su W, Gu B, Xu H, Wu C Y, Yin P, Li H T and Zhang Y Y 2019 A dual-signal colorimetric and near-infrared fluorescence probe for the detection of exogenous and endogenous hydrogen peroxide in living cells *Sens. Actuators* B **280** 120

[63] Zhou T, Su Z, Wang X, Luo M, Tu Y and Yan J 2021 Fluorescence detections of hydrogen peroxide and glucose with polyethyleneimine-capped silver nanoclusters *Spectrochim. Acta* A **244** 118881

[64] Goldstein B P 2014 Resistance to rifampicin: a review *J. Antibiot. Res.* **67** 625

[65] Bi W, Zhu L H, Jing X N, Liang Y R and Tao E X 2013 Rifampicin and Parkinson's disease *Neurol. Sci.* **34** 137

[66] Pandey P, Pant N D, Rijal K R, Shrestha B, Kattel S, Banjara M R, Maharjan B and Rajendra K C 2017 Diagnostic accuracy of GeneXpert MTB/RIF assay in comparison to conventional drug susceptibility testing method for the diagnosis of multidrug-resistant tuberculosis *PLoS One* **12** 1

[67] Nasiri M J, Zamani S, Pormohammad A, Feizabadi M M, Aslani H R, Amin M, Halabian R and Imani Fooladi A A 2018 The reliability of rifampicin resistance as a proxy for multidrug-resistant tuberculosis: a systematic review of studies from Iran *Eur. J. Clin. Microbiol. Infect. Dis.* **37** 9–14

[68] Wu X M, Zhang J H, Feng Z S, Chen W X, Zhang F and Li Y 2020 An ultra-sensitive "turn-off" fluorescent sensor for the trace detection of rifampicin based on glutathione-stabilized copper nanoclusters *Analyst* **145** 1227

[69] Zhao Y, Tang Y, He J, Xu Y, Gao R, Zhang J, Chong T, Wang L and Tang X 2018 Surface imprinted polymers based on amino-hyperbranched magnetic nanoparticles for selective extraction and detection of chlorogenic acid in honeysuckle tea *Talanta* **181** 271

[70] Hu Y, Wang Q, Zheng C, Wu L and Lv Y 2013 Recyclable decoration of amine-functionalized magnetic nanoparticles with Ni^{2+} for determination of histidine by photochemical vapor generation atomic spectrometry *Anal. Chem.* **86** 842

[71] Cai Y, Wang J, Liu C, Yang S, Zhanga Y and Liu A 2020 Histidine-triggered turning-on of gold/copper nanocluster fluorescence for the sensitive and selective detection of histidine *Chem. Commun.* **56** 11637
[72] Zhang B and Wei C 2020 The sensitive detection of ATP and ADA based on turn-on fluorescent copper/silver nanoclusters *Anal. Bioanal. Chem.* **412** 2529
[73] Mostafalou S and Abdollahi M 2017 Pesticides: an update of human exposure and toxicity *Arch. Toxicol.* **91** 549
[74] Worek F, Wille T, Koller M and Thiermann H 2016 Toxicology of organophosphorus compounds in view of an increasing terrorist threat *Arch. Toxicol.* **90** 2131
[75] Nazir K, Ahmed A, Hussain S Z, Younis M R, Zaheer Y, Ahmed M, Hussain I and Ihsan A 2020 Development of gold nanoclusters based direct fluorescence restoration approach for sensitive and selective detection of pesticide *Appl. Nanosci.* **10** 3411
[76] Sharma D, Wangoo N and Sharma R K 2021 Sensing platform for pico-molar level detection of ethyl parathion using Au–Ag nanoclusters based enzymatic strategy *Talanta* **221** 121267
[77] Chena S, Wanga Y and Feng L 2020 Specific detection and discrimination of dithiocarbamates using CTAB-encapsulated fluorescent copper nanoclusters *Talanta* **210** 120627
[78] Gu J, Qiao Z, He X, Yu Y, Lei Y, Tang J, Shi H, Heb D and Wang K 2020 Enzyme-free amplified detection of miRNA based on target-catalyzed hairpin assembly and DNA-stabilized fluorescent silver nanoclusters *Analyst* **145** 5194
[79] Tan L, Fu S, Lu J, Hu K, Liang X, Li Q, Zhao S and Tian J 2020 Detection of microRNA using enzyme-assisted amplifying and DNA-templated silver nanoclusters signal-off fluorescence bioassay *Talanta* **210** 120623
[80] Wen Q L, Peng J, Liu A Y, Wang J, Hu Y L, Ling J and Cao Q E 2020 DNA bioassays based on the fluorescence 'turn off' of silver nanocluster beacon *Luminescence* **35** 702
[81] Dang W, Luo R, Fan J, Long Y, Tong C, Xiao F, Xiong X and Liu B 2020 RNase A activity analysis and imaging using label-free DNA-templated silver nanoclusters *Talanta* **209** 120512
[82] Sych T S, Polyanichko A M, Plotnikova L V and Kononov A I 2019 Luminescent silver nanoclusters for probing immunoglobulins and serum albumins in protein mixtures *Anal. Methods* **11** 6153
[83] Ardekani L S, Moghadam T T, Thulstrup P W and Ranjbar B 2019 Design and fabrication of a silver nanocluster-based aptasensor for lysozyme detection *Plasmonics* **14** 1765
[84] Li H, Yang M, Kong D, Jin R, Zhao X, Liu F, Yan X, Lin Y and Lu G 2019 Sensitive fluorescence sensor for point-of-care detection of trypsin using glutathione-stabilized gold nanoclusters *Sens. Actuators* B **282** 366

www.ingramcontent.com/pod-product-compliance
Ingram Content Group UK Ltd.
Pitfield, Milton Keynes, MK11 3LW, UK
UKHW051307130926
8556IPUK00048B/729